HIGH-SPEED DIGITAL CIRCUITS

HIGH-SPEED DIGITAL CIRCUITS

Masakazu Shoji

ADDISON-WESLEY PUBLISHING COMPANY

Reading, Massachusetts · Menlo Park, California · New York · Don Mills, Ontario
Wokingham, England · Amsterdam · Bonn · Sydney · Singapore · Tokyo · Madrid
San Juan · Seoul · Milan · Mexico City · Taipei

Many of the designations used by manufacturers and sellers to distinguish their products are claimed as trademarks. Where those designations appear in this book and Addison-Wesley was aware of a trademark claim, the designations have been printed with initial capital letters.

The publisher offers discounts on this book when ordered in quantity for special sales. For more information, please contact:

Corporate & Professional Publishing Group
Addison-Wesley Publishing Company
One Jacob Way
Reading, Massachusetts 01867

Library of Congress Cataloging-in-Publication Data
Shoji, Masakazu, 1936–
  High-speed digital circuits / Masakazu Shoji.
    p.   cm.
  Includes bibliographical references and index.
  ISBN 0-201-63483-X (hc)
  1. Digital integrated circuits—Mathematical models.   2. Very high speed integrated circuits—Mathematical models.   3. Electric circuits, Equivalent—Mathematical models.   4. Digital computer simulation.   I. Title.
TK7874.65.S56   1996
621.3815—dc20   95-46485
                                                                      CIP

Text design by Wilson Graphics & Design (Kenneth J. Wilson)
Text printed on recycled and acid-free paper.

ISBN 0-201-63483-X
1 2 3 4 5 6 7 8 9-MA-99989796
First printing February 1996

CONTENTS

6 COMPLEXITY OF INTERCONNECTS

To execute the complex signal processing of a modern computer system, a huge number of transistors and large silicon chips are required. State-of-the-art MOS integrated circuits contain several million field-effect transistors within an area of about 1 cm × 1 cm. Active devices such as MOSFETs and BJTs scale down to the micron and to the submicron sizes, and their switching speed has increased to the limit where many new physical limits emerge. Logic gates switch from the high to the low logic states in less than 100 ps. In that time an electromagnetic signal in free space propagates only 3 cm. A signal excited on a lossy transmission line fabricated on a dielectric propagates slower than that, and much effort has been concentrated on reducing the delay to the theoretical limit. Even in the limit, however, a signal sent out from a corner of the chip does not reach another corner in zero time. The fact that an electrical signal cannot propagate instantly from the source to the destination is crucial in determining the chip performance. It means that the signal propagation must be treated accurately according to Maxwell's electromagnetic theory. This theoretical development is carried out by incorporating inductance into the conventional integrated circuit theory. When that is done, a circuit that has size Λ (cm) has signal delay Λ/c, where $c = 3 \times 10^{10}$ cm/sec is the speed of light. The unrealistic feature of the conventional theory that the signal propagates at infinite speed is removed. Thus the effect of including inductance is to build a *relativistic* circuit theory. This impressive adjective heightens our interest in investigating these new issues.

Inductance is peculiar in a circuit. There are two kinds of inductances: a shielded and an unshielded, or *naked*, inductance. The value of the shielded inductance is not related to the size of the circuit. The framework of circuit theory which includes shielded inductances is well established. Here we study the properties of a circuit that include naked inductance, which creates a magnetic field in the region of the circuit. The magnitude of the naked inductance is related to the physical dimensions of the circuit. Inclusion of naked inductance adds many new features to circuit theory.

In the operational regime where the signal propagation delay becomes the key issue, the *transient* phenomena within electron devices are unaffected by the relativistic complications. Owing to the extremely small device size achieved by the scaledown, electron devices can be understood using classical physics, even though device operation is based on such a slow phenomenon as current-carrier diffusion. In this book I intend to develop an integrated circuit theory of *classical* electron devices in a *relativistic* interconnect environment.

This text presents an opportunity to study many practical circuit problems in great depth, in terms of basic physics. Many existing circuit design aids have been produced without critically looking into the basic electrical phenomena of integrated circuits at high speeds. One issue raised in this book is quite fundamental: How far is the use of electrostatic potential, or voltage, justified in describing circuit operation? As Maxwell's theory teaches, a *conservative* electrostatic potential does not exist if the electromotive force created by a time-varying magnetic field becomes significant. An integrated circuit in the limit of high frequencies is immersed in the time-dependent magnetic field the circuit itself makes, and therefore the definition of the circuit's node voltage becomes ambiguous. Conventional integrated circuit theory, that is, the description of the charge and discharge capacitance through resistance, assumes the universal existence of voltage. A new high-speed integrated circuit theory should be built by including the effects of naked inductance into conventional integrated circuit theory, and I intend to show that self-inductance can be included consistently into integrated circuit theory. In this new theory some node voltages lose their physical meaning and we suffer from that inconvenience, but we find a pleasing general consistency, in that integration of self-inductance into the theory can be carried out quite neatly using the many peculiar features of scaled-down integrated circuits. The extension of conventional circuit theory is limited, however, by a fundamental difficulty of including mutual inductance in the theory. A study of this difficulty shows that it constitutes the natural and fundamental limit of the equivalent circuit model.

Integrated circuits are very complex devices. On an integrated circuit chip, various interactions between the circuits occur, which affect circuit performance fundamentally. This complexity becomes significant even at the lower switching speeds at which the relativistic effects are still not significant. This effect originates from nature's inability to maintain the separate identity of excitations, expressed as the thermodynamic law of irreversibility. The structures that exist on an integrated circuit chip tend to mix signals, creating complex and often undesirable effects. For our theory to be practical this complexity must be included at high frequencies. Although these complex effects are fundamental to high-speed integrated circuit theory, they have not been studied enough. Some material related to this subject is presented in this book.

Fundamental to this new high-speed integrated circuit theory is the equivalent circuit. Equivalent circuits are theoretical models of the physical objects of our present research or development. Modeling an integrated circuit by an equivalent circuit is not a trivial task, since many devices and passive components are densely packed, and no circuit is totally independent of the others. One issue addressed in this book is how to draw simple and physically accurate equivalent circuits. This issue deserves attention, since equivalent circuits are the most powerful construct of electronic circuit theory. Equivalent circuits enable us to model any physical phenomena that exist in nature, and in some cases maybe even those that are merely imagined by human minds.

Acknowledgments

The two basic ideas on which this work was founded came from two giants of modern electronics by whom I was educated in the 1960s, Professor Aldert van der Ziel and Dr. William Shockley. The first idea, that an electronic circuit contains a lot of *physics*, that must be clarified and interpreted by a complete mathematical analysis, is due to Professor van der Ziel. The origin of his viewpoint will be clear if we remember how he discovered the mechanisms of parametric amplification. The second idea, that all the semiconductor devices are variations on one fundamental device, the *electron triode*, and that semiconductor circuits using different devices are all very much alike, is my interpretation of what I learned from Dr. Shockley. I am surprised to find how these great mentors have influenced their student so deeply and so fundamentally for so long a time. I am eternally grateful to them.

Professor O. Wing agreed to review the manuscript at a very early phase. He gave me a lot of moral support in carrying out the work. Professor E. Friedman of the University of Rochester, Professor S. Kang of the University of Illinois, and Professor A. Willson of the University of California, Los Angeles, reviewed the original manuscript and gave valuable comments, suggestions for improvement, and encouragement. I am grateful to Dr. M. D. Alston, who read the entire manuscript, corrected English, improved technical interpretations, and gave valuable technical comments. Ms. H. Khorramabadi and Messrs. M. Pinto, T. Wik, V. Califano, and R. Marley reviewed an early version of the manuscript and gave me helpful comments. Ms. Khorramabadi, Mr. Pinto, and Mr. Wik have been my co-workers for many years; we share many happy memories of technical successes. While I was writing this book, I had many enlightening discussions with Mr. Kwok Ng, who was also writing a book. His book was recently published, and is referred to in this one.

My co-workers in AT&T Bell Laboratories, Cedarcrest, Pennsylvania, especially Mr. R. Krambeck and Ms. J. Sabnis, have supplied a lot of technical problems that became the nucleus of this book. The problem of Section 3.5 was one of the design problems Ms. Sabnis and I had to solve in the late 1980s. The work was supported by the management of the Computing Science Research Center of the AT&T Bell Laboratories, especially by T. G. Szymanski, R. Sethi, P. J. Weinberger, and A. G. Fraser. They have supported the fundamental theoretical study of electrical phenomena in integrated circuits for many years now. Their faith in this work was essential to its completion. I am grateful to the editors of the Addison-Wesley Publishing Company, Messrs. John Wait and Simon Yates, who kindly offered to publish this work, and especially to Ms. Avanda Peters, who worked out the difficult problems of printing a book that contains many complicated mathematical formulas and illustrations.

CHAPTER ONE

Definition of the Problems

1.1 Introduction—Circuit Complexity

The theory of electronic circuits is a branch of physical science. The theory has been constructed from the principles of physics, such as Maxwell's theory of electromagnetism and the theories of electron devices and of circuit components, which in turn are built on the principles of solid-state physics, statistical mechanics, and thermodynamics. In recent years, it has become increasingly apparent that traditional circuit theory is not adequate to deal with many of the complex problems that have emerged since the advent of the very large-scale integrated (VLSI) circuits. VLSI circuits contain so many active devices, and their components interact with each other so strongly, that traditional circuit theory is unable to adequately handle the analysis. To fill the immediate needs of integrated circuit design, numerical simulation techniques have been developed. Although numerical analysis is effective for predicting the behavior of well-specified circuits with the parameter values specified, it provides practically no general understanding of the circuits. Often, interpretation of the simulation results is the hardest part of the circuit design verification. Numerical simulation provides an extension of the experiments, but not of the theory. A strange situation is now emerging: At a time when millions and billions of circuits on silicon are fabricated every day, fundamental understanding of the circuits has not made significant progress. Many circuits fail on the first trial after the design, because of insufficient understanding.

The foregoing observation suggests that complex circuits in an integrated circuit chip require a circuit theory built on a significantly different foundation, that is, on information science, which can handle complexity, rather than on physical science, which treats everything in its simplest terms. The crucial difference is the inclusion of complexity originating from the complex connection of the large number of the circuit components. Inclusion of the interconnect complexity creates many new features

1

in circuit theory. Establishing such a circuit theory is a challenging task, which will continue into the next century.

This scientific endeavor can be illustrated in the following manner. Understanding the basic properties of about 100 distinct atoms is the task of atomic physics. Although a heavy atom has a complex internal electronic structure, its overall properties can be understood using the shell model of the electrons, which leads to the concept of *valence* of an atom. The valence determines how atoms combine to form molecules. The multitude of properties of complex molecules, including ultimately the properties of biological macromolecules, is the area covered by organic chemistry. In this regard, organic chemistry is the study of the superstructures of atoms in atomic physics. In the same manner, we begin with semiconductor device physics. Semiconductor devices are the "atoms" in the analogy. We then wish to build a field of study of superstructures of semiconductor devices (analogous to organic chemistry). In doing so we wish to go beyond traditional circuit theory, which corresponds to inorganic chemistry. This new scientific area will provide the basis for understanding complex integrated circuits that have rudimentary intelligence, in the same way as organic chemistry is fundamental in understanding the complex phenomena of life.

This is an ambitious endeavor, which will require many steps of theoretical development. The first step will be the following: Analyses of the various fundamental concepts of physics required to understand the operation of integrated circuits, to find a way of integrating these concepts with the analysis of circuits with complex connectivity. This first step is the objective I wish to accomplish in this book.

In this chapter, the prerequisites for understanding the operation of complex high-speed integrated circuits will be reviewed. In this book, I stick with the viewpoint that the interesting and useful properties of electronic circuits originate more from how the active devices and the passive components are connected together than from the detailed device and component characteristics. This viewpoint is supported by a simplification procedure, whereby the control function of the active devices can be separated from the passive parasitics. The parasitics can be lumped into the rest of the circuit. The active device thus simplified is quite featureless: Its role is limited to controlling the circuit connectivity. As the effort of scaledown and speedup of the active device progresses, properties of the semiconductor triodes approach the ideal limit, which can be represented by the very simple and universal characteristic of the *collapsible* current generator [1]. Differences in the characteristics of the various amplifying devices—BJT, MOSFET, or MESFET—then become relatively insignificant. These small differences can be included in the theory by adjusting the current-voltage characteristics or the node impedance level, or by including the leakage paths. The size of the device becomes so small that its working mechanism is still determined by the less fundamental solid-state physics principles. The device can be represented by a simple model of a switch. Complicated problems of signal propagation delay and mutual interaction of the circuit nodes are governed by the more fundamental Maxwell's electromagnetic theory. This is a very considerably different foundation from that of the conventional integrated circuit theory. To lay this foundation, we need to examine several basic physical concepts.

1.2 Complexity and High Speed

This book's subject is circuit theory in the boundary region between physics and information science. To explore the new frontier fully, I add a second specialization. Many complex circuit problems are reduced to triviality in the limit of low frequency, or low switching speed. As we observed in Section 1.1, circuit complexity originates from the device and the component interconnection, but how the complexity emerges requires some consideration. An interconnect is a featureless signal path at low frequencies. The connectivity and characteristics of devices and components determine the circuit operation. Understanding of the low-speed, *intrinsic* circuit operation is fundamental, but because of the generality of the connectivity it is quite difficult to accomplish this at this moment. There is another source of circuit complexity originating from the interconnects themselves. At high frequencies the interconnects become components, and they add to the complexity. The complexity problem of this kind is less general and is easier to study than the backbone problem. We investigate the high-speed limit, where the limit of the capability of the interconnects sets the maximum speed of operation. This limit is reached, historically, following the several steps of circuit model improvements.

If the speed of the device is low, as in BJTs of the 1950s and in MOSFETs of the 1970s, the electrical transients of the circuits can be understood from the following model. The active devices are equivalently described by controlled current generators, or by resistors, which may in general be nonlinear. The devices and the interconnects have parasitic capacitances. The delay time of the circuit is determined by computing the time required to drain the charge stored in the capacitance through the current generator or the resistor that represents the intrinsic part of the switching device. That is the theory we have now, and it is used extensively. Analyses of complex integrated circuits based on this model were popular in the 1980s. As circuit complexity increased in the 1990s a problem began to arise: The interaction between two circuits on the same chip that were not intended to interact with each other. This may be regarded as a general noise problem, aggravated by the aggressive design style of modern VLSI chips. This problem can be handled within the framework of the current-generator/resistor-capacitor theory, although the mathematical formalism must be significantly changed and the computation becomes quite extensive. Study of this problem showed an essentially new theoretical issue originating from the complexity: What is the universal voltage reference of a complex integrated circuit? Does a universal voltage reference really exist?

Resistance-capacitance modeling of circuits presents one more practical issue. The resistance of the interconnects can be included in the theory, but if that is done, a circuit node that is connected by a resistive interconnect must be taken into account separately, and the complexity of the problem increases dramatically. To mitigate this difficulty, and to improve circuit performance by reducing the interconnect RC delay, state-of-the-art CMOS circuits are wired together using several levels of metal interconnect that have high conductivity. In more ambitious quarters, the use of high-temperature superconducting wires is considered [2]. This is, above all, to simplify the analysis and design verification. Although RC delay is difficult to deal with in practice, this delay does not pose a fundamental limit on the circuit's performance, if the process cost and the design cost are not the issue. The essential limit on circuit performance is set by the inductance of the interconnects.

In the resistance-capacitance modeling two locations of a circuit can be made equivalent by using a zero-resistance interconnect between them. This simplification is no longer valid if inductance is included. Then two locations of a circuit that are connected by a perfectly conductive wire cannot be called the same node, because a location can be called a circuit node only if the node voltage can be defined. In general, the definition of node voltage loses its sense in inductive circuits. The resistance of the wire can be reduced, but the inductance cannot be reduced below a limit determined by the circuit dimensions. Therefore the problem is inevitable in high-frequency (high-speed) circuits. An immediate consequence of inductance in circuit modeling is that there will be no universal ground reference extending over the entire dimension of the circuit. Since the signals must then be transmitted as potential *differences*, a signal must be routed by a *pair* of conductors. If only one conductor wire is routed, and the voltage is referenced to the uncharacterized ground, the transmitted "signal" may or may not be a signal at all.

The maximum size L of the ground conductor is estimated as follows. Let the velocity of propagation of light through the signal conductor be c (cm/sec). If a circuit switches within time t (sec), the signal propagates over a distance ct (cm). If the ground conductor is larger than that, the potential of the ground cannot be defined. If $c = 1.5 \times 10^{10}$ cm/sec and $t = 5 \times 10^{-11}$ sec $= 50$ ps, then $ct = 0.75$ cm. This is a typical size of a VLSI chip. At present, the switching time of fast bipolar devices is in the 50 ps range [3]. As the present trend continues, scaled-down CMOS devices of $0.2\text{-}\mu m$ feature size will switch significantly faster than 100 ps. The substrate of the VLSI integrated circuit and the metallic substrate of the package can no longer be taken as the ground voltage reference. Obviously the subject of this book can be expected to have practical relevance quite soon.

As we observed in this section, there is a hierarchy of equivalent circuit representations. Namely:

1. Device resistance only (DC)

2. Device resistance and device capacitance (low frequencies)

3. Device resistance, device capacitance, and wiring capacitance (medium frequencies)

4. Device resistance, device capacitance, wiring capacitance, and wiring resistance (high frequencies)

With each level, the equivalent circuit becomes more complex. We wish to add one more level:

5. Device resistance, device capacitance, wiring resistance, wiring capacitance, and wiring inductance (close to the limit of integrated circuit operation)

We study regimes 4 and 5 in this book.

1.3 Time and Frequency Metrics in Digital and in Analog Circuits

We study high-speed analog and digital circuits. The difference between analog circuits and digital circuits may appear obvious, but in reality it is not. In my previous book, a definition of digital and analog circuits using the concepts of a circuit's configurational change was introduced [1]. This is a theoretically clean classification in CMOS and complementary bipolar circuits, but is not necessarily the only one. Perhaps the most popular definition would be that analog circuits use continuous values of voltage or current to represent the value of a variable, while digital circuits compare the continuous values with a logic threshold voltage, and determine discretized Boolean values. In conventional digital circuits, discretization of the continuous value is carried out only when the data are stored in a memory, or in a latch. The rest of the digital circuit works like an analog circuit: The analog variables of the digital circuit often require *interpretation* to determine the digital values [1]. Then the only crucial difference between the two types of circuits is the precision of the continuous value of the variables.

To attain higher precision, the voltage or current must settle close to a *final* value that is a function of the values of the input variables to the circuit. Higher precision requires closer settling, and that requires more time: The more precise circuits operate more slowly. Digital circuits may be considered as the most imprecise analog circuits. To attain voltage precision ΔV in a circuit, time t is required, and they are generally related by the logarithmic relationship

$$-\log(\Delta V / V_S) = t/t_D \qquad (\Delta V \ll V_S) \qquad (1.1)$$

where V_S is the range of voltage swing, and t_D is the delay time of the amplifying device. Here we have considered the issue of precision with regard to the circuit's transient response. There are other issues in the circuit design, namely in the mechanical construction of the components. More precise circuits require more precisely specified devices, and if this requirement is applied to the photolithography process, it means larger devices: Larger size can be specified with greater relative precision on a semiconductor chip. However, a larger device inevitably has higher capacitance if the conductance is maintained at the same level. Thus the circuits using more precisely constructed devices have a second reason for being slower. We may conclude that digital circuits utilize the technological capability for producing the smallest possible devices, but analog circuits attain the higher precision by trading off the processing speed of the technology against precision in device construction and operation.

Analog and digital circuits normally use voltage or current as the signal variable. Then time becomes a variable that has a special status. It is possible, however, for a delay time to be used to represent a continuous analog variable. In that case, the voltage level at which the delay is measured becomes the criterion of precision. Since a comparator (or quantizer) circuit has high sensitivity at the threshold, better time precision is attained by increasing the range of the node voltage swing. If V_S is the dynamic range of voltage swing as before, ΔV_{TH} is the uncertainty in the switching threshold voltage, and Δt is the switching time uncertainty and t_D is the rise/fall time of the signal, we have

$$\Delta V_{TH} / V_S = \Delta t / t_D \qquad (1.2)$$

where ΔV_{TH} is a parameter that depends on the device design, especially on the degree of matching of the devices. There are complementary relationships for a given device technology: To increase the voltage precision, the time must be extended, and to increase the time precision, the voltage range must be extended.

Any circuit, analog or digital, has a maximum operational frequency or operational speed. The maximum operation is determined by the quality of the amplifying devices used in the circuit and by the quality of the signal required. Digital signal processing is *low*-quality processing. For low-quality processing the maximum capability of the amplifying device is available. Analog signal processing is *high*-quality processing, for which the maximum device operational speed can never be exploited. To attain precision, the device design criteria must be tightened, and the circuit speed must be compromised as well. Both analog and digital circuits have their maximum frequencies of satisfactory operation. A circuit may have a minimum frequency of operation as well. The latter can be reduced to zero (DC) if certain other requirements (especially the long-term stability) are relaxed. Low-frequency stability is a different problem of circuit engineering, and we do not discuss it here. The circuits we consider here are assumed to operate from DC to a certain maximum frequency.

Criteria for the signal quality that determine the maximum operational frequency or speed are set as follows. We test the response of a circuit to a high-frequency sinusoidal signal, or alternatively, to a fast-rising step function or pulse. Both time-domain and frequency-domain analyses are used in the test, as well as in the analysis. We need to establish a simple criterion relating the time and frequency measures in analog and in digital circuits. The Fourier component $F(\omega)$ of an arbitrary time-dependent waveform $F(t)$ is given by the Fourier transform formula as [4]

$$F(\omega) = \frac{1}{\sqrt{2\pi}} \int_{-\infty}^{+\infty} F(t) \exp(-j\omega t) \, dt$$

and $\qquad (1.3)$

$$F(t) = \frac{1}{\sqrt{2\pi}} \int_{-\infty}^{+\infty} F(\omega) \exp(j\omega t) \, d\omega$$

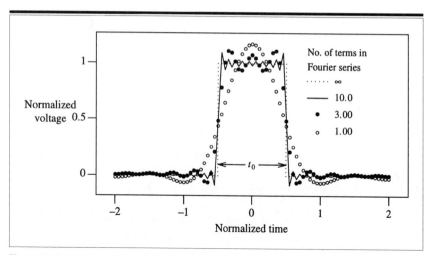

Figure 1.1(a) Fourier Synthesis of Isolated Square Pulse

In Figure 1.1(a) the dotted curve shows an isolated pulse having width t_0 and unit height. The Fourier component is calculated as

$$F(\omega) = \frac{t_0}{\sqrt{2\pi}} \frac{\sin(\omega t_0/2)}{(\omega t_0/2)} \qquad (1.4)$$

We have used $t_0 = 1$ in Figure 1.1(a). The pulse is synthesized by summing the Fourier components up to $f_0 = 1/t_0$, up to 3 times that, up to 10 times that, and all (an infinite number of) Fourier components. By summing the Fourier components up to 3 times the characteristic frequency f_0, we get a reasonably square waveform that is acceptable as a digital signal. Although this judgement is somewhat subjective, anyone who has observed digital signal waveforms in a CRT oscilloscope will agree that acceptable quality of digital pulse waveforms is obtained for sums up to $3f_0$ (closed circles) or at worst up to $10f_0$ (solid curve). The accuracy in the voltage is about 10% or better.

The waveforms of Figure 1.1(a) are not, however, accurate enough for analog signal processing, if the pulse height is to reflect an analog value. To consider analog signal processing, it is unrealistic to use an isolated pulse like that of Figure 1.1(a). If a unit step function having rise time t_0 is synthesized by summing the Fourier component

$$F(\omega) = -\frac{j}{\sqrt{2\pi}} \left[\frac{t_0}{2} \right] \frac{\sin(\omega t_0/2)}{(\omega t_0/2)^2} \qquad (1.5)$$

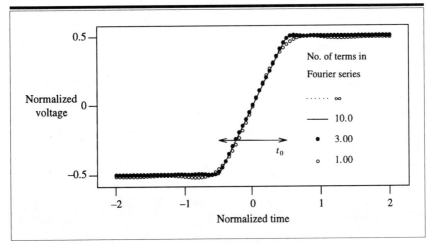

Figure 1.1(b)   Fourier Synthesis of Gradual Step Function

up to the characteristic frequency $f_0 = 1/t_0 = 1$, up to 3 times that, up to 10 times that, and all (an infinite number of) frequency components, the results are shown in Figure 1.1(b). The sum up to $3f_0$ is reasonably accurate. The sum up to $10f_0$ is adequate.

From the observations of Figures 1.1(a) and (b) we may conclude the following. In analog and in digital circuits, the type of signals that contains the information is different. In digital circuits an isolated pulse as shown in Figure 1.1(a) is the basic unit of information, since digital values 1 and 0 must be clearly distinguished. The pulse is characterized by the width t_0. In analog circuits, the instantaneous voltage level carries information. The voltage ramp sets the standard, and the time range of continuous voltage change, or the rise and fall times t_0 of the ramp in Figure 1.1(b), is the characteristic time parameter. The reciprocal of the characteristic time is the characteristic frequency. In terms of the characteristic frequency f_0 appropriate to each type of waveform, each circuit requires bandwidth from 3 to 10 times f_0.

1.4 Lumped Versus Distributed Equivalent Circuits

Traditional circuit theory focuses on lumped parameter circuits. Distributed parameter circuits like transmission lines are handled as special cases of the lumped parameter circuits. Distributed parameter circuits, however, model the real hardware better than lumped parameter circuits. Lumped parameter circuits are in essence approximations of the distributed parameter circuits. Then what are the assumptions used to draw lumped parameter equivalent circuit diagrams? This issue can be addressed in two ways: (1) how the components of a lumped parameter circuit model are identified, and (2) how discretization of the real distributed component should be made.

Electronic circuits are driven by energy. The circuit components generate, control, convert, store, and consume energy. Where the energy transfer/conversion activities take place is the key criterion in the component identification. A lumped parameter component can be identified only if the energy transfer/control activity takes place in a small, well-defined area, compared with the size of the circuit. This condition is, however, not always well satisfied. Small spatial size of a lumped component has two significances. If a certain (but not every) dimension is small, the component is *pure* (it has small parasitics): For instance, a smaller resistor has smaller associated parallel capacitive and series inductive components, and a smaller capacitor has smaller associated series resistive and inductive components. This is practically important, but is not the only consequence of small size. The more important issue is that the degree of isolation of the components is improved by small component size. For the equivalent circuit to be accurate, the mutual interaction of the components must be insignificant. If two two-terminal components interact, they must be regarded as a single four-terminal component, and this alters the circuit connectivity fundamentally. A single real piece of hardware can be represented by an infinite number of different equivalent circuit diagrams, with different degrees of approximation. A simple equivalent circuit is desirable, and such a circuit often assumes small component size.

To begin with we consider the three fundamental linear components, resistance (R), capacitance (C), and inductance (L). First we consider the methods of identifying the significant parameter values for each, and the methods of reducing the component interactions. For this we observe the differences among L, C, and R. Resistance is determined by the material resistivity and the size. Smaller-size resistors interact less. The resistance R of the small-size resistor shown in Figure 1.2(a) can still be made high by increasing the material resistivity ρ and by reducing the cross-sectional area S. The length l, however, cannot be reduced substantially below $\sqrt{S}$ for practical reasons. The resistor scaledown reaches its limit when the formula

$$R = \rho l / S = (1/q\mu N)(l/S) \tag{1.6}$$

becomes invalid, where q is the electronic charge, μ is the mobility and N is the charge carrier density. The limit is reached if the charge carrier flow becomes so tenuous that hydrodynamic carrier transport ceases, since then only a small number of carriers exist in the resistor volume Sl. In this

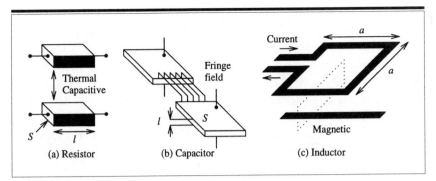

Figure 1.2 Isolation of Components

limit the resistor generates shot noise as current flows, due to the corpuscular nature of the charge. The mobility μ has its limit as well: The mean free path of the carrier cannot be less than the lattice period of the crystal. These two limits come after the practical limits, the constraint due to power density and reliability and the frequency limit due to the carrier transit-time effects. Two resistors interact via capacitive coupling or via thermal coupling. The couplings can be reduced by scaledown or by shielding, if some distance between the two resistors is allowed. Two resistors can be made practically noninteracting on a single integrated circuit chip.

The capacitance of a capacitor depends on the dielectric constant ϵ of the material in the gap, the gap l, and the electrode area S shown in Figure 1.2(b):

$$C = \epsilon S / l \qquad (1.7)$$

The electric field is practically confined within the narrow gap area, and the maximum is the breakdown field: either the field where tunneling of electrons from one plate to the other conducts current, or the avalanche breakdown field of the dielectric. The interaction of two nearby capacitors is primarily due to the *fringe* field, the field that leaks out from the edge of the capacitor. It is a dipolar field, and its characteristic length, or range of influence, is of the order of the gap, the *smallest* dimension specifying the capacitor. Because of this small characteristic length, any two capacitors are reasonably well isolated, except possibly for use as the precision capacitors of an analog circuit.

Figure 1.2(c) shows an inductor. The magnetic field lines generated by the current are closed loops surrounding the conductor, as schematically shown by the dotted lines in Figure 1.2(c). The inductance is estimated by

$$L = f\mu a \cdot N^2 \qquad (1.8)$$

where μ is the magnetic permeability of the medium (except for ferromagnetic materials, $\mu \approx \mu_0 = 4\pi \times 10^{-7}\,\text{H/m}$) and f is a numerical factor on the order of unity. If the conductor has N turns, L is proportional to N^2. A large inductor requires a large dimension a, and the magnetic field exists in a volume several times that of the $a \times a \times a$ cube surrounding the coil. The magnetic field decreases as the inverse square of the distance in the limit at infinity, and the characteristic length of the dependence is the radius a of the current loop, the *largest* dimension specifying the inductor. Within the characteristic length the magnetic field is inversely proportional to the distance, rather than to the square of the distance, because only the nearest current is effective. If another conductor loop exists within this region, it generates significant electromotive force by electromagnetic induction, and the electromagnetic interaction reacts on the inductance of its source current loop. For significant inductance values the size of the loop must be large. An inductance always has a large area around it where a magnetic field exists. Isolated inductance is therefore unrealistic, except when a shield is provided or a ferromagnetic core is used.

A lumped parameter circuit is represented by a well-defined circuit diagram. A component shown on the circuit diagram can be located in the hardware. How the real circuit is represented by a lumped equivalent circuit diagram depends on the simplification used. Traditionally, integrated circuit interconnects on semiconductor chips were represented by quite simple lumped parameter circuit diagrams. This representation becomes inaccurate as the speed of the circuit increases.

Electronic circuits work from zero to a certain maximum frequency. The operational frequency range sets significant circuit structural constraints. Inductance can be used together with resistance for bandwidth optimization (peaking), but pure inductance alone cannot be used as a load component, except for special cases. Capacitance can be a parasitic component to a resistive load, and the combined load determines the maximum frequency of operation of the circuit. Then the series impedance of interconnects is either resistive or inductive but not capacitive (capacitive series impedance appears in AC-coupled amplifiers, but we do not consider it an interconnect). Inductive impedance is proportional to frequency, and is insignificant at low frequencies. At low frequencies the frequency-independent series resistance dominates. Inclusion of interconnect series resistance splits a single circuit node into many circuit nodes, thereby increasing the complexity of the circuit representation. Figure 1.3(a) to (c) show successively less detailed representations of a resistive interconnect. How a distributed *RC* transmission line should be modeled by cascaded

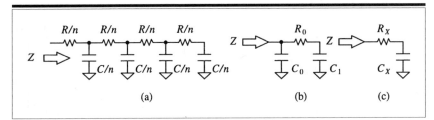

Figure 1.3 Discretization of Continuous *RC* Line

RC stages of a ladder network depends on how accurately the voltage profile on the distributed line must be modeled in the analysis. If high accuracy is required, the transmission line is split into many sections, and each joint between adjacent sections becomes an independent circuit node. If the transmission line is separated into more sections, the resistance and capacitance of each section become less, and accordingly the time step of integration of the numerical analysis becomes less. This creates a significant increase in the computation time.

Let a distributed *RC* line have total series resistance R and total parallel capacitance C. If the line is discretized into n equal sections, each section has series resistance R/n and parallel capacitance C/n as shown in Figure 1.3(a). In the simplest case the entire length is one section, and in the accurate distributed representation the length is divided into an infinite number of sections. Let us consider the driving-point impedance of the line. The driving-point impedance of the n-section discretized *RC* line of Figure 1.3(a), Z_n, is written as

$$Z_n = R \left[\frac{1}{n} + \frac{R_n(\omega/\omega_0) - jI_n(\omega/\omega_0)}{D_n(\omega/\omega_0)} \right] \qquad (1.9)$$

where $\omega_0 = 1/RC$. We have

$$D_1(\Omega) = \Omega^2 \qquad R_1(\Omega) = 0 \qquad I_1(\Omega) = \Omega$$

$$D_2(\Omega) = \Omega^4 + 2^6\Omega^2 \qquad R_2(\Omega) = 2^3\Omega^2 \qquad I_2(\Omega) = 2\Omega^3 + 2^6\Omega$$

$$D_3(\Omega) = \Omega^6 + 810\Omega^4 + 3^{10}\Omega^2$$

$$R_3(\Omega) = 3^3\Omega^4 + 10935\Omega^2$$

$$I_3(\Omega) = 3\Omega^5 + 1944\Omega^3 + 3^{10}\Omega \qquad (1.10)$$

$$D_4(\Omega) = \Omega^8 + 4096\Omega^6 + 3407872\Omega^4 + 4^{14}\Omega^2$$

$$R_4(\Omega) = 4^3\Omega^6 + 180224\Omega^4 + 58720256\Omega^2$$

$$I_4(\Omega) = 4\Omega^7 + 14336\Omega^5 + 8912896\Omega^3 + 4^{14}\Omega$$

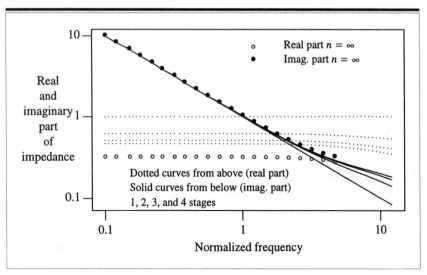

Figure 1.4 Impedance of Uniform *RC* Line Approximated by Segments

where $\Omega = \omega/\omega_0$ is the normalized angular frequency. In the limits of infinite and zero frequencies,

$$Z_n(\Omega \to 0) \to -\frac{jR}{\Omega} \qquad Z_n(\Omega \to \infty) \to (R/n) - (njR/\Omega) \quad (1.11)$$

Figure 1.4 shows the plots of the real and the imaginary parts of $Z_n(\Omega)$ versus normalized angular frequency Ω. The real part of the impedance is practically independent of the frequency up to $\Omega \approx 3$. The frequency-independent real part can be determined as follows. In the limit of zero frequency, the Π and the L equivalent circuits of Figure 1.3(b) and (c), respectively, should give the same impedance. Then the parameters R_0, C_0, C_1, R_X, and C_X are related by

$$R_X = [C_1/(C_0 + C_1)]^2 R_0 \qquad C_X = C_0 + C_1 \qquad (1.12)$$

Using these formulas as recursive relationships, the *RC* chain circuit of Figure 1.3(a) is reduced from the right end. After reduction we have a simple, one-stage *RC* chain, from which we obtain

$$\mathrm{Re}[Z_1(0)] = R$$

$$\mathrm{Re}[Z_2(0)] = (5/8)R$$

$$\mathrm{Re}[Z_3(0)] = (14/27)R \qquad (1.13)$$

and

$$\mathrm{Re}[Z_4(0)] = (15/32)R$$

In general, we obtain a formula

$$\text{Re}[Z_n(0)] = [(1/3) + (1/2n) + (1/6n^2)]R \qquad (1.14)$$

In the limit as $n \to \infty$, Re $[Z_n(0)] \to R/3$.

At high frequencies the real part decreases and approaches R/n in the limit of infinite frequency. Then only the resistance of the first link is effective, and that part becomes frequency-independent as the high-frequency limit is approached. The imaginary part of the impedance decreases inversely as Ω up to about $\Omega \approx 1$. Up to this frequency the RC line looks as if it were a frequency-independent capacitance. At the higher frequencies the effective capacitance decreases because the capacitances away from the driving point become ineffective due to the series resistance [7]. We have

$$\text{Im}(Z_n) \approx -jR/\Omega \quad (\Omega \to 0) \tag{1.15}$$

$$\text{Im}(Z_n) \approx -njR/\Omega \quad (\Omega \to \infty)$$

The impedance in the limit as $n \to \infty$ (continuous line) can be determined as follows. We write $Z_n = Rf_n(\omega/\omega_0)$ and seek $f_n(x)$ in the limit as $n \to \infty$. Suppose that the RC line was first split into n and then $n + 1$ sections, where n is a large number. Both f_n and f_{n+1} give the same impedance in the limit as $n \to \infty$. Then

$$Rf_n(\omega/\omega_0) = Rf_{n+1}(\omega/\omega_0) \qquad (n \to \infty) \tag{1.16}$$

where the right-hand side can be rewritten as

$$Rf_{n+1}(\omega/\omega_0) = \frac{R}{n+1}$$
$$+ \cfrac{1}{j\omega \cfrac{C}{n+1} + \cfrac{1}{R[n/(n+1)]f_n[(\omega/\omega_0)(n^2/(n+1)^2)]}} \qquad (1.17)$$

We expand $f_n(x)$ in the denominator into a power series and retain only the terms of the order of $1/n$. After some straightforward but tedious algebra we obtain a differential equation satisfied by the function $f(x) = \lim_{n \to \infty} f_n(x)$ as

$$2x \frac{df(x)}{dx} = 1 - f - jxf^2 \tag{1.18}$$

We seek the solution of this differential equation satisfying $Z_\infty \to 1/j\omega C$ in the limit as $\omega \to 0$. We have an infinite series solution

$$f(x) = (1/jx) + (1/3) - (j/45)x - (2/945)x^2 + \cdots \quad (1.19)$$

In the limit as $x \to \infty$ we seek for an asymptotic solution proportional to $x^{-1/2}$. We have, by substitution,

$$f(x) = (1 - j)/\sqrt{2x} \quad (1.20)$$

Z_∞ is written as

$$Z_\infty = Rf(\omega/\omega_0) \quad (1.21)$$

The real and imaginary parts of Z_∞ are plotted in Figure 1.4 by the open and closed circles, respectively.

From the analysis we conclude as follows. The imaginary part begins to deviate from the simple inverse dependence at

$$\omega = \omega_C = \sqrt{45}\omega_0 \quad (1.22)$$

This is about the frequency where the real part begins to decrease (Figure 1.4). In the limit as $\omega \to \infty$

$$|\text{Re } Z_\infty| \approx |\text{Im } Z_\infty| \approx 1/\sqrt{\omega/\omega_0} \quad (1.23)$$

If the real part at low frequency must have accuracy α, then n must satisfy

$$(3/2n) + (1/2n^2) < \alpha \quad (1.24)$$

1.5 Computational Methods of Analysis

We use closed-form analyses most of the time in this book, to understand and to interpret the problem, as well as to get the answer. Computers are used to get the values of the closed-form formula, especially to compute the Fourier transforms. These are mathematically well-defined operations. After setting our direction this way, we ask a question: Couldn't we use computers for the purpose of thinking, rather than just to get results? I think that there are ways to use a computer directly as a help in understanding a circuit problem. The following example is associated with the discretization issue of the last section.

If a circuit simulation program has a *run-time* problem (as distinct from the problems originating from the logic flow and the program control), does it mean that the program has a bug? If that has happened, should we perhaps consider that a new and not yet completely understood physical mechanism exists and the program's inefficiency suggests this? This is related to the fundamental philosophical issue: Does nature compute before revealing physical phenomena? I think the answer is affirmative.

Let us consider the following example: If the circuit shown in Figure 1.3(a) is simulated, the increment of time, Δt, must be chosen significantly less than the time constant of each link:

$$T_{RC} = (R/n) \cdot (C/n) = RC/n^2 \gg \Delta t \qquad (1.25)$$

This is because within time T_{RC} the voltage developed across one RC link decreases by a large factor (about $e = 2.718$). If $\Delta t \approx T_{RC}$ is used as the step of integration, the computed numbers become grossly inaccurate. If Δt is more than a certain limit, the inaccuracy can be detected by a casual glance at the numerical results. The step of time progression becomes quite small if a modest increase in the spatial discretization is made, so that spatial precision costs a lot in computing time. If the simulation program gives a rational result for properly chosen Δt but if it gives an obviously wrong result for other choices, that indicates the existence of a time constant shorter than the simulation time increment. This idea can be generalized: First we look for numerical analysis difficulties, and then we identify the underlying physical mechanisms. This is a good way to use a computer as a thinking aid.

To detect a hidden physical mechanism, the program must not have any *coverup* provisions. A simpler program is better for this purpose. If the time increment is held constant all through a simulation, the existence of the smaller time constant will be revealed. If a variable time increment is used in a single simulation run, then small time constants can be hidden. Since detecting the existence of a small time constant is a useful capability, we show what happens if the time increment is chosen differently in the simulation program of the RC chain circuit of Figure 1.5.

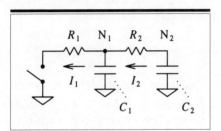

Figure 1.5 Two-Stage *RC* Chain

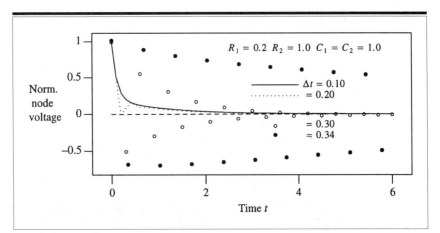

Figure 1.6(a) Node 1 Waveforms Computed at Various Time Increments

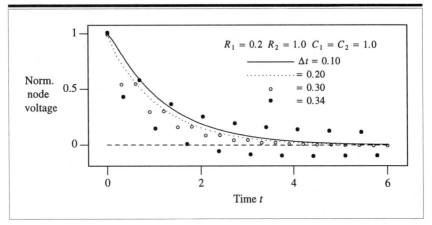

Figure 1.6(b) Node 2 Waveforms Computed at Various Time Increments

In Figures 1.6(a) and (b) node N_1 and N_2 waveforms computed for several values of Δt are shown. The values of R_1, R_2, C_1, and C_2 are given in the figures. The two-stage RC chain is characterized by the two time constants T_1 and T_2 given by [1]

$$T_1 = R_1(C_1 + C_2) \qquad T_2 = R_2 C_1 C_2 / (C_1 + C_2) \qquad (1.26)$$

where T_1 is the time constant of the *joint* discharge of C_1 and C_2 through R_1, and T_2 is the time constant of equalization of node N_1 and node N_2 voltages through R_2. In the first example $T_1 < T_2$ because $R_1 \ll R_2$. The capacitances C_1 and C_2 are charged to unit potential, and the discharge is

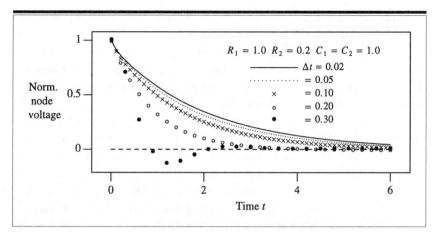

Figure 1.7(a) Node 1 Waveforms Computed at Various Incremental Times

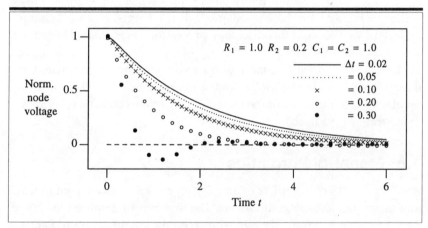

Figure 1.7(b) Node 2 Waveforms Computed at Various Incremental Times

begun at $t = 0$ by closing a switch. If time increment Δt is small ($\Delta t \leq 0.1$), the computed node waveform is reasonable (solid curves). If Δt is increased, the N_1 voltage decreases during the period from 0 to Δt by an excessive amount, because it is computed using the current I_1 at time 0. Consequently the N_1 voltage becomes negative. This is of course an unrealistic result, but the computation proceeds nonetheless. The negative N_1 voltage at Δt creates negative I_1, which countercharges the node N_1, and the voltage at the next time point $2\,\Delta t$ becomes positive. The node N_1 is then driven back to a negative voltage at $3\,\Delta t$ by the same mechanism. The N_1 voltage swings back and forth around zero at every time step. By reducing Δt, the N_1 voltage can be made to lead to a more accurate computation of I_1. The value of I_1 determines the rate of reduction of the node voltage accurately, and overshoot of the N_1 voltage does not take place. If Δt is less than a certain limit, the node voltage waveform is practically independent of Δt. This stability indicates that numerical simulation gives meaningful results. This may be rephrased as follows: A physical conclusion is invariant to the modification of the mathematics, as long as the modification is mathematically valid. In Figure 1.6(b), if the N_1 voltage overshoots, the N_2 voltage swings as well, but the voltage excursion is less, since $R_2 \gg R_1$. A large $R_2 C_2$ time constant smooths out the voltage at N_2, but the result is still unreasonable. If $T_1 < T_2$, increase in the time step results in a *high-frequency* oscillation of the numerical result. The N_1 voltage oscillates at a high frequency, with period 2 Δt, which is obviously unrealistic in that the period would vary with Δt.

   If $T_2 \ll T_1$, oscillation occurs, but at a lower frequency. In this case the N_2 voltage changes only if the two node voltages are different. If the N_2 voltage is higher than the N_1 voltage, C_2 is discharged and C_1 is charged by I_2, and vice versa. If this discharge process is overestimated because of a large Δt, that affects the joint discharge of the two nodes. As Δt increases, the effects of I_1 on the discharge of the two nodes are overestimated and the node voltages fall below ground, an unrealistic result. The period of oscillation is of the order of the joint discharge time constant T_1, as shown in Figures 1.7(a) and (b).

1.6 Essential Parasitics

An electronic circuit consists of amplifying devices, passive components, and unwanted *parasitic* components. The first two are required to synthesize the circuit function. They are shown on the circuit diagram. Parasitic components usually are not shown. This has been the practice from the time of discrete component electronics. In integrated circuits, the passive components and the parasitics are often indistinguishable: They have about the same magnitude, and have the same effects. A circuit diagram not showing the parasitics is unrealistic. Parasitics of discrete component circuits are always passive. In integrated circuits, however, parasitic amplifying devices may exist. An example is the mechanism called latchup in CMOS ICs, caused by parasitic BJTs.

   Parasitics in integrated circuits have different effects on analog versus digital circuits. In analog circuits, parasitics must be reduced so that circuit precision is not compromised. Analog circuits are designed not to be sensitive to the absolute values of the precision passive components: The precision depends on their ratios only. To achieve the ratio matching, the size of the precision components must be sufficiently large so that the size uncertainties are reduced to insignificance. In precision analog circuits, the desired functions are created by the amplifying devices, the passive component value ratios and by the architecture.

In digital circuits, the designer's intention is not to add any parasitics at all. The parasitics of digital circuits are *inevitable* parasitics of the active devices and of the interconnects, which cannot be reduced by ordinary techniques. This never means that the parasitics have no role in digital circuits. Digital circuits must have high reliability. High reliability in a noisy environment is achieved practically by accumulating the signal effects over a period of time. Parasitic capacitance is used to accumulate signal. More accumulation of the signal creates more delay, but the circuit becomes less sensitive to noise. To minimize delay, we use the parasitic capacitance for accumulation.

Active devices are described by a controlled current generator or a controlled conductance, which is inevitably associated with a parasitic capacitance. If the controlled current generator has transconductance g_m (this is the change of the device current per unit input voltage change) and the parasitic input capacitance of the device looking into the input terminal (or gate or base) is C_1, the ratio of the two, T_D, has the dimensions of time and characterizes the switching delay time of the device in a circuit:

$$T_D = C_1/g_m \qquad (1.27)$$

This is the *intrinsic* delay time of the signal from the input terminal of the device to its output terminal, which is loaded by the same device accepting the processed signal. It is not realistic to characterize a device delay in an unloaded condition. Basic information-theoretical issues are involved: Any gate that is externally observed is loaded, and the observation *cost* is the capacitance C_1. For semiconductor triodes T_D is independent of the device scaling. A semiconductor triode has an active region (like the channel of an FET, or the base of a BJT) through which the device current flows. If more current carriers are generated there, more current flows. The current carriers are created in the active region by the control electrode or gate, which attracts the mobile carriers from the source or from the emitter by the electrostatic attraction. The electrostatic attraction of the control gate for the current carriers in the active region is governed by the same physics as a capacitor: More carriers for same voltage means more capacitance. Therefore the device current and the input capacitance are proportional to each other. On paralleling two devices the *intrinsic* delay time of the device does not change. T_D is determined primarily by the processing technology of the device.

In the traditional integrated circuit theory, only resistances and capacitances are involved. This is essentially a low-frequency theory. In it, active devices are modeled using a switched resistor or a controlled current generator. At the low frequencies where the metal interconnects can be approximated by their capacitance to ground only, the switching of a digital circuit is a process of charging and discharging the capacitance (device and wiring) through the resistor or the current generator representing the device. At moderately high frequencies, the series resistances of the interconnects add to the circuit delay. As the frequency increases still further, the effects of the inductance becomes significant, as we observed in Section 1.2.

At the highest frequencies reachable by the present integrated circuit technology, the effects of inductance of the metal wire become comparable to the effects of the series resistance. If high-temperature superconductor wires are used in the future, the effects of inductance will dominate. Any of C, R, or L can be the dominant parasitics to determine the operation limit. Then it is of interest to ask the following question: Among the parasitic components of a circuit, which of resistance, capacitance, and inductance is more (or less) fundamental in the high-frequency circuits?

The **first** hint is that resistance can be *synthesized* as the characteristic impedance of a transmission line, using only inductance and capacitance. Figure 1.8 shows the equivalent circuit of a lossless inductance-capacitance (LC) transmission line. The inductance L and capacitance C are per unit length of the LC line. The velocity of propagation of the electromagnetic wave along the line, c, is given by [5]

$$c = 1/\sqrt{LC} \qquad (1.28)$$

Suppose that the input of the LC line is pulled up from voltage zero to V_0 within a negligibly short time. The step-function wavefront propagates a

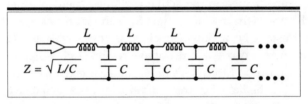

Figure 1.8 Characteristic Impedance of Transmission Line

distance ct in time t, and the capacitance $C \cdot (ct) = \sqrt{C/L}\,t$ is charged to voltage V_0. This derivation assumes that the transmission line is infinitely long. The total charge supplied by the voltage source is $V_0\sqrt{C/L}\,t$. The current at the input is then $V_0/\sqrt{L/C}$. The impedance looking into the transmission line is the ratio of voltage to current, and it is a pure resistance having value

$$Z_0 = \sqrt{L/C} \tag{1.29}$$

The observation, that resistance can be synthesized from inductance and capacitance, seems to indicate that resistance is less fundamental than inductance and capacitance. Although this argument sounds convincing, there is one point that requires attention: The line must be infinitely long. If the line is short, the reflected wave returns to the driving point some time later. After that time the line is no longer equivalent to a resistance. What a finite-length transmission line does is to carry the electrical energy away from the driving point and set it aside for some time, so that the energy does not get involved in the circuit operations during that time.

Let us look at this problem from a **second** angle. Among the three passive components (L, C, and R) of linear circuit theory, resistance is unique in that it dissipates electrical energy unconditionally. Resistors dissipate electrical energy by converting it to heat. Since heat cannot be converted back to electrical energy by the resistor itself, the energy is lost. This is a clean way of removing the energy. If electrical energy is converted to some other form of energy that can still be converted back to electrical energy, the component may appear like L, C, or a finite-length transmission line. An example is a DC motor, whose terminal characteristics is similar to those of a resistor when the mechanically loaded rotor begins to rotate and draw energy from the source, but which acts like a capacitor in the unloaded and rotating state. These observations show that electrical energy can be either converted to another form, or carried away to a distant dumping site. "Infinity" functions as an energy dumping site. Physically this means that an infinite space is required. Then is the characteristic impedance of a lossless transmission line really a resistance? If we keep throwing energy away to the infinity, that may affect the system's history, by modifying the boundary conditions to the physical problem.

These two distinct mechanisms of expending energy in a circuit, by a resistor and by a transmission line, have an interesting consequence when one generalizes circuit theory. A resistor shown in a circuit diagram is completely specified by its resistance: In this sense resistors are integral parts of circuit theory. The free space surrounding the circuit works as a transmission line. The characteristics of the free space as a transmission line are never specified by the circuit diagram. Its characteristics are therefore unknown, since the locations and the shapes of all the reflecting and the absorbing bodies outside the circuit can never be specified. If the circuit theory is extended to include the free outside space, it becomes undecidable, because of the lack of information on the outside space. Conventional circuit theory includes the first kind of energy loss mechanism, but not the second kind, except for special cases.

The **third** reason why resistance is less fundamental is that its value depends on the material, and there are even resistanceless materials (superconductors) in nature. Any attempt to create negative capacitance or negative inductance to compensate for the existing capacitance or inductance ends up with active circuits or devices that consume energy [6]. While resistance can be substantially reduced, capacitance and inductance are, in a sense, inescapable.

The inductance and capacitance of an interconnect are determined primarily by its size and not by the material, and their values are not independent. This is the **fourth** reason why they are more fundamental. If the conductor is thinner, the capacitance is less and the inductance is more. If conductor gap is smaller, the capacitance is more and the inductance is less. Their product, however, remains the same. From Equation (1.28) the product determines the wave velocity. In many multiconductor structures, the wave velocity is determined by the medium material alone, and not by the details of the conductors' structure. Figure 1.9 shows the two conductors A and B that make a transmission line; the cross section of the conductors is arbitrary. Suppose that an electromagnetic wave propagates perpendicular to the page. The electric field $\vec{E}$ and magnetic field $\vec{H}$ at a point P a long distance away from the conductor pair are shown in the figure. They are perpendicular to each other, and the wave field is closely similar to the ideal transverse electromagnetic (TEM) wave, which is also the mode of propagation of an ordinary plane light wave. This conclusion is justified as long as the wave is supported by a pair of conductors. The velocity of the wave is given by $v = 1/\sqrt{\epsilon\mu}$, where ϵ and μ are the dielec-

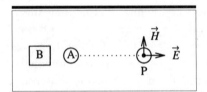

Figure 1.9 Wave Velocity of Transmission Line

tric constant and the magnetic permeability of the medium, respectively. Since the TEM part of the wave far away and the distorted wave close to the conductors must have the same velocity, we conclude that

$$LC = \epsilon\mu \tag{1.30}$$

and thus L and C are related to each other. Inductance and capacitance come in pairs, and their product cannot be reduced below a limit set by the signal propagation delay. This is the **fifth** of five reasons why inductance and capacitance are more fundamental than resistance.

1.7 Estimation of Inductance Effects

The effects of wiring inductance show up at high switching speeds. It is important to know at what speed they are observable, and how significant they are. The inductance delay in IC interconnects shows up as a lossless (LC), or as a lossy (LCR) transmission-line delay. This delay is as much an inductance effect as a resistance effect. To understand the transmission-line delay, the wire must be modeled by an equivalent circuit including inductance, shown in Figure 1.10(a) and (b). We compare the LC transmission-line delay with device charge-discharge delay, and study how much the series resistance increases the former.

In the equivalent circuit of Fig. 1.10(a), the active device is represented by a resistor R_D, the load capacitance by C_L, and the interconnects by L and C. The transmission-line section is shown by the dotted box. At low switching speed the effect of L is negligible, and the delay is estimated by the time constant $T_D = R_D(C_L + C)$: If C_L and C are charged to unit voltage, it takes time T_D to reduce the capacitance voltage by a factor of $1/e$. The values of L and C determine the *transmission delay* of the interconnect, $T_{TR} = \sqrt{LC}$. Since L and C are both proportional to the length of the transmission line, T_{TR} is proportional to the line length. If $T_{TR} > T_D$, the LC delay effects determine the signal delay. Let the velocity of the electro-

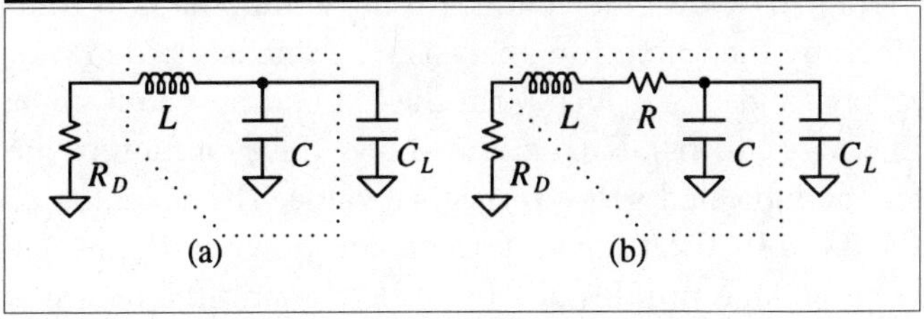

Figure 1.10 Effects of Intrinsic and Wiring Delay

magnetic wave be c; then $c = \Lambda / T_{TR}$ or $T_{TR} = \Lambda / c$, where Λ is the length of the transmission line. The propagation velocity of the wave in the TEM mode in a uniform nonmagnetic medium is given by $c = c_0 / \sqrt{\epsilon / \epsilon_0}$, where ϵ and ϵ_0 are the dielectric constants of the material and of vacuum, respectively, and $c_0 = 3 \times 10^{10}$ cm/sec is the velocity of light in vacuum. In silicon integrated circuits, silicon dioxide, whose relative dielectric constant $\epsilon / \epsilon_0 \approx 4$, is the most commonly used dielectric. We have $c \approx 1.5 \times 10^{10}$ cm/sec in the dielectric. For many practical reasons (complexity of circuits, packaging, chip production yield, and thermal considerations) most silicon chips are designed to be less than 1 cm square [7]. In such a chip, 1.5-cm-long critical-signal-path wires exist [7]. The delay time of a 1.5-cm-long resistanceless wire is 100 ps. This is the lower bound of the critical-signal-path wire delay of a VLSI circuit. This delay time is quite significant in comparison with the switching delay time of fast bipolar transistors, which ranges from 10 to 20 ps at present. The interconnect signal delay due to the electromagnetic wave propagation time is expected to be the predominant circuit delay in future high-speed ICs.

The inductance-capacitance delay is the ultimate limit on the speed of an integrated circuit, but is not generally recognized as a serious design issue at present. In the present silicon IC (MOS, or bipolar) design style, the delay due to the series resistance of the long wires in the densely packed fine interconnects dominates. *RC* delay and *LC* delay do not simply add, and that makes the *LC* delay less obvious than the *RC* delay. But their coexistence has a significant effect on the transmitted waveform as well, as shown in the following analysis. We make an estimate using the equivalent circuit including the series resistance and the inductance, shown in Figure 1.10(b). Since the resistance-capacitance delay is *RC*, in the present technology we have the condition

$$RC > \sqrt{LC} \quad \text{or} \quad R > \sqrt{L/C} \tag{1.31}$$

that is, the series resistance over the entire wire length is higher than the characteristic impedance of the resistanceless *LC* transmission line. Figure 1.11(a)–(d) shows the voltage profile within the *LCR* transmission line obtained by the numerical analysis. The line is divided into $N = 50$ equal sections, and it is driven from the left end by a unit voltage ramp (ramp time $t_0 = 0.01$ ns). Each graph shows the evolution of the voltage profile for the indicated series resistance value. The potential profiles at times 0.05, 0.1, 0.15, 0.2, 0.25, and 0.3 ns are shown in Figure 1.11(a) and (b), and the voltage profiles at 0.05, 0.1, 0.15, and 0.2 ns are shown in Figure 1.11(c) and (d). Figure 1.11(a) shows that the line is effectively an

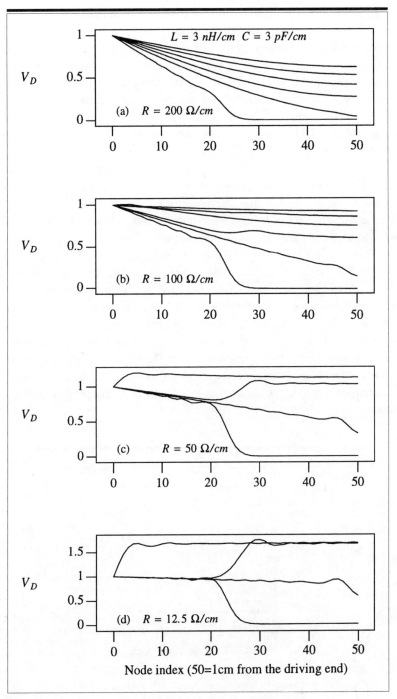

Figure 1.11 Potential Profile Within an *LCR* Transmission Line

RC transmission line. This is approximately the case of the minimum-width wire in the 1-μm CMOS technology.

Until the initial voltage step arrives at the right end of the line, the right-end voltage remains at 0. There is no *early warning* signal observed as an artifact in the *RC* transmission-line simulation. The *LCR* circuit model conforms with special relativity, and no signal propagates faster than the velocity of the electromagnetic wave. The initial voltage step in the first profiles [Figure 1.11(a)] decays before arriving at the right end. We define the transmission delay time T_{TR} of the *LCR* line by the interval from the input transition to the time when the voltage of the right-end node reaches $\frac{1}{2}$ for the first time. Figure 1.12 shows T_{TR} versus the series resistance per section, R/N. We use the characteristic impedance $Z_0 = \sqrt{L/C} = 31 \ \Omega$ as the reference. Since the transmission line is divided into 50 sections, the total resistance is 50 times the value plotted on the horizontal axis [4 for $R = 200 \ \Omega$ in Figure 1.11(a)]. The delay time is close to 0.1 ns until about $R/N = 1.5$, or $R = 75 \ \Omega$. This is the *LC* delay time, estimated at $\sqrt{3 \ \text{pF} \times 3 \ \text{nH}} = \sqrt{0.9 \times 10^{-20}} \approx 0.1$ ns. If R is increased beyond the limit, the delay time increases rapidly with increasing R. This is the transition from the *LC* to the *RC* transmission-line regime. In the *LC* transmission line the delay is proportional to the length. In the *RC* transmission line it is proportional to the length squared. To improve the IC delay characteristic, the *RC* interconnect delay must be reduced. Improvement of the active-device delay has become secondary in high-speed digital VLSI chips.

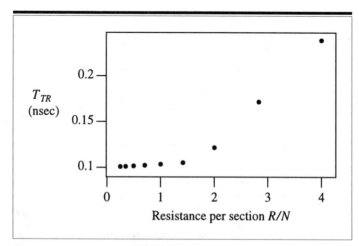

Figure 1.12 Effects of Series Resistance on Transmission-Line Delay

Scaledown of the technology feature size decreases the device delay but not the interconnect delay, and ultimately the scaledown will not improve the VLSI circuit speed any more. The series resistance of the wires determines the delay. It is necessary to reduce the series resistance, while maintaining the parallel capacitance and the series inductance unchanged. The simplest way is to increase the cross-sectional area of the metal wire. The series resistance decreases inversely as the cross-sectional area, but the parallel capacitance can be maintained if the distances to the nearby conductors are increased proportionally. The series inductance can be maintained as well, by scaling the conductor size and the interconductor distance simultaneously. Although this method is the simplest, it leads to many problems in the wafer processing. For example, if the thickness of the insulating dielectric is increased, the aspect ratio of the contact hole decreases. It becomes very hard to open small contact holes and fill them up with metal. Thus there is a limit to the applicability of this method.

Let us consider the design problem of transmitting a critical signal through a VLSI interconnect 1.5 cm long. Suppose that the transmission delay must be reduced by any method available. This requirement is often practicable, since the performance limit is usually set by several critical signals carried by long wires, and not by the many local signals on short wires. The wire must be a metal wire to reduce the dominant resistance-capacitance delay, and it must be on an upper level, such as the second or third metal level, so that the capacitance to the substrate will be small. It is better not to have any features underneath the wire. The capacitance of a wide wire is given by

$$C = \epsilon_{\text{ox}}(W\Lambda)/T_{\text{ox}} = 0.345(W/T_{\text{ox}})\Lambda = 5.18 \text{ pF} \qquad (1.32)$$

where W and Λ are the width and length of the wire, and T_{ox} is the thickness of the silicon dioxide insulator. We use $W = 15 \ \mu\text{m}$, $T_{\text{ox}} = 1.5 \ \mu\text{m}$, and $\Lambda = 1.5$ cm. If the metal is 1-μm-thick aluminum, the sheet resistance is $0.04 \ \Omega/\text{square}$. The series resistance is $(1.5 \text{ cm}) \times (0.04 \ \Omega/\text{square})/ (15 \ \mu\text{m}) = 40 \ \Omega$. The RC delay of the unloaded wire is

$$T_{\text{D}}(RC) = (40/2) \times 5.18 = 102 \text{ ps} \qquad (1.33)$$

If the wire is perfectly conductive, the LC delay is given by $T_{\text{D}}(LC) = \Lambda/c = 100$ ps. The LC delay is independent of the wire width. Then we have

$$T_{\text{D}}(RC) \approx T_{\text{D}}(LC) \qquad (1.34)$$

If $T_D(RC) \geq T_D(LC)$, the inductance is often neglected in the interconnect modeling. Then there is an unrealistic *early warning* signal as a simulation artifact, which propagates instantly from the input to the output. The voltage at the output begins to increase at the same time as the input transition. The early warning signal never exists in the real hardware, or in the more complete *LCR* model. The output voltage stays at 0 until a small step function, which is an attenuated version of the input drive waveform, arrives at time $T_D(LC)$ later. It is necessary to recharacterize the delays of CMOS dynamic gates to the waveforms delivered from an *LCR* transmission line, which are significantly different from the waveform from the *RC* transmission line. Inclusion of inductance creates as many waveform issues as delay issues. $T_D(LC)$ is the ultimate lower limit of the signal delay.

In fact 15 μm is unusually wide for a wire in a submicron-feature-size CMOS chip; such a width is used only for special purposes, such as transmitting clock signals. If the width is reduced to 5 μm, the series resistance triples, but the capacitance is not reduced to one-third, since fringing capacitance effects show up. Assuming realistically that the capacitance is reduced to two-thirds, $T_D(RC) = 204$ ps. Then the inductance delay is about half the resistance delay. In an ordinary 1.5-μm-wide wire, the series resistance is 400 Ω, and the capacitance is about 2 pF. Then $T_D(RC) = 800$ ps, and the inductance delay is only about 10% of that. For an accurate delay analysis of a long wire, inclusion of inductance is still desirable, but in many cases it is not critical. For automatically designed or interconnected ICs the conventional resistance-capacitance delay model can be used, at least for some time. The recent trend toward using the scaled-down CMOS technology at reduced speed at a large scale of integration to accomplish certain business objectives, such as field-programmable gate arrays, will prolong the usefulness of the conventional delay model. With design-intensive high-performance ICs it is different. The operational regime of integrated circuits where inductance effects are significant has already arrived at the high-speed limit. This is the area of application of the theory of this book.

The estimate of inductance delay in this section gives a lower bound. Since the silicon substrate is never a perfect conductor, the magnetic field penetrates into it. If the substrate has low conductivity, the entire thickness of the chip may be permeated by the magnetic field. Thus the space occupied by the magnetic field expands, and the effective inductance becomes higher than the simple estimate assuming that the field exists only between the wire and the substrate. The capacitance remains the same, since the dielectric relaxation time of the substrate is less than 1 ps, and this means that the electric field does not penetrate into the substrate. Then the electromagnetic wave propagation velocity is reduced by the ratio of the square roots of the inductances [8]. The effect may increase the inductance delay by a factor of 2 or more. This peculiar substrate effect, called a slow-wave mode, may not exist if there is a bulky conductor structure such as a power bus beside or below the signal wire. Such a structure, however, is often avoided in present-day design style to reduce capacitance.

There are several techniques to reduce the series resistance of the interconnects. The material used for interconnects on CMOS VLSI chips is aluminum. Silver and copper have higher conductivity, but using these materials has not been considered yet. If the dielectric constant of the insulator is reduced, the gap between the interconnect wires can be reduced while maintaining the same sidewall capacitance value, and therefore the cross section of the metal wire can be increased. Multilevel metal with larger gaps between the wires, insulated by low-dielectric-constant material, would be the best practical option. If the temperature of the integrated circuit is reduced, the resistance of the metal interconnect decreases faster than proportionally to the absolute temperature. Reducing the device temperature is advantageous in CMOS ICs, since the FET characteristics improve, and there are other advantages. Operation of silicon CMOS devices at 77 K (liquid nitrogen temperature) has been studied [9]. By that means, a 2-μm-wide and 1-μm-thick metal wire that is 1 cm long can be brought to the range where the series resistance is less important than the series inductance. Reducing the temperature of the electronic system, however, has met with serious resistance and skepticism. It is not an option for bipolar circuits, because the characteristics of bipolar transistors deteriorate quite significantly at 77 K.

A high-temperature ceramic superconductor would be the ultimate choice for the interconnect material, but critical temperatures are still not high enough to allow room-temperature operation. For liquid nitrogen temperature, that problem is already solved, and limited use of superconductive wire is expected [2]. A problem with high-temperature superconductors is that the maximum current density of the material is not very high. IC interconnects carry currents of the order of milliamperes and have cross-sectional areas of 10^{-8} cm^2. The resulting current density of 10^5 A/cm^2 is too high to use superconductors reliably at present. This problem, however, will be solved as the mechanism of high-temperature superconductivity becomes better understood.

1.8 Lumped Versus Distributed Reactive Components

In this section we consider the following problem: When can an IC interconnect be represented by a lumped inductance and when must it be represented by a more complex distributed inductance-capacitance circuit such as a transmission line? The impedance of the transmission-line model should be similar to that of the interconnect at low frequencies. A real interconnect is a nonuniform transmission line. We neglect, for simplicity, the nonuniformity. We also assume the line is lossless. We study the driving-point impedance of the transmission line shown in Figure 1.13(a) and (b). The driving-point impedance should be capacitive in the open-ended transmission line of Figure 1.13(a), and inductive if the end is short-circuited as in Figure 1.13(b). How do they behave on various time scales? We analyze this problem both in the frequency and in the time domain.

We choose a coordinate x along the transmission line as shown in Figure 1.13(a) and (b). The length of the line is Λ, and the parallel capacitance is C per unit length. As for the inductance, we may assign L_1 and L_2 per unit length for the pair of lines as shown in Figure 1.13(c), but only the sum $L = L_1 + L_2$ has physical meaning, as we discuss in the next section. The differential voltage $V(x, t)$ developed at location x at time t and the current $I(x, t)$ at the same location and time satisfy the following set of equations:

$$\frac{\partial V(x, t)}{\partial x} = -L\frac{\partial I(x, t)}{\partial t} \qquad \frac{\partial I(x, t)}{\partial x} = -C\frac{\partial V(x, t)}{\partial t} \qquad (1.35)$$

If $I(x,t)$ is eliminated between the two equations, we obtain the following wave equation:

$$\frac{\partial^2 V(x, t)}{\partial x^2} = LC\frac{\partial^2 V(x, t)}{\partial t^2} \qquad (1.36)$$

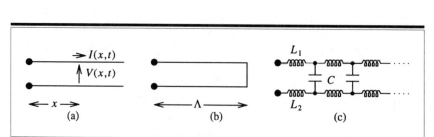

Figure 1.13 Transmission Line and Reactive Components

1.8.1 Frequency-Domain Analysis

If $V(x, t)$ and $I(x, t)$ depend on time as $\exp(j\omega t)$, we have

$$\frac{\partial^2 V}{\partial x^2} = -LC\omega^2 V \tag{1.37}$$

We consider the shorted-end line of Figure 1.13(b). By applying the boundary condition $V(\Lambda, \omega) = 0$ of Figure 1.13(b) we have

$$V(x,\omega) = A \exp(j\omega t) [\exp(-j\omega\sqrt{LC}\,x)$$
$$- \exp(-2j\omega\sqrt{LC}\Lambda + j\omega\sqrt{LC}\,x)] \tag{1.38}$$

$$I(x,\omega) = \sqrt{C/L}\,A \exp(j\omega t) [\exp(-j\omega\sqrt{LC}\,x)$$
$$+ \exp(-2j\omega\sqrt{LC}\Lambda + j\omega\sqrt{LC}\,x)]$$

The driving-point impedance defined at $x = 0$ is computed as

$$Z_0(\omega) = V(0, \omega)/I(0, \omega) = j\sqrt{L/C}\,\tan(\omega\sqrt{LC}\Lambda)$$

If ω is small,

$$Z_0(\omega) = j\omega(L\Lambda)[1 + (1/3)\omega^2 LC\Lambda^2] \qquad (\omega \to 0)$$

where $L\Lambda$ is the inductance of the entire closed loop made by the shorted-end transmission line. The *effective* value of the inductance presented by the loop increases with increasing frequency. At the higher frequencies the shorted-end transmission line goes into a parallel resonance at $\omega = \omega_1$, where $\omega_1 = (\pi/2)\omega_0$, and where $\omega_0 = 1/(\sqrt{LC}\Lambda)$.

1.8.2 Time-Domain Analysis

The wave equation has the general solution of the form

$$V(x,t) = F[x + (t/\sqrt{LC})] + G[x - (t/\sqrt{LC})] \tag{1.39}$$
$$I(x,t) = -\sqrt{C/L}\,[F[x + (t/\sqrt{LC})] - G[x - (t/\sqrt{LC})]]$$

Suppose that the left end of the transmission line is driven by a voltage source $V_{IN}(t)$, and $V_{IN}(t) = 0$ if $t < 0$. If $0 \le t \le t_1 = \sqrt{LC}\Lambda$, then $F = 0$, since there is no reflected wave. Then by requiring that $V(0, t) = V_{IN}(t)$ we get

$$V(x,t) = V_{IN}(t - \sqrt{LC}\,x) \tag{1.40}$$

At $t = t_1$ the incident wave arrives at the shorted end of the transmission line. After that time,

$$V(x,t) = V_{IN}(t - \sqrt{LC}\,x) + F[x + (t/\sqrt{LC}\,)] \qquad (1.41)$$

and we determine the function F. By requiring $V(\Lambda, t) = 0$ at the shorted end we have

$$0 = V_{IN}(t - \sqrt{LC}\,\Lambda) + F[\Lambda + (t/\sqrt{LC}\,)] \qquad (1.42)$$

On denoting the argument of F by $\Lambda + (t/\sqrt{LC}) = Y$, we find $F(Y) = -V_{IN}[\sqrt{LC}\,(Y - 2\Lambda)]$. Then we have

$$V(x,t) = V_{IN}(t - \sqrt{LC}\,x) - V_{IN}(t + \sqrt{LC}\,x - 2\sqrt{LC}\,\Lambda) \quad (1.43)$$

At time $t_2 = 2\sqrt{LC}\,\Lambda$ the reflected wave arrives at the driving point. After that the second forward wave is launched, so that

$$\begin{aligned} V(x,t) = {}& V_{IN}(t - \sqrt{LC}\,x) - V_{IN}(t + \sqrt{LC}\,x \\ & - 2\sqrt{LC}\,\Lambda) + G[x - (t/\sqrt{LC}\,)] \end{aligned} \qquad (1.44)$$

and the new G is determined by requiring that $V(0, t) = V_{IN}(t)$. We have

$$\begin{aligned} V(x,t) = {}& V_{IN}(t - \sqrt{LC}\,x) - V_{IN}(t + \sqrt{LC}\,x \\ & - 2\sqrt{LC}\,\Lambda) + V_{IN}(t - \sqrt{LC}\,x - 2\sqrt{LC}\,\Lambda) \end{aligned} \qquad (1.45)$$

This procedure is repeated. The mathematical expression of $V(x,t)$ acquires one extra term in every passage of time $\sqrt{LC}\,\Lambda$. The result is summarized as follows. Let $t_i = i \times \sqrt{LC}\,\Lambda$. Then

$$\begin{aligned} V(x, t) = {}& V_{IN}(t - \sqrt{LC}\,x) & (t < t_1) \\ & - V_{IN}(t + \sqrt{LC}\,x - 2\sqrt{LC}\,\Lambda) & (t_1 < t < t_2) \\ & + V_{IN}(t - \sqrt{LC}\,x - 2\sqrt{LC}\,\Lambda) & (t_2 < t < t_3) \\ & - V_{IN}(t + \sqrt{LC}\,x - 4\sqrt{LC}\,\Lambda) & (t_3 < t < t_4) \quad (1.46) \\ & + V_{IN}(t - \sqrt{LC}\,x - 4\sqrt{LC}\,\Lambda) & (t_4 < t < t_5) \\ & - V_{IN}(t + \sqrt{LC}\,x - 6\sqrt{LC}\,\Lambda) & (t_5 < t < t_6) \\ & + \cdots & \end{aligned}$$

Let us consider the interval $0 < t < t_1$. We have $V(x, t) = V_{IN}(t - \sqrt{LC}\,x)$. By integrating Equation (1.35) with respect to t we have

$$I = -(1/L) \int_0^t [\partial V(x,t)/\partial x]\, dt = \sqrt{C/L}\, V_{IN}(t - \sqrt{LC}\,x) \quad (1.47)$$

The driving-point impedance Z_0 is defined as the ratio of $V(0, t)$ and $I(0, t)$:

$$Z_0 = \frac{V(0, t)}{I(0, t)} = \sqrt{L/C} \tag{1.48}$$

the transmission line looks as if it were a pure resistance Z_0. In general, using the formula $-L(\partial I/\partial t) = (\partial V/\partial x)$, we have

$$
\begin{aligned}
I(x,t) = {} & \sqrt{C/L}\, V_{IN}(t - \sqrt{LC}\,x) & (t < t_1) \\
& + \sqrt{C/L}\, V_{IN}(t + \sqrt{LC}\,x - 2\sqrt{LC}\,\Lambda) & (t_1 < t < t_2) \\
& + \sqrt{C/L}\, V_{IN}(t - \sqrt{LC}\,x - 2\sqrt{LC}\,\Lambda) & (t_2 < t < t_3) \\
& + \sqrt{C/L}\, V_{IN}(t + \sqrt{LC}\,x - 4\sqrt{LC}\,\Lambda) & (t_3 < t < t_4) \quad (1.49) \\
& + \sqrt{C/L}\, V_{IN}(t - \sqrt{LC}\,x - 4\sqrt{LC}\,\Lambda) & (t_4 < t < t_5) \\
& + \sqrt{C/L}\, V_{IN}(t + \sqrt{LC}\,x - 6\sqrt{LC}\,\Lambda) & (t_5 < t < t_6) \\
& + \cdots
\end{aligned}
$$

At the driving point we set $x = 0$ and obtain

$$I(0,t) = \sqrt{C/L} \left[V_{IN}(t) + 2 \sum_{k=1}^m V_{IN}(t - 2k\sqrt{LC}) \right] \quad (1.50)$$

where m is the maximum integer that satisfies $t/(2\sqrt{LC}) > m$. This formula shows the nature of the inductance of a shorted-end transmission line quite well. If the variation of $V_{IN}(t)$ within the interval $2\sqrt{LC}$ is small, the formula may be rewritten as

$$
\begin{aligned}
I(0, t) = {} & \sqrt{C/L} \left[V_{IN}(t) + 2 \sum_{k=1}^m V_{IN}(t - 2k\sqrt{LC}) \right] \\
& \to (1/L) \int_0^t V_{IN}(t)\, dt
\end{aligned}
\quad (1.51)
$$

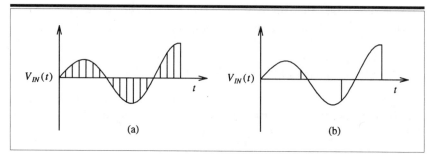

Figure 1.14 Approximation of Integral by a Discrete Sum

The last relationship is the usual formula for an inductance: the current in it is proportional to the time integral of the voltage applied to it.

For a slowly varying driving voltage, the response of the shorted-end line is similar to that of a pure inductance. The deviation from this inductance characteristic is represented mathematically by the difference between the integral of $V_{IN}(t)$ and the discrete sum of it. The difference becomes significant if the variation of $V_{IN}(t)$ within one integration step is significant. In Figure 1.14(a) the integration step is much smaller than that required to create large variation of $V_{IN}(t)$. The discrete sum represents the integral closely, and therefore the transmission line behaves like an inductance. In Figure 1.14(b) the integration step is much larger than that, and the discrete sum is a poor approximation of the integral. As a special case of this, if $V_{IN}(t)$ is a sinusoid and if the step of integration is taken at points that have the same phase, the integral diverges to either plus or minus infinity. This is the case of a series resonance. If the step of integration is taken at points that are alternately in and out of phase, then the discrete sum is zero. This is the case of a parallel resonance. In either case, the effects of the input voltage long before the present time determine the current at present. At the other extreme, immediately after a step-function voltage waveform occurs, the contribution from the past is insignificant. Then the transmission line works effectively as a resistance. The relationship between the sum and the integral clarifies many confusing points of modeling a distributed parameter circuit by a discrete component.

1.9 Voltage and Current in a Circuit

An electronic circuit is built by connecting devices and components with conducting wires. Currents flow through wires and voltages are developed across components and devices. The voltages and the currents carry information, and are called *signals*. The intensity of a signal must be measurable. Electronic measurements have a curious feature, the instrument that measures a circuit's voltage or current is also a circuit. If the measured values are used to exercise control, the signal user is a circuit as well. This hierarchy may extend to any level. If we stretch our imagination, a human observer who supervises the electronic system is yet another complex circuit. This endless hierarchy confuses the issue. A voltmeter and an ammeter are defined as a convenient termination of the hierarchy. The readings of the meters are processed by a human mind that is assumed to have unlimited capability to use or interpret the data.

To establish a circuit theory, it is necessary to examine a circuit's operations in great detail. To examine a circuit, we set up a model of it and carry out experiments, whose results are determined by theoretical deductions, using the basic principles of physics. This theoretical exercise is called a *gedanken*experiment. In this gedankenexperiment, we choose the physical principles we adhere to, and we construct an experimental procedure consistent with the principles and determine the experimental results. Other than that, we do not have to worry about the practical details of the experimental setup, as long as the basic principles are not violated. Our gedankenexperiments are most often to measure currents and voltages. Such gedankenexperiments are indispensable in constructing a circuit theory including inductance.

The current $\vec{J}$ in a resistive conductor is the sum of the conduction current $(\sigma \vec{E})S$ and the displacement current $(\epsilon \, \partial \vec{E}/\partial t)S$, where σ and ϵ are the conductivity and the dielectric constant of the resistive conductor, respectively, S is the cross-sectional area, and $\vec{E}$ is the electric field. In a resistive conductor the conduction and the displacement currents coexist, but their ratio depends on σ and ϵ. If the conductor is broken open at the location of a capacitor, all the current is carried there by the displacement current. Within a perfect conductor there is no electric field, and all the current is carried by the conduction current. Both the conduction and the displacement currents make a magnetic field around the current path.

The current $\vec{J}$ is determined by measuring the magnetic field created by it. In Figure 1.15(a) the magnetic field $\vec{H}$ is related to current $\vec{J}$ carried by the conductor as dictated by Ampere's law:

$$\int_C \vec{H} \, \vec{ds} = \int_S \vec{I} \cdot \vec{N} \, dS \quad \text{where} \quad \vec{I} = \sigma \vec{E} + \epsilon(\partial \vec{E}/\partial t) \quad (1.52)$$

is the total current density dS is the element of the surface enclosed by the closed path C, $\vec{N}$ is the unit normal to the surface, and $\vec{ds}$ is the line-element vector taken along the closed path C. The line element $\vec{ds}$ and $\vec{N}$

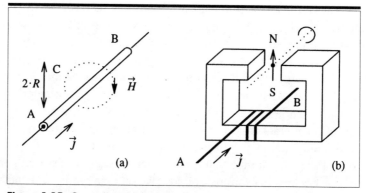

Figure 1.15 Operational Principle of Ammeter

follow the right-hand rule: The vector $d\vec{s}$ rotates in the direction to drive a conventional right-hand screw in the direction $\vec{N}$. If $\vec{J}$ is confined within the narrow conductor $\overline{AB}$ of Figure 1.15 and if C is a circle having radius R, we have

$$|\vec{J}| = 2\pi R |\vec{H}| \tag{1.53}$$

and therefore the current is determined by measuring the magnetic field. Here the magnetic field measurement is carried out not by the other circuit but by a human experimenter, so that the details of the measurement techniques are not required. Still it is informative to observe how the measurement is carried out. The section $\overline{AB}$ of the current path belongs to the ammeter. It is not obvious what impedance the ammeter presents to the rest of the circuit. In a modified, more realistic ammeter shown in Figure 1.15(b) a magnetic core is used to concentrate the magnetic field at the magnetic needle indicator $\overline{SN}$. It rotates around the dotted axis to indicate the current. The magnetic force is balanced by the mechanical spring, which restores the needle to the zero if current ceases to flow. From this setup it is obvious that an ammeter presents an inductive impedance to the rest of the circuit. In either setup, the size of the ammeter can be made as small as we wish. Miniaturization of the ammeter is accomplished by making $\overline{AB}$ short and R small in Figure 1.15(a). To connect the ammeter to the circuit, a narrow gap is cut to the current path and the meter is inserted there. That the ammeter can be made small and requires no connection wire means that current can be measured at any location: it can be measured *locally*.

The voltage of a node is always measured relative to another node, most often to ground. The two nodes are spatially separated, and an elec-

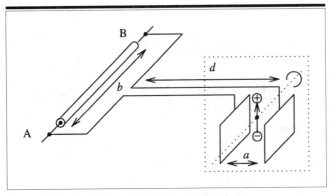

Figure 1.16 Operational Principle of Voltmeter

tric field exists in the space between the pair of nodes. The voltage difference between the two nodes A and B of Figure 1.16 is given by

$$V_{BA} = - \int_A^B \vec{E} \, d\vec{s} \qquad (1.54)$$

where $\vec{E}$ is the electric field and $d\vec{s}$ is the line element of an arbitrary path
connecting points A and B. Voltage is measured by connecting nodes A
and B to a voltmeter with a pair of conducting wires, as shown in
Figure 1.16. The voltmeter consists of a pair of electrodes a distance a
apart, where the electric field $|\vec{E}| = V_{BA}/a$ is created. We note that the pair
of wires is analogous to the magnetic core of the current measurement
setup of Figure 1.15(b), in that it creates a strong field in the gap. To measure the electric field, an electric dipole moment is placed as shown in
Figure 1.16 and the torque exerted on the moment is measured. The ideal
DC voltmeter is an electrometer as described above, consisting of a static
electrode and a rotatable indicator that moves by the Coulomb force. As is
observed from this example, an ideal voltmeter has a capacitive internal
impedance. The wire pair connects the voltmeter to the circuit, and the
horizontal section having length d contributes to the capacitance of the
voltmeter as well.

If the voltmeter is connected to a circuit, it is allowed to draw energy to
measure voltage in the short time within which the reading is recorded.
Thus the pair of host's nodes must be able to provide the energy in a short
time. The pair of nodes must have a parallel capacitance C_N that holds
the energy, as shown in Figure 1.17(a). The capacitance of the voltmeter,
C_M, must satisfy $C_N \gg C_M$. If $C_N = 0$, there is no way to satisfy the in-
equality. Therefore we must say that the voltage of a node that has no

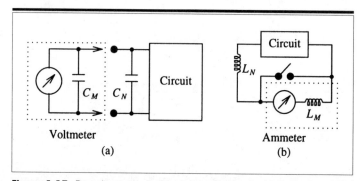

Figure 1.17   Requirements for Nodes and Loops for Voltage-Current Measurements

capacitance is, strictly speaking, unmeasurable. In the limit of high speed in a common-ground circuit, the voltage is measurable only if the node has capacitance to ground. The equivalent circuit must be drawn conforming to this requirement. In the current measurement a similar issue exists, as shown in Figure 1.17(b). The ammeter has internal inductance L_M. It draws energy from the host circuit to move the indicator. To keep the current undisturbed, series inductance is required. For the meter being able to draw energy from the circuit, the inductance in the circuit, L_N, must be able to force the same current through the meter. We must have $L_N \gg L_M$. This means that the ammeter must be much smaller than the host. The current in an inductanceless loop, or a loop much smaller than the ammeter, is not measurable.

The voltmeter enclosed in the dotted box of Figure 1.16 can be made as small as we wish. To connect the voltmeter to the circuit, however, two types of wire pairs are necessary. One pair brings the two node voltage signals to a small locality, which in Figure 1.16 has length b. The other is to connect the voltmeter, and it has length d. The second wire pair is often made of twisted wires that couple to each other strongly, thereby forming a transmission line. The second pair is unnecessary if the voltmeter is connected directly to the end of the first pair. The first pair is always necessary for a voltage measurement, since nodes A and B are some distance apart. The voltmeter connection must have nonzero length, and this is a crucial difference from the ammeter connection. Thus it may be said that current is a locally measurable variable, but voltage is not.

The currents of a circuit satisfy the fundamental laws of conservation of the loop current. Figure 1.18 shows that a capacitor is discharged through a device. In the highly conductive wire the current is carried by the conduction current. The device is less conductive than the wire,

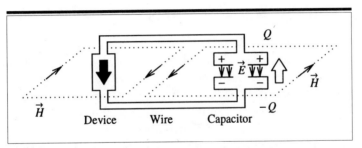

Figure 1.18 Discharge of a Capacitor Through a Device

where the conduction and the displacement currents coexist. This is one reason why a device is always associated with a parasitic capacitance. The current in the capacitor is the displacement current. The capacitor loses charge when the device turns on, and the device and the wire carry current dQ/dt. The capacitor has area S, and the insulator has dielectric constant ϵ. The field in the capacitor gap is

$$E = (1/\epsilon)(Q/S) \tag{1.55}$$

Then the displacement current of the capacitor is given by

$$I_D = \epsilon S \frac{dE}{dt} = \frac{dQ}{dt} \tag{1.56}$$

and therefore the current is the same everywhere in the loop.

At the electrode of the capacitor, the conduction current vanishes at the charge, and all the current is carried by the displacement current in the dielectric. Since the current loop appears broken open at the location of the capacitor, it might be thought that the definition of inductance would become unclear. This is not the case. The inductance of the loop is defined by the magnetic flux that links the loop current. The magnetic flux is created by both conduction and displacement currents. In creating magnetic flux they are absolutely equivalent, as dictated by Maxwell's equation. At the location of the capacitor of Figure 1.18 the displacement current may spread out over the area. This effect is the same as for the inductance of a resistive loop that has significant cross-sectional area, over which the current is distributed. The concept of the *internal* and *external* inductances needed to account for the flux linkage can be used to compute the loop inductance [10]. The existence of the components such as devices and capacitors in the closed current loop does not alter the definition of the inductance.

1.10 Voltage Measurements

In Section 1.9 we showed a voltage measurement setup in Figure 1.16. In the setup the voltmeter and the horizontal connecting wires (having length d) may be considered together as a voltmeter, and the wires directly to the two nodes A and B of the circuit under test (the vertical wire having length b) are the voltmeter connection wires. In this section we consider the problems of the voltmeter connection wires.

In Figure 1.19(a) the inductance has a magnetic core. The core is a closed space in which all the magnetic flux generated by the winding is confined. The inductance is shown in the dotted box. No magnetic field leaks out. The inductance is driven by a time-dependent current source, and the voltage developed across the inductance is measured. In this arrangement the voltmeter can be connected to nodes A and B either like V by a twisted pair of wires, or like V' by an open pair of wires, and the reading is independent of the way the voltmeter is connected. This means that an inductance having a magnetic core is a true two-terminal *lumped* component: Its characteristic is completely specified by the terminal (A-B) characteristic.

In Figure 1.19(b) the inductance shown by a thick circle (the almost closed arc ABC) has no magnetic core, and therefore the magnetic field spreads out in the vicinity of the winding. The voltmeter connection wires must go through the magnetic field. Current $\vec{J}$ flows in the direction indicated by the arrow. The inductance current path follows O-L-F-A-B-C-G-

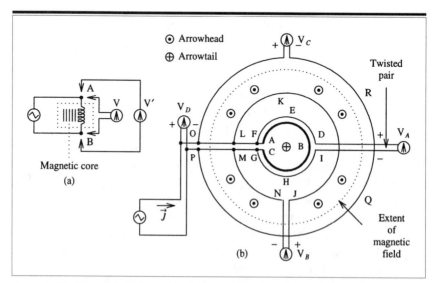

Figure 1.19 Uncertain Voltage Measurement in a Circuit Having a Naked Inductance

M-P. The directions of the magnetic field are shown by the arrowtail and the arrowhead symbols. The downgoing flux within the circle *A-B-C*, shown by a single arrowtail symbol, equals the upgoing flux outside the circle, shown by the eight arrowhead symbols. The outside upgoing flux is confined within the dotted circle. If the current changes, then the magnetic flux changes, and a voltage is developed around any closed conductor loop wholly or partly within the dotted circle. The voltage equals the rate of change of the magnetic flux linked to the closed loop. The pair of wires *O-L-F-A* and *C-G-M-P* are *twisted* wires. A pair of twisted wires are close together so that there is no space between where the magnetic flux can creep in. In the figure, the minimum-gap pairs of wires are meant to be twisted pairs.

Suppose that $\tilde{J}$ increases. The downgoing magnetic flux Φ in the loop *A-B-C* increases as well. The rate of increase is Φ' (prime means time derivative). Voltmeter V_D is in a closed loop V_D-*O-L-F-A-B-C-G-M-P*-V_D, and except for the portion *A-B-C* it consists of a twisted pair of wires. Then the reading V_D is given by

$$V_D = \Phi' \tag{1.57}$$

The polarity of the voltmeters is shown by $+$ and $-$ signs. The voltmeter V_D reads positive. The voltmeter V_A is in a closed loop V_A-*D-E-F-A-B-C-G-H-I*-V_A, and it consists of a twisted pair of wires only. Then the reading V_A is

$$V_A = 0 \tag{1.58}$$

The voltmeter V_C is in a closed loop V_C-*O-L-F-A-B-C-G-M-P-Q-R*-V_C. The magnetic flux in the loop is everywhere upgoing, and since the outer loop is outside the range of the magnetic field, the reading V_C is

$$V_C = \Phi' \tag{1.59}$$

The voltmeter V_B is in a closed loop V_B-*J-K-L-F-A-B-C-G-M-N*-V_B. The magnetic flux in the loop is upgoing, but is less than the total magnetic flux Φ. Thus

$$V_B < \Phi' \tag{1.60}$$

The readings of the voltmeters disagree. Which voltmeter connection provides a meaningful result?

A physical measurement should provide a result that is independent of any test setup modification that maintains the same measurement principle and that does not affect the test object. Voltmeter connection wires are a part of the test setup. The gedankenexperiment showed that the routing of the connection wire affects the measured result. We note, however, that in some cases the result depends sensitively on the route of the wire, in some cases much less sensitively, and in some cases not at all (for small changes). The reading of V_B depends sensitively on the routing: The reading is linearly dependent on the radius of the loop K-L-M-N-V_B-J. The reading of V_C is independent of small variation of the radius of the outside loop P-Q-R-V_C, since the loop is outside the range of the magnetic field. The reading of V_D is independent of the location of the voltmeter connection (O and P). The connecting points can be moved to L and M, and the reading remains the same. The reading of V_A depends weakly on the loop radius. Since the magnetic field is zero on the loop A-B-C, the variation is proportional to the square of the difference of the radii of the loops A-B-C and D-E-F-G-H-I. We consider that a *stationary* reading, independent of the routing, reflects something physically real. An experimental physicist would consider the *reproducibility* of the data as one of the key criteria of reality of the measurement. We take the same line in the gedankenexperiment. The voltmeters V_D and V_C give the real reading Φ'. Its reading is the voltage developed across the *naked* inductance, and the inductance value is defined as the ratio of the voltage to the rate of the current change. As for the voltmeter V_A, we must explain what its peculiar reading, 0, means.

Figure 1.20(a) shows a side view of a short section of an interconnect wire, which carries current $\vec{J}$. The current generates a magnetic field, whose directions are shown by the arrowtail and the arrowhead symbols. Suppose that a voltmeter is connected to points A and B. If $\vec{J}$ increases, the magnetic field in the voltmeter connection loop (an arrowhead) increases,

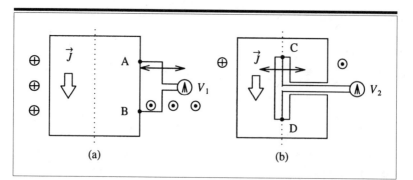

(a) (b)

Figure 1.20 Measurement of Inductance of a Straight Section

and the voltmeter measures a positive voltage V_1. The magnitude of the voltage, however, depends on how large the loop is. If the voltmeter and the doglegs of the connection wire are moved to right and left as shown by the double-headed arrow in Figure 1.20(a), the reading changes linearly with the movement. The measured value is meaningless. If we compute the inductance of section A-B, the reading V_1 is divided by the rate of change of the current, $|d\vec{J}/dt|$. Since V_1 does not have a meaningful value, the "inductance" computed this way is an arbitrary number. It is impossible to define an inductance of a section of a wire. Inductance can be defined only if the voltage reading is stationary, as in the loop of Figure 1.19(b). An immediate consequence is that the inductance of a closed loop cannot be broken down into components.

In Figure 1.20(a) we assumed that the voltmeter connection point was on the outside surface of the conductor. If the connection point can be moved inside by cutting a small hole as shown in Figure 1.20(b), the reading of V_2 can be made stationary at the center of the conductor against the movement of the contact points C and D. The stationary reading is zero. What is the physical meaning of the stationary value zero? This question is best answered if we consider how to make this value not zero. We assumed that the conductor section has zero resistance. If the section has nonzero resistance R, the stationary voltage will be $R|\vec{J}|$.

An interesting question is: If resistance exists in a closed loop, can it be measured separately from the inductance? In the measurement setup of Figure 1.21(a) the wire section $\overline{AB}$ has resistance as well as being a segment of the loop's inductance. Suppose that the current in the wire increases. We take the voltmeter reading at the time when the conductor current I has reached a fixed value I_0, and plot the voltmeter reading V_M versus voltmeter distance D, as shown in Figure 1.21(b). The voltage reading V_M is

$$V_M = RI_0 + (d\Phi/dt) \tag{1.61}$$

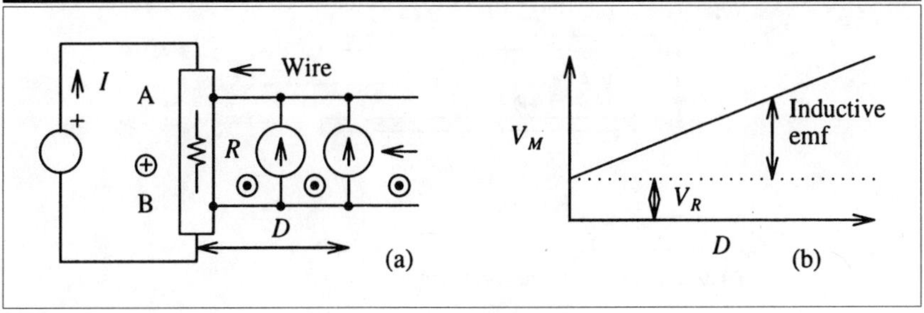

Figure 1.21 Coexistence of Inductance and Resistance

where R is the resistance of the conductor section and Φ is the magnetic flux linked to the voltmeter connection loop. We have $\Phi \to 0$ if $D \to 0$. If the voltmeter is pushed to the center of the wire, there is a nonzero voltage reading, which equals the resistance voltage drop. In this way the resistance part can be determined separately. Indeed, the *profile* of resistance along the loop can be determined. This means that resistance is *locatable* on a closed loop, but inductance is not. Inductance is an attribute of the loop as a whole. This peculiarity of inductance raises very considerable new issues in modeling integrated circuits that include naked inductances.

Figure 1.22 shows a capacitor to which a voltmeter is connected. We note that a capacitor has large capacitance if the electrode area is large and the electrode gap is small. Current is carried by the displacement current in the dielectric. The displacement current is spread over the large electrode area. A capacitor is the part of a closed current path where the cross-sectional area is the largest. Since the current spreads out over the area, it may not be easy in practice to locate the center of the displacement current distribution where the magnetic field becomes zero, but still such a center exists. Since the magnetic field at the center is zero, the voltmeter reading there becomes stationary. This stationary voltage is the capacitor voltage. Its measurement has no ambiguity, since the voltmeter is directly connected to the capacitor electrodes, which are only small distance d apart.

The ambiguity in the voltage measurement can be interpreted, alternatively, as follows. The naked inductance and the voltmeter connection wires make a pair of magnetically coupled transformer windings, and the voltage induced in the connection wires acting as the secondary adds to the inductance voltage. Such an interaction happens because a naked inductance is not a two-terminal component. This explanation is obvious in the cases shown in Figure 1.23(a)–(c). In Figure 1.23(a) the connection wire adds to the winding of the naked inductance, and the voltmeter reads a

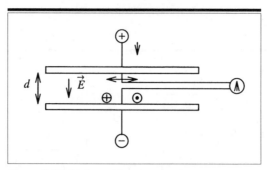

Figure 1.22   Voltage Measurement on a
Capacitance

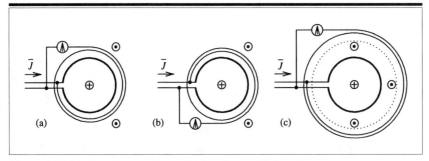

Figure 1.23 Various Connections of Voltmeter and Its Readings

higher voltage. A variable AC power transformer used to step up the AC line voltage operates by this mode. In Figure 1.23(b) the connection wire voltage subtracts from the inductance voltage. In Figure 1.23(c) the voltage reading is the same, since the wire outside the magnetic field range does not produce any electromotive force (the wire loop encircles equal upgoing and downgoing magnetic flux).

The ambiguity originating from the multiple turns of the connection wires gives a set of multiple stationary voltages that are apart by $|\Phi'|$. This ambiguity can be eased by selecting the voltage that has the smallest absolute value, but this criterion is not watertight. There can still be an ambiguity of $\pm|\Phi'|$. Observation of the voltmeter connection wire routing is the way to resolve this ambiguity.

What has been discussed in this section can be seen from the perspective of Maxwell's theory of electromagnetism as follows. The electromagnetic field of a conductor system is described by the vector and the scalar potential, $\vec{A}$ and ϕ, respectively, as

$$\vec{B} = \text{curl } \vec{A} \qquad \vec{E} = -\nabla\phi - \frac{\partial \vec{A}}{\partial t} \tag{1.62}$$

but there is arbitrariness in selecting $\vec{A}$ and ϕ. A set of new potentials

$$\vec{A}' = \vec{A} - \nabla\psi \qquad \phi' = \phi + \frac{\partial\psi}{\partial t} \tag{1.63}$$

where ψ is an arbitrary function, gives the same electromagnetic field, as can be proven by direct substitution (note that operator identity curl $\nabla = 0$) [11]. The components of the electromagnetic fields, $\vec{E}$, $\vec{H}$, etc., are the variables that can be measured *locally* (the electric and the magnetic fields at a point can be determined in principle by placing a small electric charge or a magnetic monopole there and measuring the force on it). This is consistent with the fact that the electromagnetic field is independent of the selection of function ψ. This property is called gauge invariance of the electromagnetic field. Suppose that $\psi = \text{const} \times t$. Then we have

$$\phi' = \phi + \text{const} \tag{1.64}$$

This means that the electrostatic potential, or the circuit node voltage, becomes arbitrary. This is consistent with the results of the gedanken-experiment that the node voltage becomes unmeasurable in general. Only the electrostatic potential *difference* developed across certain components that have small dimension in the direction of the current flow, to which the voltmeter can be connected and readings can be taken unambiguously, retains physical meaning. Circuit theory must be constructed using only such voltages as variables.

1.11 Equivalent Circuit and Distinctive Circuit Nodes

We observed that node voltages generally cease to exist in a circuit including a naked inductance. This means that the circuit theory may not have voltages as variables. Circuit theory without voltage sounds bizzare. However, I show an example. In a closed resistive conductor loop shown in Figure 1.24(a), the current established in the loop decays. The current decay provides an example of a voltageless circuit theory. The loop current $\vec{I}$ shown in Figure 1.24(a) is established as follows: The loop is broken open, and a magnetic field directed perpendicular to the page [the arrow-tail mark in Figure 1.24(a)] is applied. The loop is closed, and then the field is turned off. The conductor loop attempts to maintain the magnetic flux linked by it. As a consequence the current indicated by the arrows is

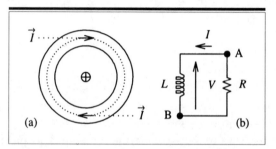

Figure 1.24 Example of Voltageless Circuit Theory

induced in the conductor. This is the same mechanism as that of a transformer winding. The current decay is usually calculated by using the equivalent circuit of the loop shown in Figure 1.24(b). The inductance L of the loop and its total resistance R can be defined clearly. In the simple equivalent circuit a single L and a single R are connected in series. Since $I = V/R$ and $V = L\, dI/dt$, the current I satisfies the equation

$$L\frac{dI}{dt} = -RI \quad \text{or} \quad I = I_0 \exp[-(R/L)t] \qquad (1.65)$$

where I_0 is the loop current established at $t = 0$. In this analysis the physical meaning of current I is clear. I is measurable anywhere on the loop. How about the physical meaning of V? Where are the nodes A and B of Figure 1.24(b)? V is defined only in the equivalent circuit of Figure 1.24(b), that is, a lumped parameter equivalent circuit. V is a variable that does *not* have any physical meaning in the real hardware of Figure 1.24(a).

When V loses its physical meaning, many confusions can arise from its use. An example is the following gedankenexperiment. Suppose that the conductor loop is perfectly circular and uniform. Instead of using the one-resistor, one-inductor equivalent circuit, we may use a fine-grained equivalent circuit. A single two-terminal component L or R of Figure 1.24(b) is split into N pieces. Since the ring has circular symmetry, this appears reasonable at the first sight. We redraw the equivalent circuit so that L and R are divided into N equal pieces and connected alternately. With the substitution $L \to L/N$ and $R \to R/N$ for each section, the current I remains unchanged, since the circuit equation written using I remains unchanged. In both the coarse-grained and in the fine-grained equivalent circuits, the voltages developed across the inductance and the resistance are equal in magnitude and opposite in sign. Their sum is zero. This voltage cancellation takes place on the *microscopic* level, since N can be arbitrarily large. Then on connecting a voltmeter to any section of the loop, the voltage reading should be zero. Is this a right conclusion? It is not. If the voltmeter is connected to a section of the ring, and the connecting wire is routed so that the area enclosed by the connecting wire and the center of the ring is zero (the wire may have to be routed inside the ring's conductor as we saw in Figure 1.20(b)), the voltmeter measures voltage developed across the resistance of that section, and that is *not* zero.

This contradiction occurred because dividing the loop into N *equal* pieces and connecting the resistive and inductive parasitics alternately in series in the equivalent circuit is a procedure that is not allowed. The inductance of a section need not be L/N. It can have any value, as long as the sum of all is L. This is true even if the ring has high structural symmetry and material uniformity. This is a logical conclusion, but is certainly against intuition and is confusing, as symmetry is much respected as a clue to problem solving in physics. The only measurement that is allowed is the measurement of the component voltage developed across the resistance of the section. To do so, the voltmeter is connected so that the connection loop has a negligible magnetic flux linkage. If the current is distributed over the conductor ring cross section, there is a point at the geometrical center of the current distribution where the magnetic field vanishes. If these points at all the cross sections of the ring are connected, they make a closed path. The voltmeter connection wires are routed along that path. Then the voltmeter reading divided by the current I gives R/N of the section, the *local* resistance. This is true even if the ring does not have uniform resistivity.

A second confusing example is as follows. Figure 1.25(a) shows a setup to measure the resistance of the short section $\overline{AB}$ located on the right side of the conducting ring. The voltmeter connection goes through the closed path inside the ring where the magnetic field is zero. The reading of the voltmeter M, divided by the current I, gives the resistance of the short section. Now suppose that the points A and B are moved to the left side of the ring, as shown in Figure 1.25(b). Does the voltmeter read the voltage developed across the resistance of the short section $\overline{AB}$, or the long section $\overline{ACB}$? The voltmeter connection wires follow the zero-magnetic-field path that exists inside the conducting ring, and the ring and the loop are closely coupled: All the magnetic flux linked by the conductor ring links the connection wires as well. As the conductor ring current decays with the time constant L/R, the magnetic flux that links the connection loop changes, and a voltage is induced in the connection loop that equals the voltage

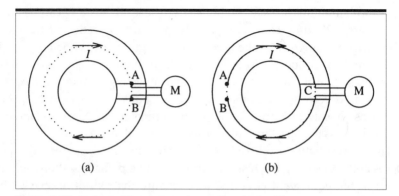

(a)   (b)

Figure 1.25   Voltage Measurement on a Conductor Ring

developed across the inductance of the conductor ring. That voltage in turn equals the voltage developed across the total resistance of the ring. Thus the voltmeter should read the voltage developed across the resistance of the entire ring. The setup of Figure 1.25(b) gives the resistance of the long section $\overline{ACB}$.

It might be thought that measuring the voltage developed across the resistance of a resistive ring is a violation of the rule of voltage measurement of Section 1.9. The pair of nodes to which a voltmeter is connected must have a capacitance so that the voltmeter does not load the circuit. We are not violating the rule in this problem because we connect the voltmeter to real hardware (a conductor ring) and not an equivalent circuit model. In a real conductor ring, a section that has resistance always has a parallel capacitance associated with it. The product of the resistance and the capacitance is the dielectric relaxation time of the resistive material. In an equivalent circuit *model*, however, the voltage difference between a pair of nodes that do not have a capacitance between them is not measurable. In modeling a resistive ring, this has the following consequence. The nodes A and B of one conductor ring indicated on the equivalent circuit diagram of Figure 1.24(a) and (b) have no capacitance between them, because such nodes never exist in the real hardware ring, and accordingly there is no way to assign a capacitance to them. Accordingly, their voltage difference cannot be measured. The inability to determine the capacitance and the inability to measure the voltage are consistent. The voltage difference between two nodes that are the two terminals of a real resistor can be measured, since the real resistor is always associated with a capacitance by the dielectric relaxation mechanism. The resistance in the real hardware and the resistor in the equivalent circuit model are different, and this is a crucial point of the theory.

In the real hardware on an integrated circuit chip, several issues related to this criterion must be clarified. An integrated circuit consists of amplifying devices that always have parallel capacitance. An electron triode has a drain capacitance to the substrate or to the source. If the capacitance is small, however, the equivalent circuit analysis can be simplified by neglecting it. We need to consider the drain voltage as measurable. If a current generator or a resistor is used as the model of the amplifying device, its drain-source voltage is measurable as well, because the model is considered to have a parallel capacitance. The load circuit connected to the output of a circuit may be considered as a voltmeter. In digital circuits the input capacitance of the load circuit is significant or often dominant, and it acts as the node capacitance, since the load circuit is never disconnected. In this sense a load on a digital circuit is always a voltmeter.

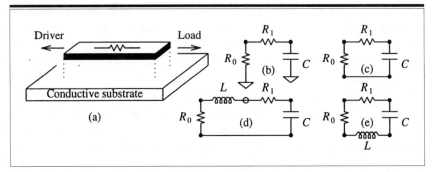

Figure 1.26 Modeling a Section of an Interconnect

We consider the problems of drawing an equivalent circuit using a few examples. Figure 1.26(a) shows a section of a resistive transmission line fabricated on a perfectly conducting substrate. The driver is a conducting triode, which is equivalently represented by the resistor R_0. This resistor has a parasitic capacitance, and even though that is not shown on the circuit diagram, the voltage developed across R_0 is measurable. The load is a capacitance C, and the voltage developed across C is always measurable. This capacitance is the lumped capacitance of the entire transmission line. The *backbone RC* equivalent circuit is shown in Figure 1.26(b). We note that the connectivity of the backbone circuit determines which of the node voltages are measurable (carry valid signal or information). This circuit must be redrawn as in Figure 1.26(c) to explicitly indicate the ground connection. A loop inductance L can be added to Figure 1.26(c), either in series with R_1, or in the ground connection, as shown in Figure 1.26(d) and (e), respectively. In the circuit including inductance, the voltage developed across the series connection of L and R_1 in Figure 1.26(d), or across L of Figure 1.26(e), is *not* measurable. In the redrawn circuit including inductance, many voltages lose their physical meanings. The node shown by the open circle in Figure 1.26(d) is a node that exists only in the equivalent circuit, and not in the real hardware. There is no proper capacitance value to be assigned to this node, and therefore the node voltage is considered unmeasurable. If we try to measure the voltage developed across the series connection, we obtain only the voltage developed across R_1, if the voltmeter connection wire is adjusted to give the stationary reading.

Figure 1.27 shows the equivalent circuits of a resistive transmission line, fabricated on a resistive substrate. R_1 and R_s, representing the series resistance of the line and the substrate, respectively, are locatable. The line capacitance can be lumped into a single capacitance C or into two capacitances C_1 and C_2. Depending on the decision, the equivalent circuit con-

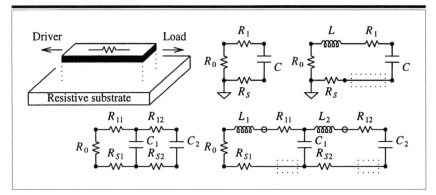

Figure 1.27 Modeling a Section of Interconnect on a Resistive Substrate

tains a single loop or two loops. For each loop an inductance is added. The voltages developed across R_0 and across C or C_1 and C_2 are the only measurable voltages.

The requirement of voltage measurability restricts the types of equivalent circuits that model a digital circuit. Figure 1.28(a) shows an interconnect fabricated on a perfectly conducting substrate that has a branch point B. The capacitances that the conductor wire sections $\overline{AB}$, $\overline{BC}$, and $\overline{BD}$ make to the substrate are lumped into C_{AB}, C_{BC}, and C_{BD} in the equivalent circuit of Figure 1.28(b), respectively. The resistances of these sections are likewise lumped into R_{AB}, R_{BC}, and R_{BD}. At high frequencies the inductances of the loops $AA'B'BA$, $BB'C'CB$, and $BB'D'DB$ of Figure 1.28(a)—L_{AB}, L_{BC}, and L_{BD}, respectively—are added to the equivalent circuit. The loop $AA'B'BA$ is closed at the left end (AA') by a switching device and its output capacitance, and at the left end (BB') by the lumped capacitance C_{AB}.

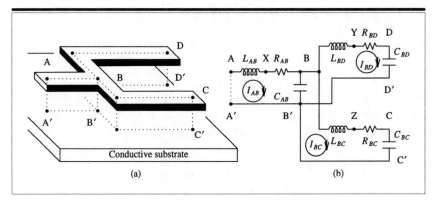

Figure 1.28 Inductance of the Interconnect on the Substrate

The other loops are closed at their left ends by C_{AB}, and at their right ends by C_{BC} and C_{BD}. The equivalent circuit of the digital interconnect circuit consists of loops that are closed by the capacitors. The nodes X, Y, and Z of Figure 1.28(b) emerged when the equivalent circuit was drawn. These nodes do not actually exist, and their voltages are never definable. The circuit equations must be written using only the meaningful voltages. In the equivalent circuit of Figure 1.28(b) the voltages developed across the terminals of the capacitances C_{AB}, C_{BC}, and C_{BD} are V_{AB}, V_{BC}, and V_{BD}, respectively, and they are the measurable and physically meaningful voltages. Summarizing the observations on the equivalent circuit, the circuit equations of a digital circuit must be written subject to the following requirements:

1. Use the voltages developed across the capacitors as the dependent variables.

2. The inductances are defined only for the loops closed by the capacitances.

We define the loop currents I_{AB}, I_{BC}, and I_{BD} that circulate in the loops AB, BC, and BD, respectively. We have

$$C_{AB} \frac{dV_{AB}}{dt} = I_{AB} - I_{BC} - I_{BD}$$

$$C_{BC} \frac{dV_{BC}}{dt} = I_{BC} \qquad C_{BD} \frac{dV_{BD}}{dt} = I_{BD}$$

$$\tag{1.66}$$

$$L_{BC} \frac{dI_{BC}}{dt} + R_{BC}I_{BC} = V_{AB} - V_{BC}$$

$$L_{BD} \frac{dI_{BD}}{dt} + R_{BD}I_{BD} = V_{AB} - V_{BD}$$

Among the five equations the two variables I_{BC} and I_{BD} are eliminated. We need one more equation, specifying the driver. If a voltage source $V_0(t)$ is connected across AA', for instance, we get

$$V_0(t) = L_{AB} \frac{dI_{AB}}{dt} + R_{AB}I_{AB} + V_{AB} \tag{1.67}$$

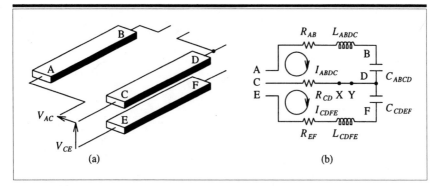

Figure 1.29   Equivalent Circuit Including Two Inductance Loops

and the set of equations then determines the voltage waveforms of V_{AB}, V_{BC}, and V_{BD}, if solved subject to the initial conditions.

Figure 1.29(a) shows three interconnects $\overline{AB}$, $\overline{CD}$, and $\overline{EF}$. The first signal voltage source V_{AC} is connected to A and C, and the second signal voltage source V_{CE} is connected to C and E. The signals are received at BD and DF, respectively. The capacitances of the pair of lines $\overline{AB}$ and $\overline{CD}$ and the pair $\overline{CD}$ and $\overline{EF}$ are lumped into C_{ABCD} and C_{CDEF}, respectively, in the equivalent circuit of Figure 1.29(b). The resistances of the three lines are lumped into R_{AB}, R_{CD}, and R_{EF}. The driving ends, AC and CE, are connected to the driver device. The pair of lines $\overline{AB}$, $\overline{CD}$, the capacitor C_{ABCD}, and the device close a loop $ABDC$, which has inductance L_{ABDC}. Similarly, the pair of the lines $\overline{CD}$, $\overline{EF}$ close a loop, whose inductance is L_{CDFE}. L_{ABDC} is the inductance of a horizontal current loop, and L_{CDFE} is the inductance of a vertical loop. We assume that they are not magnetically coupled. The equivalent circuit of this interconnect circuit is drawn as shown in Figure 1.29(b). The voltages developed across AC, CE, BD, and DF are measurable, and they are V_{AC}, V_{CE}, V_{BD}, and V_{DF}. The resistances R_{AB}, R_{CD}, and R_{EF} are the locatable resistances, and they are placed on the branches AB, CD, and EF of the equivalent circuit, respectively. If the inductances are neglected, Figure 1.29(b) is a conventional RC equivalent circuit of the interconnect. Inductances L_{ABDC} and L_{CDFE} are placed on the branches $\overline{AB}$ and $\overline{EF}$, respectively. They are *not* placed between the points X and Y of branch $\overline{CD}$. This is because the two inductances are independent: L_{ABDC} affects V_{BD} only, while L_{CDFE} affects V_{DF} only. If any inductance is placed between X and Y, it affects both V_{BD} and V_{DF}.

The circuit equations are written, using the two loop currents I_{ABDC} and I_{CDFE}, as

$$V_{AC}(t) = R_{AB}I_{ABDC} + R_{CD}(I_{ABDC} - I_{CDFE}) + L_{ABDC}\frac{dI_{ABDC}}{dt} + (X) \qquad (1.68)$$

$$V_{CE}(t) = R_{EF}I_{CDFE} + R_{CD}(I_{CDFE} - I_{ABDC}) + L_{CDFE}\frac{dI_{CDFE}}{dt} + (Y) \qquad (1.69)$$

and

$$I_{ABDC} = C_{ABCD}\frac{dV_{BD}}{dt} \qquad I_{CDFE} = C_{CDEF}\frac{dV_{DF}}{dt} \qquad (1.70)$$

In Equations (1.68) and (1.69) the effects of the mutual inductance between the two loops can be included. To do so, one includes terms (X) and (Y) of the general form

$$(X): \quad M\frac{dI_{CDFE}}{dt} \qquad (Y): \quad M\frac{dI_{ABDC}}{dt} \qquad (1.71)$$

By eliminating the loop currents I_{ABDC} and I_{CDFE}, the equations satisfied by the two measurable voltages V_{BD} and V_{DF} are derived. Figure 1.30 shows a third example, which has perfectly conducting interconnect fabricated on a resistive substrate, and which has three current loops.

The steps to draw the equivalent circuit including inductances can be summarized as follows. First the *RC*-only backbone equivalent circuit is drawn. The distributed capacitances are lumped at the nodes where the desired voltages become measurable. The voltages developed across the capacitances and the voltage sources (the latter may be regarded as having infinite internal capacitance) are measurable, and they can be used to write the circuit equations. Inductances are added in each loop, such that the common part of two or more loops does not have inductance. This is

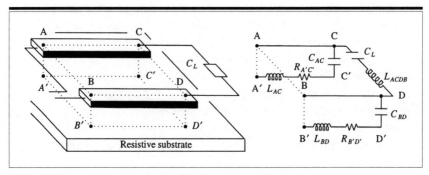

Figure 1.30 Inductance of the Parallel Interconnects

because if it did, an inseparable loop inductance would be separated into two parts, and that is not allowed. Only resistance can be placed in the common part of the loops.

1.12 Modes and the Equivalent Circuit

More than one equivalent circuit can model a single piece of hardware. To draw an equivalent circuit that reflects the transient of the hardware with reasonable correctness, we need some understanding of the transient itself. To understand the transient at a rudimentary level, the concept of *mode* is convenient. A mode is a physically realizable pattern of current flow in the hardware. Since current can be measured locally, a mode can be identified without relying on problematical voltage measurements. For any physically consistent pattern of current flow in the hardware, there is a mode and an equivalent circuit. Equivalent circuit and mode are in one-to-one correspondence. Since a mode is a specific current flow pattern in the hardware, it can be determined from the basic set of equations describing the hardware, often the complete set of Maxwell equations with boundary conditions, by eliminating the time dependence. The mathematical procedure amounts to replacing the time derivative $\partial/\partial t$ by the complex frequency $j\omega$. A mode has a spatial current distribution $\vec{J}(\vec{r})$ multiplied by the time dependence, $\exp(j\omega t)$. The frequency ω is in general a complex number.

In a digital circuit in which the switching devices are able to control the circuit connectivity, a mode is a substructure of a *microstate*, discussed in my previous book [1] (also see Section 2.5). In an analog circuit that does not change connectivity, a mode reflects the current flow pattern in the hardware directly, and this point is clearer in *LC* circuits than in *RC* circuits. Figure 1.31(a) shows a cross section of a reentrant cavity resonator,

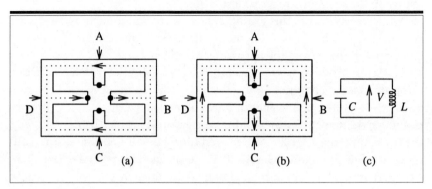

Figure 1.31 Equivalent Circuit of a Resonant Cavity

similar to those used in a microwave magnetron. The cavity resonator has many resonant modes, of which the current flow patterns for the two lowest-frequency modes are shown in Figure 1.31(a) and (b).

Modes are identified by understanding the hardware operation as a whole, and this is often as difficult as understanding the equivalent circuit operation. There are several clues to identifying the modes. A gap between two conductors like *AC* or *BD* of Figure 1.31(a) and (b) works as a capacitance, and a long conducting path connecting the gap like BAD or BCD of Figure 1.31(a) works as an inductance. A parallel combination of a capacitor and an inductor makes a resonant oscillator. There can be many such oscillators in real hardware, especially if the hardware has a structural symmetry. In the mode of Figure 1.31(a) the BD gap functions as the capacitor, and the current paths $\overline{BADB}$ and $\overline{BCDB}$ function as the inductor. In the mode of Figure 1.31(b) the *AC* gap functions as the capacitor and the current paths $\overline{ABCA}$ and $\overline{ADCA}$ function as the inductor. Both modes have the equivalent circuit of Figure 1.31(c), but the physical meanings and the values of *C* and *L* are different. Since resistance is locatable, where to place them is obvious, and often it is important to consider how effective the resistors are in dissipating the energy of the resonant oscillator. Any resistance in series with the inductance attenuates the oscillation. In the structure of Figure 1.31(a) and (b), yet another resonant mode exists, in which two of the four gaps *AB*, *BC*, *CD*, and *DA* (say AB and CD) function as the capacitor, and the alternative quadrant current paths (say, $\overline{AB}$ and $\overline{CD}$) function as the inductor. In this mode the current paths $\overline{BC}$ and $\overline{DA}$ are not active. The hardware is effectively disconnected into two pieces. The two independent resonant pieces do not interact in the mode, since the transients in the symmetrical structure proceed exactly in phase. In the mode, not all the circuit connections that are shown in the diagram or are obvious in the hardware structure are effective. The connections shown in the diagram are a maximum set.

As is observed from this example, modes are determined by hardware symmetry. Whether or not two modes have the same resonant frequency can be determined from the hardware symmetry alone.

In symmetrical hardware the voltage measurability is often confusing. Suppose that the circuit of Figure 1.32 is oscillating in the mode shown in Figure 1.31(a). The points *A* and *C* are some distance apart. If a voltmeter is connected to them, is the reading uncertain? It is not. If the voltmeter connection points A and C are moved along the dotted line α, the voltmeter reading remains constant at 0. If the voltmeter is moved in the direction perpendicular to that, the reading is also unchanged. This is because

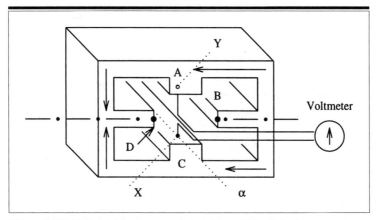

Figure 1.32 Voltage Measurement on a Symmetrical Mode

the circuit is symmetrical. If the hardware is not symmetrical with respect to the dot-dash symmetry plane, the reading depends on the location of the voltmeter connection. In the equivalent circuit of this mode, the capacitance between B and D is C in the circuit diagram of Figure 1.31(c), but there is no capacitance between the voltmeter connection points A and C. Then why is voltage measurement possible? It is because the measured voltage is always zero, and the voltmeter draws no energy from the circuit.

The modes of an *LC* circuit have real frequency, and the electrical activity of the modes continues forever. The modes of an LCR circuit have, in general, a complex frequency that represents attenuation of the electrical activity. The modes of an *RC* circuit are important because the modes of the backbone RC circuit are modified to give the modes of a circuit including inductances. This is convenient, and perhaps the only rational route to establishing an IC theory including inductance. By adding inductance to the backbone circuit we are able to get good insight into the LCR circuit, especially if the inductance effects are small. Understanding the modes of the *RC* backbone circuit is therefore important, and this is the subject of Chapter 2.

RC circuits often used in digital-circuit modeling have the general structure shown in Figure 1.33(a). A single resistor R_0 to ground represents the switching device pulling down the circuit, and the rectangular box contains an arbitrary resistance network. Each node of the network is connected to a capacitor, whose other terminal is grounded. In the mode of this circuit that is essential to its operation, all the capacitors discharge jointly through R_0. In this mode the voltages of all the capacitors decrease.

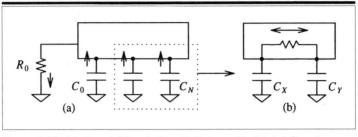

Figure 1.33   Modes of *RC*-only Circuits Relevant to Digital-Circuit Modeling

This mode has the longest time constant. There are many other modes, having shorter time constants. The action of the other modes is to equalize the voltages of the capacitors, shown in Figure 1.33(b).

1.13   Isolated Versus Nonisolated Inductances

Inductance arises on assigning a nonzero impedance to an impedanceless connection wire. Since a connecting wire does not have any identity in the circuit diagram, assigning impedance to such unidentified hardware creates a confusion. If a connecting wire acquires the status of a circuit component, an immediate question is the internal structure of the inductance itself: If inductances are connected by wires, the connecting wires of the inductances have inductance as well. This logical complication is resolved in the last section: Inductance is an attribute of a complete loop of wire, not a section.

An inductance as a practical circuit component is almost always provided with measures for confining the magnetic flux, such as an iron or ferrite core and a shield can. If such inductances are connected together, each inductance behaves independently. This kind of inductance has been an integral part of circuit theory. Our problem is that when real high-speed hardware is modeled by an equivalent circuit, some inductances can be isolated, but most cannot. An isolated inductance can be treated like a resistor. For example, it is possible to connect two 1-H choke coils in series to replace one 2-H choke coil. However, if there are two current loops in an integrated circuit that interact with each other, the simple inductance addition does not hold. It is therefore convenient to set up a notation for isolated and nonisolated inductances. In this book, isolated inductance is surrounded by a dotted box, and nonisolated inductance is represented by the bare symbol, as shown in Figure 1.34(a) and (b), respectively.

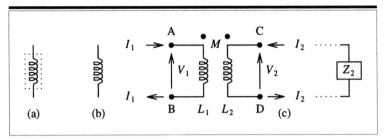

Figure 1.34 Coupled Inductances

The formula to describe the characteristics of the two interacting inductances shown in Figure 1.34(c) is [12]

$$V_1 = L_1 I_1' + M I_2' \tag{1.72}$$

$$V_2 = M I_1' + L_2 I_2' \tag{1.73}$$

where L_1 and L_2 are the self-inductances and M is the mutual (or coupling) inductance. The primes show the time derivatives. The first term of Equation (1.72) is the time derivative of the magnetic flux $L_1 I_1$ that links the first inductance (or primary) if its current is I_1. Likewise, $M I_2$ is the flux the secondary inductance makes and that links the first inductance if the secondary current is I_2. There is a relationship

$$M^2 \leq L_1 L_2 \tag{1.74}$$

that sets the maximum mutual flux linkage. This relationship is derived from the requirement that the magnetic energy of the coupled inductances given by

$$E_M = (1/2) L_1 I_1^2 + (1/2) L_2 I_2^2 + M I_1 I_2 \tag{1.75}$$

be positive definite (positive for any values of I_1 and I_2). If $M^2 = L_1 L_2$, the two inductances are perfectly coupled (close-coupled) and there is no *leakage flux*. If not, some magnetic flux leaks out from the two-inductance system.

If terminals B and C are connected and terminal A and D are used as a pair of the terminals, we have from Equations (1.72) and (1.73)

$$V_1 = (L_1 + M) I' \qquad V_2 = (M + L_2) I' \qquad I_1 = I_2 = I \tag{1.76}$$

and therefore we obtain a relationship

$$V = V_1 + V_2 = (L_1 + L_2 + 2M)I' \qquad (1.77)$$

The effective inductance is $L_1 + L_2 + 2M$. If terminals B and D are connected and terminals A and C are used as a pair, the effective inductance is $L_1 + L_2 - 2M$, and similarly, if the terminal pair A and C and the pair B and D are tied together, the *paralleled* inductances have an effective inductance value $(L_1L_2 - M^2)/(L_1 + L_2 - 2M)$. In these formulas we note that $L_1 + L_2 \geq 2\sqrt{L_1L_2} \geq 2M$ and all the *composite* inductances are positive.

If the secondary of the coupled inductances is loaded by an impedance Z_2, the impedance Z_1 looking into the primary is affected by that. We have

$$V_1 = L_1 \frac{dI_1}{dt} + M \frac{dI_2}{dt}$$

$$V_2 = M \frac{dI_1}{dt} + L_2 \frac{dI_2}{dt} \qquad (1.78)$$

and

$$V_2 = -Z_2 I_2$$

By assuming a sinusoidal time dependence, the primary impedance at the angular frequency ω is given by

$$Z_1 = \frac{V_1}{I_1} = \frac{j\omega L_1 Z_2 - (L_1L_2 - M^2)\omega^2}{Z_2 + j\omega L_2} \qquad (1.79)$$

We note that

$$Z_1 \rightarrow j\omega L_1 \quad (Z_2 \rightarrow \infty \text{ or } M \rightarrow 0)$$

$$Z_1 \rightarrow j\omega \frac{L_1L_2 - M^2}{L_2} \quad (Z_2 \rightarrow 0) \qquad (1.80)$$

The limit as $Z_2 \rightarrow \infty$ or $M \rightarrow 0$ means that the secondary does not exist. Since $L_1L_2 \geq M^2$, in the other limit $Z_1(Z_2 \rightarrow 0)$ is an ordinary positive inductive impedance. This is the case where leakage magnetic flux exists. In this case only the leakage flux contributes to the effective inductance of the primary. If $L_1L_2 = M^2$ (perfect coupling), the magnetic flux the primary makes is exactly canceled by the flux the secondary current makes. The primary appears as a perfect short circuit.

If $Z = R$ (resistance),

$$Z_1 = \frac{j\omega L_1 R - (L_1 L_2 - M^2)\omega^2}{R + j\omega L_2}$$

$$Z_1 \rightarrow j\omega L_1 + (M^2/R)\omega^2 \quad (R \rightarrow \infty) \tag{1.81}$$

and

$$Z_1 \rightarrow j\omega[(L_1 L_2 - M^2)/L_2] + (M^2/L_2^2)R \quad (R \rightarrow 0)$$

We note that the real part of Z_1 is proportional to R in the limit as $R \rightarrow 0$, but is inversely proportional in the limit as $R \rightarrow \infty$. If $Z_2 = 1/j\omega C$ (capacitance loading),

$$Z_1 = \frac{j\omega L_1[1 - (C/L_1)(L_1 L_2 - M^2)\omega^2]}{1 - \omega^2 L_2 C} \tag{1.82}$$

and there is a parallel resonance at the resonant frequency of the secondary, $\omega_{\text{RESONANCE}} = 1/\sqrt{L_2 C}$.

Figure 1.35 shows two resonant circuits that consist of L_1, C_1 and L_2, C_2. If they are not coupled ($M = 0$), their resonant frequencies are given, respectively, by

$$\omega_1 = 1/\sqrt{L_1 C_1} \qquad \omega_2 = 1/\sqrt{L_2 C_2} \tag{1.83}$$

If they are coupled, the circuit is described by the set of the equations

$$V_1 = L_1 \frac{dI_1}{dt} + M \frac{dI_2}{dt} \qquad V_2 = M \frac{dI_1}{dt} + L_2 \frac{dI_2}{dt} \tag{1.84}$$

and

$$I_1 = -C_1 \frac{dV_1}{dt} \qquad I_2 = -C_2 \frac{dV_2}{dt} \tag{1.85}$$

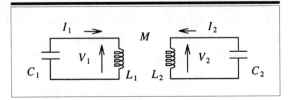

Figure 1.35 Transformer-Coupled Tuned Circuit

If a sinusoidal time dependence, $\exp(j\omega t)$, is assumed, the resonant angular frequency ω is determined from the secular equation

$$(L_1 C_1 \omega^2 - 1)(L_2 C_2 \omega^2 - 1) - M^2 C_1 C_2 \omega^4 = 0 \qquad (1.86)$$

We write $\omega = \omega_1 + \Delta_1$ and $\omega = \omega_2 + \Delta_2$, and seek the first-order correction to the resonant frequency when M is small. If $\omega_1 \neq \omega_2$ we have

$$\Delta_1 = \frac{M^2}{2L_1 L_2} \frac{\omega_1}{1 - (\omega_2/\omega_1)^2} \qquad \Delta_2 = \frac{M^2}{2L_1 L_2} \frac{\omega_2}{1 - (\omega_1/\omega_2)^2} \qquad (1.87)$$

If $\omega_1 > \omega_2$, then $\Delta_1 > 0$ and $\Delta_2 < 0$. Mutual inductance coupling between the resonant circuits shifts the higher resonant frequency higher and the lower resonant frequency lower. If $\omega_1 = \omega_2 = \omega_0$, we have

$$\omega_{1,2} = \omega_0[1 \pm (M\sqrt{C_1 C_2}/2)] \qquad (1.88)$$

The resonant frequency shift due to mutual inductance coupling is used for broadbanding the interstage transformers of an intermediate-frequency amplifier (staggered tuning).

Let us consider a current loop having size Λ. If current I flows in the loop, the magnetic field at the center of the loop is estimated at $H = I/\pi\Lambda$. The magnetic flux Φ that links the loop is estimated by $\Phi = \mu_0 \Lambda I$, and the self-inductance of the loop is $\mu_0 \Lambda$, where μ_0 is the magnetic permeability of free space, $\mu_0 = 4\pi/10^{-9}$ H/cm. A factor of the order of unity is multiplied by the estimated value, but to simplify the following argument we neglect that. The value of the self-inductance L reflects the size of the circuit, Λ, in an obvious way, like $\mu_0 \Lambda$. The value of the capacitance is also related to the size, like $\epsilon_0 \Lambda$. The delay time of the signal, that is, the time required for an electromagnetic wave to propagate through the circuit, Λ/c (where c is the velocity of the electromagnetic wave), is included in the circuit theory through the value of the inductance and the capacitance associated with it. Let us clarify this point by considering the following example.

In Figure 1.36(a) and (b), the voltage source V_1 drives a loop having dimension Λ. In Figure 1.36(a), the capacitance is placed at the far end of the loop, and in Figure 1.36(b) it is near the driving voltage source. It is easy to see that the step-function voltage signal takes time Λ/c to reach C in Figure 1.36(a), but how about Figure 1.36(b)? Both circuits are represented by the same equivalent circuit of Figure 1.36(c). The capacitance

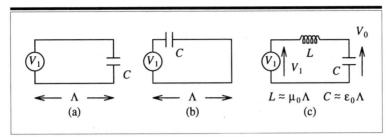

Figure 1.36 Size Dependence of Delay Time of an *LC* Circuit

voltage V_0 is found from the equation

$$V_1 = V_0 + LC \frac{d^2 V_0}{dt^2} \tag{1.89}$$

as

$$V_0 = V_1(t/\sqrt{2LC})^2 \approx (t/\Lambda \sqrt{\epsilon_0 \mu_0})^2 \tag{1.90}$$

This means that even if the voltage source is nearby, the loop current requires time Λ/c, where $c = 1/\sqrt{\epsilon_0 \mu_0}$, to build up, and indeed, the signal takes time Λ/c. A self-inductance reflects the circuit size in an obvious way.

If the circuit includes mutual inductance, a new issue emerges. Conventional mutual inductance is a single real parameter. Small mutual inductance can be due to poor coupling between two nearby loops, but it can be due to the long distance between the two loops in a plane. The propagation delay time of the signal between the two loops is not reflected in the parameter. How is a signal delay included in the mutual inductance coupling between loops that have some distance between them?

Let us consider the two interacting current loops, having currents I_1 and I_2, and the voltages developed across the loops, $V_1(t)$ and $V_2(t)$. They are related by the pair of linear equations

$$V_1(t) = L_1 I_1'(t) + M I_2'(t) \qquad V_2(t) = M I_1'(t) + L_2 I_2'(t) \tag{1.91}$$

where $I_{1,2}'(t)$ means the time derivative of current as before, L_1 and L_2 are the self-inductances of loops 1 and 2, respectively, and M is the mutual inductance. This relationship holds if the variation of voltages and currents is so slow that an electromagnetic wave is able to travel the distance

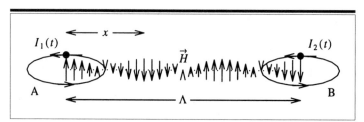

Figure 1.37 Time Delay Involved in Mutual Inductance

between the two loops within much shorter time than the characteristic time of the transient.

We noted that the energy of inductance is stored outside the conductor. Inclusion of space outside the circuit is the essential new feature created by including inductance in the circuit theory. In mutual inductance, the outside is even more significant than in self-inductance. If propagation of the electromagnetic wave outside the loop is considered, the two loops effectively form a pair of transmitting and receiving antennas, as schematically shown in Figure 1.37. The first current loop (left side) is driven by a signal source that creates a current $I_1(t)$. The magnetic field the current loop makes is like a magnetic dipole field that propagates from the current loop. The magnetic field is polarized perpendicular to the plane of the current loops. If the distance between the two loops is much larger than the loop radius, the electromagnetic wave between the two loops is a transverse wave whose electric field is in the plane containing the two loops. The wave takes time Λ/c to travel from one loop to the other, where c is its velocity. The wave arrives at the second loop that amount of time later, and generates a voltage proportional to the rate of change of the magnetic flux that links the second loop. Then the voltage developed in the second loop is given by

$$V_2(t) = MI_1'[t - (\Lambda/c)] + L_2 I_2'(t) \tag{1.92}$$

where M is a real number characterizing the magnitude of the mutual inductance coupling in the limit of a slow transient, and the argument of I_1', $t - (\Lambda/c)$, characterizes the time delay due to the distance between the two loops.

The effect of the time delay of mutual inductance is now considered separately. The coupling coefficient M decreases with increasing distance, but it also depends on the angle the two loops make. The reverse coupling from the second to the first loop is the same, except that the voltage induced in the first loop is a delayed function of the current in the second loop—delayed by a time Λ/c.

The formula that includes delay of the mutual inductance coupling must be used with care. Since mutual inductance coupling brings the conditions outside of the circuit into the circuit, unrealistic conclusions may be reached by using this formula. The following shows the point. If we use the formula for small delay time Δt, we obtain

$$V_1(t) = L_1 \frac{d}{dt} I_1(t) + M \frac{d}{dt} I_2(t - \Delta t)$$

$$= L_1 \frac{d}{dt} I_1(t) + M \frac{d}{dt} I_2(t) - M \Delta t \frac{d^2}{dt^2} I_2(t) \qquad (1.93)$$

$$V_2(t) = M \frac{d}{dt} I_1(t - \Delta t) + L_2 \frac{d}{dt} I_2(t)$$

$$= M \frac{d}{dt} I_1(t) - M \Delta t \frac{d^2}{dt^2} I_1(t) + L_2 \frac{d}{dt} I_2(t)$$

If the secondary loop is loaded by an impedance Z_2 and if sinusoidal time dependence is assumed,

$$V_2(\omega) = -Z_2 I_2(\omega)$$

$$V_1(\omega) = j\omega L_1 I_1(\omega) + (j\omega M + M \Delta t\, \omega^2) I_2(\omega) \qquad (1.94)$$

$$V_2(\omega) = (j\omega M + M \Delta t\, \omega^2) I_1(\omega) + j\omega L_2 I_2(\omega)$$

The impedance looking into the primary is given by

$$Z_1(\omega) = \frac{V_1(\omega)}{I_1(\omega)} = j\omega L_1 - \frac{(j\omega M + M \Delta t\, \omega^2)^2}{Z_2 + j\omega L_2} \qquad (1.95)$$

whence $Z_1(\omega) \to j\omega L_1$ if $M \to 0$ or $Z_2 \to \infty$. In the limit as $Z_2 \to 0$ this formula gives a strange result. Then

$$Z_1 = j\omega[(L_1 L_2 - M^2)/L_2] - 2[(M^2 \Delta t)/L_2]\omega^2 + o(\omega^3) \qquad (1.96)$$

or

$$\frac{1}{Z_1} = \frac{L_2}{j\omega(L_1 L_2 - M^2)} - \frac{2M^2 L_2\, \Delta t}{(L_1 L_2 - M^2)^2} \qquad (1.97)$$

The real part of $Z_1(\omega)$ is negative. If the real part of $Z_1(\omega)$ were real, the two coupled loops would work as an active, amplifying device. This is obviously an absurd conclusion.

The reason why this bizzare conclusion was reached is that the energy loss from the primary was not properly accounted for. From the primary, an electromagnetic wave is emitted into space. Only a fraction of the emitted wave energy is captured by the secondary to generate the current there. The secondary sends some energy back to the primary. The energy reflected from the secondary creates the negative resistance of Equation (1.95), but in reality this is only a reduction of the primary energy loss. If the primary radiation energy loss is included, the total energy loss becomes positive and the primary impedance has a positive real part. The following explains this point in detail.

The magnetic flux produced by the current in the circuit exists outside the circuit, and the structure of the outside world, especially of the path of the electromagnetic wave, is not specified to the same detail as the electrically connected part of the circuit, on which the circuit diagram provides information. It might be thought that the outside magnetic field could be described as a *magnetic circuit* [13]. The notion of a magnetic circuit is conventionally used in the design of iron-core transformers. This model is indeed useful in some cases, but it is valid only for slow variations where the magnetic flux of a closed path is generated by the magnetomotive force without delay. The magnetic circuit has a resistance-like parameter that relates magnetomotive force and magnetic flux, but has no parameter like inductance and capacitance, and that is an essential difference.

Magnetic circuit theory provides, however, an insight into the transformer model. Figure 1.38(a) shows the structure of a transformer. The magnetic flux created by the primary may go through path P_1 as well as paths P_2 and P_3. Only the magnetic flux going through path P_3 links the secondary, and the flux variation generates the secondary voltage. The inductance looking into the primary is, however, the sum of the self-inductances contributed by paths P_1, P_2, and P_3. The parts contributed from the flux of P_1 and P_2 are separately indicated in the equivalent circuit of Figure 1.38(b) as L_1'. Addition of side paths like P_1 and P_2 increases the volume for magnetic energy storage, and the inductance increases. If the components are separated out, the rest of the self- and the mutual inductances make a *close-coupled* transformer, where

$$(L_1 - L_1')(L_2 - L_2') = M^2 \qquad (1.98)$$

holds. An implicit assumption of the transformer model is that the mag-

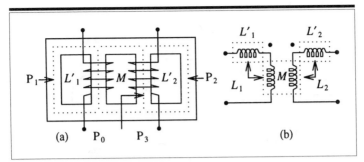

Figure 1.38 Conventional Coupled Inductance Model

netic flux is confined within the *core* region, and never spreads out indefinitely into the surrounding space. The energy that is transferred from the winding to the core returns to the winding. If the transformer secondary is loaded by lossless components, the impedance looking into the primary has no real part representing a loss. The transformer model is consistent with energy conservation, and it can be used freely in the circuit modeling and analysis.

If delay is included in the transformer model, the model changes fundamentally. Certain basic assumptions break down, and this issue can be understood from the point of view of physics as well as information science. Instantaneous interaction and delayed interaction differ in the following point: In the instantaneous interaction the present state of the dependent variable is determined by the present state of the independent variables, but in delayed interaction it is determined by all the past states of the independent variables. Instantaneous interaction is described by a function, and delayed interaction by a *functional*. Even if the delay is small, the qualitative difference in the mathematical description distinguishes the two interactions. The mathematical difference reflects the physics. Since the remote past determines the present state, and since the electromagnetic wave propagates at the velocity of light, the signal that left the primary a time T before travels a distance cT from the primary. The dimension of the space swept by the wave that holds information increases with T. The volume of the core of the transformer diverges.

Information-theoretically the space around the circuit works effectively as the *memory* that stores the information of the past state of the circuit. There is a certain limit on the density of information that can be packed in the space. If the infinitely distant past has bearing on the present, the volume of the required memory space is infinite as well. According to physics and information science, inclusion of the mutual inductance delay

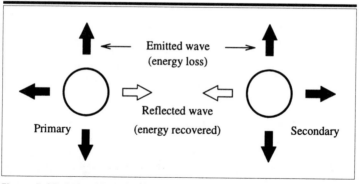

Figure 1.39 Loss of Energy by Radiation

involves inclusion of an infinite volume of the space around the inductance. The transformer model is consistent with instantaneous interaction assuming the existence of a magnetic core, but if the delay is included, the volume of the core must be extended to the entire space. As a consequence, escape of energy from the primary or the secondary of the transformer is inevitable. The primary and the secondary look like a pair of magnetic antennas, from which the electromagnetic wave is emitted into infinite space, as schematically shown in Figure 1.39. Some energy is exchanged between the primary and the secondary (open arrows), but some or even most of the energy emitted to space will never come back and is lost forever.

The secondary has some effect on the energy loss from the primary, and vice versa. Suppose, for simplicity, that the secondary has no loss. Then the current induced in the secondary by the electromagnetic wave from the primary generates a reflected wave that sends some energy back to the primary. The role of the secondary is to reduce the energy loss of the primary. If the secondary surrounds the primary in all directions, all the energy is sent back to the primary, and it appears as if the primary had no loss. In Figure 1.39 the arrows between the primary and the secondary show the interaction of the loops by sending and receiving energy. The positive contribution is, however, overcome by the loss of the energy that is not captured by the secondary, and this is more significant than the reflected energy: If the reflected and the lost energies are added, the total energy loss is positive.

Although the mechanism of the unrealistic negative real part of the impedance in the delayed transformer model is clear, it creates significant difficulties in the development of the circuit theory. It is difficult, however, to abandon the transformer model, since the alternative is to solve for the

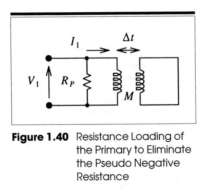

Figure 1.40 Resistance Loading of the Primary to Eliminate the Pseudo Negative Resistance

electromagnetic field from Maxwell's equations. It is desirable that the negative real part be suppressed so that physically meaningful results are obtained.

To eliminate the negative real part of the impedance in the delayed transformer model let us connect a resistor R_P in parallel with the primary (Figure 1.40). The impedance looking into the primary terminal, Z, is given by

$$(1/Z) = (1/R_P) + (1/Z_1) \tag{1.99}$$

where

$$Z_1 = j\omega L_{eff} - \Delta R$$

where $\tag{1.100}$

$$L_{eff} = \frac{L_1 L_2 - M^2}{L_2}$$

$$\Delta R = \frac{2M^2 \, \Delta t}{L_2} \omega^2$$

Here the power-series expansion of Z_1 was derived in Equation (1.96), and it is correct up to the second order of ω. After some algebra, we obtain

$$Z(\omega) = j\omega L_{eff} + [(\omega^2 L_{eff}^2 / R_P) - \Delta R] \tag{1.101}$$

The value of R_P required to obtain a positive real part is

$$R_P \le R_{PC} \qquad R_{PC} = \frac{1}{2} \left[\frac{L_{eff}}{M} \right]^2 \frac{L_2}{\Delta t} \tag{1.102}$$

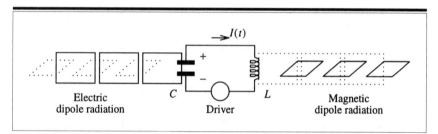

Figure 1.41 Circuit Loss by Electromagnetic Radiation

Once R_P is connected, the real part becomes positive, and we have physically meaningful results. R_{PC} tends to infinity in the limit as $\Delta t \to 0$. In the limit as $M \to 0$, $R_{PC} \to \infty$: This is the case where L_1 alone exists and its magnetic flux is confined. In the limit as $L_{\text{eff}} \to 0$, R_{PC} vanishes. This is the case of a close-coupled transformer whose secondary is short-circuited.

R_{PC} is the *maximum* resistance required to eliminate the positive real part. Its dependence on the mutual inductance is reasonable. Let us assume that $|M|$ is small, so that $L_{\text{eff}} \approx L_1 L_2$. Then increasing the mutual inductance decreases R_{PC}. This means that if there is a big secondary that surrounds the primary, the energy loss from the circuit is less.

Interaction between the components of a circuit by emitting and receiving radiation is, however, hard to characterize. If the electromagnetic field is the primary mechanism of the signal transfer, the radiation signal path becomes a waveguide (Figure 1.41). We considered the other limit, where the radiation plays a minor role: Our interest is to save the simple equivalent circuit representation when the radiation field energy is still not dominant. Still, this modeling method has its own strident limitation. Although adding parallel resistance eliminates the negative real part of the impedance, it does not avoid the inherent limitation of an equivalent circuit model as such: the equivalent circuit becomes open to the outside, and since the outside is not represented in the equivalent circuit, the latter does not contain sufficient information to make the circuit theory deterministic. This is obviously the limit of the generalization of conventional circuit theory.

CHAPTER TWO

Models of Devices and of Low-Frequency Circuits

2.1 Introduction

Integrated circuit technology is an offspring of the semiconductor device technology of the 1950s. As a consequence, ICs have been considered more as semiconductor devices than as electronic circuits. According to this view, the characteristics of ICs are determined by the amplifying devices, and every detail of the device characteristics is directly relevant. This viewpoint was valid in the early days of the IC technology, when the device quality was the key issue. As the technology matured, however, the viewpoint became too restrictive and unproductive. The device-centered bias delayed development of the circuit technology, especially the most important digital circuit technology. During the entire history of IC technology, new circuit developments were essentially limited to analog ICs. Practically no addition has been made to the list of basic digital circuits since the 1960s. The new viewpoint, that the characteristics of circuits are determined by the connection of the devices rather than by the characteristics of each device, emerged only recently, and it has not yet been universally accepted. This book is based on that viewpoint.

The new viewpoint encourages, above all, the search for better device connectivity to execute the desired signal processing. That is the activity that has been missing in the traditional digital circuit technology. The new viewpoint gains further support from the current status of semiconductor device technology, in which there exists only one type of amplifying device, a semiconductor triode. The basic structure and characteristics of BJTs, MOSFETs, and JFETs are very much alike. Furthermore, in the scaled-down VLSI technology, the amplifying devices are so small that they may still be described by a lumped circuit model even at high speed, but their interconnects have become extensive, and they must be regarded as complex circuits. According to the new viewpoint, the central subject of IC theory should be shifted from devices to interconnects. This shift is inevitable, since the components that have the largest size determine the ultimate high-frequency performance of the circuit. Before going into this subject, however, the *secondary* role played by the amplifying devices in high-frequency ICs is summarized in this chapter. Together with the device issues, I summarize the conventional IC theory, which is the theory of the charging and discharging of capacitance through resistance. We call the resistance-capacitance circuit the *backbone* circuit. In the later chapters inductances are added to the backbone circuits.

2.2 Electron Triodes

Amplifying devices used in integrated circuits—MOSFETs, BJTs, and MESFETs—are very much alike. They share the following properties [14–18]: If a device is used for voltage signal amplification, the amplifier circuit must have a pair of the input terminals and a pair of the output terminals, but one of the input terminals and the one of the output terminals are common. This common terminal is the *source* of the device. The other input terminal is the *gate*, and the other output terminal is the *drain* of the device. Here we use the most descriptive names of the electrodes, traditionally used for FETs and for all electron triodes. The terms used for BJTs—emitter, base, and collector—are less descriptive, since "base" does not imply a control function. The terms used for vacuum tubes—cathode, grid, and anode—are based on the bias polarity and are inconvenient. They correspond to the bias polarity of an *n*-type triode. In an *n*-type triode, positive voltage applied to the gate pulls electrons from the source. The number of electrons pulled out depends on the positive gate voltage. Because the mechanism of the electron extraction is the attractive force between the dissimilar charges, the higher the gate voltage, the more electrons are pulled from the source. Once electrons are pulled out, they travel under the influence of the field created by the positively biased drain. The electrons proceed from the source to the drain, and establish the device current directed from the drain to the source. Since the gate is positively biased, electrons may flow into the gate, thereby creating useless device input current. The device must be designed so that the electrons flow predominantly to the drain and not to the gate. Thus electron triodes can be classified by the means utilized in their design to prevent the electrons flowing into the gate. This classification has been discussed in my previous monograph [1] and is not repeated here.

I have explained the operation of an *n*-type triode. A *p*-type triode works in the same way, with the bias polarity reversed.

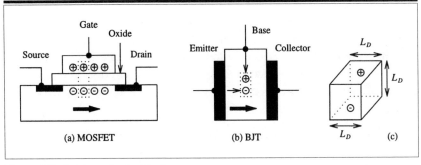

Figure 2.1 Semiconductor Triodes

In spite of the structural differences among various electron triodes, they have the same mechanisms of operation described above, and have closely similar current-voltage characteristics. I wish to show their close similarities by comparing a MOSFET and a BJT, shown schematically in Figure 2.1(a) and (b), respectively. In both devices, the mobile electrons that are created in the channel (MOSFET) or in the base (BJT) conduct the device current. To create the mobile electrons and sustain them by electrostatic attraction, positive charge must exist close to the negative mobile electrons and their source. In MOSFETs the positive charge is on the gate electrode, which is insulated from the conducting channel by the gate oxide [Figure 2.1(a)]. In BJTs the positive charge and the mobile electrons occupy the same base region [Figure 2.1(b)]. Either way, more positive charge attracts more negative mobile electrons in the channel, and that creates more device current. Positive charge is brought into the device's channel region from the gate circuit through the gate contact. Under ideal device operating conditions the positive charge moves in and out only through the gate circuit. The positive charge should not be consumed within the device. In real devices a small fraction of the positive charge is consumed by various imperfections. If positive charge is not lost, the gate input impedance is infinite at DC, like a capacitor.

Since the positive and the negative charges create a voltage difference between them, there is a capacitance between the pair of the charges. If more current is to be conducted for the same input voltage, more positive charge is required, and the gate capacitance becomes higher. This capacitance is physically obvious in MOSFETs. If gate area facing the conducting channel is S_G, and the thickness of the gate oxide is T_{OX}, the gate capacitance C_G is given by

$$C_G = \epsilon_{OX} S_G / T_{OX} \tag{2.1}$$

The base capacitance C_B of the BJT is less obvious. The area of the base (practically the area of the emitter facing the collector) is S_B, and the emitter-collector distance is L. If the base-emitter voltage is V_{BE}, the electron density at the emitter side edge of the base is $n_i \exp(q V_{BE}/kT)$, where n_i is the intrinsic carrier concentration of silicon [15]. The electron concentration decreases linearly with distance from the emitter, and becomes practically zero at the collector edge of the base. The total negative electron charge in the base is given by

$$Q_{\text{BASE}} = (1/2)q(S_B L)n_i \exp(q V_{BE}/kT) \tag{2.2}$$

Since the base capacitance C_B is the derivative of Q_{BASE} with respect to V_{BE},

$$C_B = \frac{\partial Q_{\text{BASE}}}{\partial V_{BE}} = \frac{q}{kT} Q_{\text{BASE}} \tag{2.3}$$

This relationship can be explained as follows. Inside the base the electrons and the holes are mixed. They do not recombine, however, because in a group IV semiconductor electron-hole recombination is a slow process, catalyzed by the scarce crystal imperfections. How is the capacitance between totally mixed electrons and holes defined? Because a mobile carrier in thermal equilibrium has thermal energy $kT/2$ per degree of freedom by the equipartition law [19], the carriers maintain, on the average, a mutual distance called the Debye length L_D, given by [20]

$$L_D = \sqrt{\epsilon_s kT/q^2 N_D} \tag{2.4}$$

where N_D is the density of the carriers. L_D is the distance to which two same-sign elementary charges are able to approach, and the distance from which two different-sign elementary charges are able to fly away, by the thermal energy (this is the same mathematical problem as for a comet and the sun). Consider a cube having size $L_D \times L_D \times L_D$ as shown in Figure 2.1(c). Suppose that the electron charge is smeared on the bottom surface of the cube, and the hole charge is smeared on the top surface. The capacitance between the pair of the charges is given by $\epsilon_s L_D^2/L_D = \epsilon_s L_D$. Since this is the capacitance per volume L_D^3 and the charges having different sign are completely mixed, the capacitance per unit volume is $\epsilon_s L_D^{-2}$. Since the volume of the transistor base is $S_B L$, the capacitance looking into the base is given by

$$C_B = \frac{\epsilon_s}{L_D^2}(S_B L) = \frac{q}{kT} Q_{\text{BASE}} \tag{2.5}$$

This is the same as Equation (2.3). This analysis shows how the base capacitance of a BJT is defined by a simple and intuitive model. The essential base capacitance (excluding the parasitics to the emitter and to the collector) and the true gate capacitance of a MOSFET are physically the same thing.

Although MOSFET and BJT are similar in many respects, not everything is the same. One difference is that with increasing gate-to-source voltage, the BJT current increases exponentially and the FET current linearly. This difference is due to the mechanism of injection of the current carriers into the device's working region (the channel). The following is an analogy. The source is like a sea, electrons are like seawater, and the channel is like land. To transfer water from the sea to the land, the water may be evaporated and then condensed as rain, or the land may be pushed down below the sea level. Evaporation from sea is an exponential function of temperature. This is the case of the BJT. The amount of water inundating the land is proportional to its subsidence. This is the case of FET. The level of the land and the temperature can both be measured as energies.

For a second difference, consider the concentration of the device current in the narrow peripheral region. In a MOSFET, current flows in a thin surface region of the silicon, whose thickness is of the order of the Debye length. This thickness is determined by the balance between the electrostatic attraction from the positive gate charge, which confines the induced electrons closely to the surface, and the thermal energy of the electrons, which pushes the electrons out from the surface region. In a BJT, there is an effect called *emitter crowding*. Since the base contact of a BJT is always on the side of the active base region, the current in the base is concentrated at the emitter edge directly facing the base contact. This effect creates an electron beam in the base region close to the base contact. For this reason, state-of-the-art BJTs have a narrow and long emitter contact, and two base contacts on both sides of the emitter. This is apparently quite similar to the surface conduction in a MOSFET. The two effects are, however, due to different mechanisms. Current crowding is due to the voltage drop within the base region, which reduces the effective base-emitter voltage and the minority electron injection.

A semiconductor triode structure is characterized by two size parameters corresponding to different processing techniques. Planar semiconductor processing consists of

1. growth, or alteration of semiconductor layer in the direction perpendicular to the wafer (epitaxial growth, ion implantation, or diffusion processes), and

2. definition of the features within the plane of the wafer.

Process 2 is carried out by photolithography, and has successively smaller feature size Λ as the technology scales down. Process 1 can be a relatively slow chemical process. By controlling the time and the temperature, the dimension of the vertical structure, λ, can be made very small. The thickness of the layer can be orders of magnitude smaller than the feature size:

$$\Lambda \gg \lambda \tag{2.6}$$

An interesting issue is, which of Λ or λ determines which of the crucial dimensions of the device. In BJTs λ determines the emitter-collector distance, and Λ determines the emitter width. In MOSFETs Λ determines the drain-source distance, and λ determines the gate oxide thickness.

Devices have contacts. The structure of the three contacts to the source, to the gate, and to the drain of a BJT and an MOSFET are schematically shown in Figure 2.2(a) and (b), respectively. Contacts contribute primarily to the device parasitic capacitances. The device active region is shown in the dotted box. The contacts have approximately the same size and shape, roughly a $\Lambda \times \Lambda \times \Lambda$ cube. Contacts are made by photolithography, and therefore their horizontal dimension (size measured within the plane of wafer) is dictated by the photolithography feature size Λ. The space between the contacts is also dictated by Λ. The vertical dimension of the contact structure is dictated by photolithography as well, since the size / depth ratio of the cuts from one level to the level below must be of the order of unity. ICs must have this mechanical-structural consistency, so if one dimension is scaled down, the other dimensions need to be scaled down as well.

Since device contacts have volume $\Lambda \times \Lambda \times \Lambda$ and they are separated by the distance Λ, capacitances between the contacts of the order of $\epsilon \Lambda$ are added to the minimum-size device. There are also capacitances to the common substrate, which are proportional to $\Lambda \times \Lambda$ (the drain area for

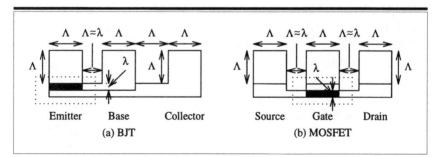

Figure 2.2 Contact Structure of Semiconductor Triodes

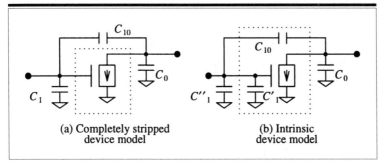

(a) Completely stripped
device model

(b) Intrinsic
device model

Figure 2.3 Simplified Device Models

the minimum-size device). The device capacitances can be separated out and credited to the interconnects, thereby *stripping out* the nonessential parameters from the device. There are two ways of removing a parasitic capacitance from the device.

1. Remove all the capacitance from the device, and represent the device by a controlled current generator or a controlled resistor alone, as shown in Figure 2.3(a). In this figure, C_0 is the triode drain capacitance, C_{10} is the gate-to-drain capacitance (often called a Miller capacitance), and C_1 is the input capacitance, which includes the gate capacitance of the triode. Only the controlled current generator within the dotted box is represented by the triode symbol.

2. Remove all capacitive parasitics except for the gate and Miller capacitances, which is essential for device operation. A device whose nonessential parasitics are thus removed is called an *intrinsic* device.

Devices simplified as in alternative 1 behave very much like switches. This simplification is convenient for studying the effects of the circuit connectivity [1]. Simplification 2 separates the physically irrelevant parasitics from the device, and it is convenient for studying the high-frequency device performance. This simplification is useful for introducing the concept of the gain-bandwidth limit of a device [21].

Semiconductor triodes are closely similar to each other. A semiconductor triode converts input voltage V_1 to control charge $-Q_C$. The control charge attracts an equal amount of mobile charge having the opposite sign, Q_C, that conducts current in the channel. Conversion from the voltage to the charge to the current is the fundamental mechanism of electron triode operation, and it imposes directionality on the signal, since the reverse conversion in the same device structure is impossible [1]. The conversion asymmetry originates from the triode structure: The charge in the channel is able to move under the influence of the drain-source field, and it can be recycled through the closed loop of the load circuit and the power supply. The charge on the gate stays there.

A semiconductor triode amplifies a signal by the following mechanisms:

1. The mobile charge Q_C in the channel is the charge deposited on the input capacitance C_1 of the triode. The charge is proportional to the input capacitance and to the input voltage.

2. The device current I_D generated in the channel is proportional to the charge. The reciprocal of the proportionality factor is the time required to move charge Q_C through the space, T_t, which is called the transit time.

3. The device current is integrated into the capacitance of the output node C_0 to regenerate the output voltage. The capacitance that inevitably exists at the output node is the input capacitance of the following amplifier stage C_1 that accepts the signal. The output voltage V_0 generated at the terminals of this capacitance is proportional to the time interval of integration, t_I.

From items 1–3 we may conclude as follows:

$$Q_C = C_1 V_1 \qquad I_D = Q_C / T_1$$
$$V_0 = I_D t_I / C_0 = V_1(t_I / T_t) \qquad \text{or} \quad V_0 / V_1 = G = t_I / T_t \tag{2.7}$$

where G is the gain of the triode as a voltage amplifier. This may be rewritten as

$$G \cdot f_{\max} = 1 / T_t \tag{2.8}$$

where $f_{\max} = 1/t_I$ is the maximum frequency at which the device is able to generate a voltage gain; it equals the *bandwidth* of the device as an amplifier. This relationship is called the gain-bandwidth product limitation. What has been shown here is the most fundamental limitation; there are many other practical limitations that push the limit lower. The gain-bandwidth product is still the most fundamental relationship that can be derived from the basic working mechanisms of the triode alone.

The ultimate speed limit of a triode is set by the transit time of the carriers traveling from the source to the drain. By technology scaledown the transit time scales down as well. In silicon the saturation electron drift velocity is 5×10^6 cm/sec. The transit time of a 0.2-μm-long FET channel is 4 ps. This transit time is much shorter than the interconnect delay time, which is of the order of 100 ps or more. The disparity between the device speed and the interconnect speed grow as the technology scales down and the system complexity increases.

2.3 Device Characteristics

At the high speeds the device I-V characteristic is described by the following simple model. The structure of a semiconductor triode is shown in Figure 2.4(a). The source-drain space is the channel region where the high field created by the drain voltage drives the carriers from the source to the drain. In a semiconductor triode the channel region consists of the space-charge layer of the reverse-biased drain-channel *pn* junction. The electric field in most of the channel is high if the triode is biased to the operating point where it has gain. The gate acts directly on the source; it pulls charge Q_C from the source, and injects it into the space-charge region.

In the channel, the charge Q_C travels with the saturation drift velocity v_∞, which depends on the semiconductor material alone, if the drain voltage V_D is high. The current I_D is given by

$$I_D = I_{D\infty} = Q_C v_\infty / \Lambda \tag{2.9}$$

where Λ is the source-drain distance, and Λ / v_∞ is the transit time T_t. The value $I_D = I_{D\infty}$ is reached if V_D is higher than the minimum V_{min} required to create the drift-velocity saturation field E_{SAT} in the channel region. Then

$$\begin{aligned} I_D &= I_{D\infty} \quad \text{if} \quad V_D \geq V_{min} = \Lambda E_{SAT} \\ &= I_{D\infty}(V_D / V_{min}) \quad \text{if} \quad V_D \leq V_{min} \end{aligned} \tag{2.10}$$

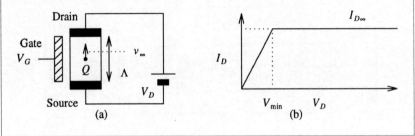

Figure 2.4 Semiconductor Triode Characteristics

This characteristic is schematically shown in Figure 2.4(b). $I_{D\infty}$ depends on the gate voltage V_G in two different ways, because the characteristic of Q_C versus V_G can be either of the two kinds discussed in Section 2.2. If the charge Q_C is created by evapolating the carriers from the source (as in BJTs), then

$$Q_C = Q_0 \exp(\alpha V_G) \tag{2.11}$$

and if it is created by electrostatic induction (as in MOSFETs and MESFETs), then

$$Q_C = Q_0(\beta V_G + \gamma) \tag{2.12}$$

where $\alpha, \beta, \gamma,$ and Q_0 are constants.

From this simple theory two models of a semiconductor triode are derived. They are for the limits of high and low channel field. The one is based on a controlled current generator, and the other on a controlled resistor. The source of the device is grounded, and the gate and the drain are biased to V_G and V_D relative to ground, respectively, as shown in Figure 2.5(a). Let us consider the device current I_D versus the drain voltage V_D. In a device modeled using a controlled current generator, I_D is determined by V_G alone, and it is independent on V_D. In real devices the model breaks down if V_D approaches zero. To make a reasonable approximation in this limit we set an exception for $V_D = 0$: If I_D forced through the device

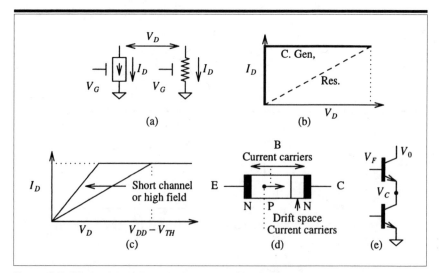

Figure 2.5 Device Modeling and Active and Passive Devices

is less than the constant current determined by V_G, then $V_D = 0$ as shown in Figure 2.5(b) (thick curve). This *collapsible* current generator model was studied in my last monograph [1].

If the device is modeled by a controlled resistor, then I_D is proportional to V_D, and the resistance is determined by the input voltage V_G, as shown in Figure 2.5(b) (dashed curve). If we compare the two models, I_D in the resistor model yields to V_D variation, but I_D in the collapsible current generator model does not. Using an anthropomorphism, the device represented by the current generator model has a strong "will" to maintain itself: We may say that the current generator model represents a more *active* device than the resistor model. The active and the passive device models are consistent with the semiconductor device physics. Figure 2.5(c) schematically shows the channel current-drain voltage characteristic of a MOSFET. A long-channel MOSFET operating at low channel field is almost an ideal controlled resistor, a very *passive* triode. As the maximum drain or gate voltage V_{DD} and the threshold voltage V_{TH} are kept constant and the channel length is reduced, MOSFET saturates at successively lower voltages than the low-field saturation voltage $V_{DD} - V_{TH}$ [16, 17]. At higher channel fields the device becomes more *active*, because of the carrier velocity saturation. BJTs are very active devices, since a very high field exists in the base-collector reverse-biased *pn* junction. The collector-base space-charge layer of Figure 2.5(d) prevents the collector voltage from influencing the device current. The BJT becomes inactive when the device goes into *saturation* (pseudo short-circuit condition at high base current and low collector voltage). A drain-channel space-charge layer in a MOSFET is created only if the drain voltage is higher than the gate voltage minus the threshold voltage. This is the reason why MOSFETs are less active than BJTs.

We need to clarify the relationship between the terms *active* and *amplifying*. *Active* means that an amplifying triode has high internal resistance, and the circuit operation is controlled by the triode and not significantly affected by the load and the drain bias. If we used this definition alone, however, smaller devices might be thought as more active. That is not what the definition means. The shape of the device's current-voltage characteristic, rather than its scale factor, is the issue. In Figure 2.5(c), a device that saturates at the same current level at a smaller drain voltage is more active. The term is best clarified using anthropomorphism again: A device or circuit (in analogy, a human) is said to be active to the extent that it can determine its output current without regard to the load impedance (in analogy, environmental conditions). Obviously, application of this term to two-terminal devices does not make sense: "Active" refers to electron triodes.

It is possible to create a highly active device by integrating less active devices into a circuit. An electron triode can be made more active by reducing the influence of the drain voltage on the device current. Figure 2.5(e) shows a BJT cascode circuit, where the collector voltage of the lower BJT is held effectively at $V_F - V_{BE}$, where V_F is the constant DC bias voltage of the upper BJT. Then the current becomes practically independent of the anode voltage V_0. The cascode circuit has been used since the days of vacuum tubes. A vacuum pentode (a cascode circuit integrated into a single vacuum tube) is a more active device than a vacuum triode. In Figure 2.5(e) the base of the upper BJT works like the screen grid of a vacuum pentode, thereby maintaining the lower BJT at the constant current. A more active device can be synthesized using less active devices and circuit techniques. This makes the boundary between complex device and circuit quite ambiguous. My view is that the boundary does not exist at all. We discuss this view in the next section.

2.4 Amplification Mechanisms

There are two distinct mechanisms of electrical signal amplification: amplification by a unidirectional triode, and amplification by a bidirectional negative-resistance diode. The triode amplification mechanism was discussed in Section 2.2. Negative-resistance diode amplification is used to reduce the net loss of an amplifier, including the loss of power delivered to the subsequent stage. In the equivalent circuit of Figure 2.6 let the internal impedance of the signal source be R_S, that of the load be R_L, and the negative resistance be $-R_N$ ($R_N > 0$). The current that flows through the load, I_L, is given by

$$I_L = I_S \frac{R_S}{R_L - R_N + R_S} \quad (> I_S) \qquad \text{if} \quad R_N > R_L \qquad (2.13)$$

Figure 2.6 Negative-Resistance Diode Amplifier

There are two different types of negative-resistance diodes: current-controlled (S-shaped *I-V* characteristics), and voltage-controlled (N-shaped *I-V* characteristics). *pnpn* switches and Esaki (tunnel) diodes are respective examples. A voltage-controlled negative-resistance diode can be biased stably in the negative-resistance region if it is connected to a low-impedance voltage source. A current-controlled negative-resistance diode can be biased stably in the negative-resistance region if it is driven by a high-impedance current source. Let us consider a simple example of a negative-resistance mechanism. A thermistor is a slab of oxide-semiconductor material having two terminals. The specific resistance of the material decreases strongly with increasing temperature. If the thermistor current is increased, its temperature increases, and the resistance decreases. As the consequence the terminal voltage decreases. Suppose that the thermistor is set at a bias point in the negative-resistance region by a current generator. If, for any reason, the thermistor voltage becomes higher, then the power consumption increases, the resistance decreases, and the thermistor voltage decreases. Accordingly, the bias point is stable. The thermistor is a *current-controlled* negative-resistance device.

The feedback mechanism creating the thermistor negative resistance suggests that a negative-resistance diode could be considered as a *circuit* in a black box, which creates negative resistance at the terminals by an internal feedback mechanism. A negative-resistance diode can be synthesized using triodes in circuits. Then the two amplification mechanisms are not independent. It is impossible, however, to synthesize an electron triode using negative-resistance diodes alone, because we need a unidirectional device. From this observation, electron triodes appear more fundamental and elementary than negative-resistance diodes. I expect that this point may not be accepted by some readers, but observation of the two types of amplification mechanism inevitably leads to this conclusion.

The mechanism of a negative-resistance diode can be modeled by a circuit, and its general characteristics can be understood from its circuit model. An example of a circuit, called a λ-diode, that presents a voltage-controlled negative resistance is shown in Figure 2.7(a) [22]. Suppose that the n- and p-type triodes in the circuit are modeled using the collapsible current generator model, with transconductances g_{mN} and g_{mP}, and threshold voltages V_{THN} and V_{THP}, respectively. The current in the n-type triode is given by $g_{mN}(V_1 - V_{THN})$. The voltage V_1 and current I_1 are related by

$$
\begin{aligned}
I_1 &= V_1/R_N \quad (V_1 \leq V_{10}) \\
I_1 &= [(1/R_N) - g_{mP}g_{mN}R_P]V_1 \\
&\quad + g_{mP}(V_{THP} + R_P g_{mN}V_{THN}) \quad (V_1 > V_{10})
\end{aligned}
\tag{2.14}
$$

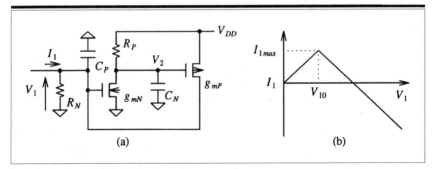

Figure 2.7 Negative-Resistance Circuit

where $V_{10} = V_{THN} + (V_{THP}/R_P g_{mN})$ is the input voltage where the p-type triode begins to conduct. We consider the DC characteristics. I_1 versus V_1 is plotted in Figure 2.7(b). The current maximum, $I_{1\,max} = V_{10}/R_N$, occurs at $V_1 = V_{10}$. The I_1-V_1 characteristic shows negative resistance in the range $V_{10} < V_1 < V_{DD}$ if $g_{mP} g_{mN} R_N R_P > 1$, where V_{DD} is the power supply voltage.

We consider the frequency characteristics of the negative resistance. Suppose that V_1 is held at the DC bias point V_{1S} within the negative-resistance region by an input voltage source of negligibly low internal resistance. At $V_1 = V_{1S}$ we have $I_1 = I_{1S}$. If a small variation of V_1 takes place around V_{1S} such that

$$V_1 = V_{1S} + v_1 \qquad V_2 = V_{2S} + v_2 \qquad I_1 = I_{1S} + i_1 \quad (2.15)$$

then the small variations v_1, v_2, and i_1 satisfy the circuit equations

$$C_N \frac{dv_2}{dt} + g_{mN} v_1 = -\frac{v_2}{R_P} \qquad C_P \frac{dv_1}{dt} + \frac{v_1}{R_N} = -g_{mP} v_2 + i_1 \quad (2.16)$$

If the small variations depend on time like $\exp(j\omega t)$, then

$$
\begin{aligned}
i_1 &= \left(j\omega C_P + \frac{1}{R_N} - \frac{g_{mP} g_{mN} R_P}{1 + j\omega C_N R_P} \right) v_1 \\
&= \left[\frac{1}{R_N} - \frac{g_{mP} g_{mN} R_P}{1 + (\omega C_N R_P)^2} \right] v_1 \\
&\quad + j\omega \left[C_P + \frac{R_P^2 g_{mP} g_{mN}}{1 + (\omega C_N R_P)^2} C_N \right] v_1
\end{aligned}
\quad (2.17)
$$

The small-signal negative conductance vanishes at the angular frequency ω_0 defined by

$$\omega_0 = \sqrt{G_N G_P - 1}/C_N R_P \qquad (2.18)$$

where $G_N = g_{mN}R_P$ and $G_P = g_{mP}R_N$ are the DC gains of the two grounded-source amplifier stages. If $G_N G_P \gg 1$,

$$\omega_0 \approx (g_{mN}/C_N)(G_P/G_N) \qquad (2.19)$$

where g_{mN}/C_N is the unit-gain angular frequency of the amplifier first stage when $R_P \to \infty$. The negative resistance vanishes at the frequency where the amplifier loses gain. What we have shown here is that the gain stages of the negative-feedback system have a frequency characteristic, and the negative resistance vanishes at the frequency where the circuit's gain is lost. It is always possible to make an equivalent circuit of a negative-resistance diode. Then the electron triode amplifier used in the equivalent circuit is always faster than or at the least equal to the amplifier using the synthesized negative resistance, because it requires two stages of amplification. Isn't it possible to conjecture that ultimately no device is faster than an electron triode? By generalizing this conclusion we ask the following question: What determines the ultimate frequency limits of negative-resistance diodes, such as *pnpn* switches and tunnel diodes? Are they really faster than triodes? I think that they are not faster than an ultimately fast triode.

Negative resistance of a *pnpn* switch is modeled by a bipolar circuit, shown in Figure 2.8(a), that closely reflects the device's working mechanisms. Negative resistance shows up between terminal P and ground. Figure 2.8(a) shows the I-V characteristic as well. A *pnpn* switch has been considered as an elementary electron device, but it is rather a circuit consisting of the more fundamental BJTs. Then can the other negative-resistance diodes be described by circuits? In particular, how about tunnel diodes? In our circuit theory we regard electron devices and passive components as being like atoms having certain valences, and circuits as being like molecules constructed from the atoms. In this analogy, "Are some conventional electron devices really as elementary as atoms are in chemistry?" is a logical and nonacademic question.

The structure of a tunnel diode is shown in Figure 2.8(b). In its simplicity and in its small size, the device appears elementary. This is a *pn* junction diode, except that the transition region between the n- and the p-type semiconductors is extremely thin, less than 100 Å and both p- and

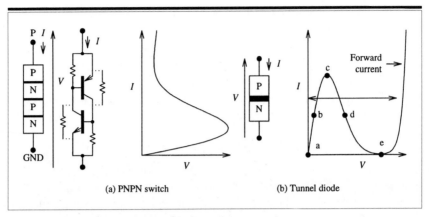

(a) PNPN switch   (b) Tunnel diode

Figure 2.8 Negative-Resistance Diodes

n-type semiconductors are doped to degeneration. The electrons in the n-type region are able to tunnel into the p-type region by the quantum-mechanical tunnel effect. Due to the tunnel effect combined with the band structure of degenerate semiconductors, a negative-resistance characteristic as shown in Figure 2.8(b) is created. The thickness of the transition region between the n- and p-type regions, W, is a crucial parameter; it is estimated from

$$W \approx \sqrt{2\epsilon_s V_{EG}/q}\ \sqrt{(1/N_A) + (1/N_D)} \qquad (2.20)$$
$$= 1.203 \times 10^{-6}\sqrt{(1/N_A) + (1/N_D)}$$

where V_{EG} is the forbidden energy gap of the semiconductor (1.11 eV for silicon), and N_A and N_D in the second line are measured in units of $10^{19}/cm^3$. To make a thin transition region, N_A and N_D must be higher than $10^{19}/cm^3$. If $N_A = N_D = 8\ (\times 10^{19}/cm^3)$, $W = 60$ Å. Electrons are able to tunnel a barrier of this width and height. The first significant effect of high doping is that the forward diffusion current of the pn junction begins to flow at a higher forward bias voltage than for conventional diodes that are doped less. A wide voltage range (500-800 mV) of low forward current is created, where the second significant effect creates a negative resistance.

Electrons in highly doped semiconductors are in *degenerate* states, which are described by Fermi-Dirac statistics in thermal equilibrium. With reference to the equilibrium energy band diagram of Figure 2.9(a), the conduction band of the n-type region and the valence band of the p-type region are both filled up to the Fermi energy E_F, and practically no electrons exist in either band whose energy is higher than E_F plus the thermal

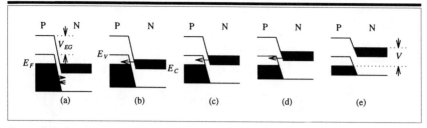

Figure 2.9 Energy Band Diagram of Tunnel Diode

energy kT/q (energies are measured in electron volts). Electrons are exchanged across the barrier by thermal excitation and by the tunnel effect, but the time-averaged current across the barrier is zero. This is the point a on the *I-V* characteristic of Figure 2.8(b).

If the forward bias voltage is applied to the *pn* junction (the *p*-type side positive and the *n*-type side negative), the Fermi level of the electrons on the *n*-type side increases by the bias voltage V as shown in Figure 2.9(b)–(e). Electrons in the *n*-type region move into the *p*-type region by the two mechanisms: a thermal excitation to the conduction band of the *p*-type region followed by recombination with the holes, and tunneling to the unoccupied valence band energy levels of the *p*-type region. The first type of transition is insignificant where tunneling transitions occur.

For a tunneling transition to take place from a conduction band state to a valence band state, the conduction band state must be occupied by an electron, and the valence band state must be empty. The probability of occupation by electrons of a single conduction or valence band energy state in thermal equilibrium is given by the Fermi-Dirac factor [23]. The pair of states between which the tunneling transition takes place must have the same energy, and must exist in the energy range $E_C \leq E \leq E_V$ shown in Figure 2.9(b). The conduction and the valence bands have densities of states per unit energy and per unit volume given, respectively, by

$$\rho_C(E) = 4\pi(2m_C^*/h^2)^{3/2}\sqrt{E - E_C} \quad (E > E_C)$$
$$= 0 \quad (E \leq E_C)$$
$$\rho_V(E) = 4\pi(2m_V^*/h^2)^{3/2}\sqrt{E_V - E} \quad (E < E_V)$$
$$= 0 \quad (E \geq E_V)$$

(2.21)

where m_C^* and m_V^* are the density-of-states masses of the conduction and the valence band, respectively, and h is Planck's constant [23]. The tunneling current from the conduction band to the valence band and that from the valence band to the conduction band are given, respectively, by Esaki's formula [24]

$$I(C \rightarrow V) = \text{const.} \int_{E_C}^{E_V} Z(E)\rho_C(E)\rho_V(E)$$
$$f_C(E)[1 - f_V(E)] \, dE \qquad (E_V \geq E_C) \qquad (2.22)$$

$$I(V \rightarrow C) = \text{const.} \int_{E_C}^{E_V} Z(E)\rho_C(E)\rho_V(E)$$
$$f_V(E)[1 - f_C(E)] \, dE \qquad (E_V \geq E_C)$$

where $f_C(E)$ and $f_V(E)$ are the probabilities that a single energy level in the conduction or the valence bands, respectively, is occupied by an electron. They must be determined by solving the electron transport equations consistently, including the tunneling. $Z(E)$ is the probability of tunneling between the conduction and the valence band states. The net current is the difference of the two, given by

$$I_D = I(C \rightarrow V) - I(V \rightarrow C)$$

$$\qquad (2.23)$$

$$= \text{const.} \cdot Z \cdot \int_{E_C}^{E_V} \rho_C(E)\rho_V(E)[f_C(E) - f_V(E)] \, dE \qquad (E_V \geq E_C)$$

neglecting the energy dependence of the tunneling probability. Z is estimated from the quantum-mechanical tunnel effect analysis as

$$Z = \exp\left[-\frac{16\pi}{3} \cdot \frac{\sqrt{m^*\epsilon_s}}{he} \cdot V_{EG} \cdot \left(\frac{1}{N_A} + \frac{1}{N_D} \right)^{1/2} \right] \qquad (2.24)$$

Equation (2.23) gives the current-voltage characteristic of the tunnel current only. The *pn* junction forward current must be added to it to compute the diode current. In the state of Figure 2.9(b) the tunneling current begins to flow [*b* of Figure 2.8(b)]. In the state of Figure 2.9(c) the forward current reaches its maximum [*c* of Figure 2.8(b)]. Upon further increase in the voltage, the number of the pairs of states in the conduction and the valence band that have the same energy decreases with increasing applied voltage, and the tunnel current decreases. In the state of Figure 2.9(e) the tunneling current vanishes.

Let us examine the physical meaning of the Esaki formula. First, the electron in the conduction band must exist in some state. The electron must have moved into the device from the outside, and then scattered into the energy state from which the tunneling takes place. The electron that has tunneled to a valence band state must be scattered out of the state and carried away from the contact to the *p*-type region. The tunneling is preceded and followed by thermodynamic scattering processes. Until the electron is scattered out of the valence band state and the conduction band state is occupied again, no further tunneling takes place between the two states. If the tunneling current is small and the thermal equilibrium is not disturbed significantly, $f_C(E)$ and $f_V(E)$ can be approximated by the Fermi-Dirac factor in thermal equilibrium:

$$f(E) = 1/[1 + \exp[(E - E_F)/kT]] \tag{2.25}$$

In the limit of high tunnel current the Fermi-Dirac factor shifts from its thermal equilibrium value. This is the case where the ultimate high-speed limit is reached.

From Equations (2.23) and (2.24) we see that the time constant of the tunnel diode depends on the transition-region width W as

$$\text{time constant} \approx \exp(-\text{const} \times W)/W \tag{2.26}$$

where the exponential factor is proportional to the diode equivalent resistance, and $1/W$ to the junction capacitance. The high-speed limit is approached by decreasing W. In the limit, however, the tunnel current should be determined by the time required for the electrons to scatter into the conduction band energy level and to scatter out of the valence band energy level. The scattering time depends on many factors that are not physically fundamental, such as lattice thermal vibrations. There are several mechanisms working together in the diode. First, if an electron tunnels from the *n*- to the *p*-type region, the junction is temporarily polarized so as to reduce the potential applied to the *pn* junction transition region. If the diode is in the negative-resistance region, the effect increases the number of filled energy levels in the conduction band and the number of empty energy levels in the valence band that energetically overlap. This effect increases the tunnel current. The junction polarization increases the effective barrier height for tunneling, thereby reducing the tunnel current. Until the electron deficiency in the conduction band is replenished by electron scattering and the occupied energy levels of the valence band are vacated, tunneling from the conduction band level to the valence band level is forbidden. This is the effect that reduces the tunnel current at high tunnel current density. We see that the mechanism of tunnel diode operation at the high frequencies is a complex web of negative and positive

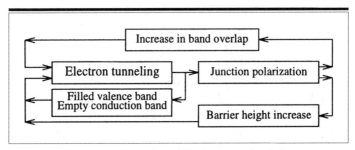

Figure 2.10 Mechanisms of Tunnel Diode Operation in the Negative-Resistance Region

feedbacks, each of which has its own time constant, as shown schematically in Figure 2.10. The most essential effect creating negative resistance, the voltage dependence of the overlap of the two bands, is a quantum-mechanical effect. The wave function of an electron in the conduction band of the *n*-type region spreads out to the *p*-type region and senses whether or not the terminal energy level is available for a tunneling transition. This is a feedback mechanism, carrying the information from downstream to upstream. The tunneling transition takes place instantly, but when it takes place is determined by the other, less fundamental processes. From this observation I judge that a tunnel diode is more like a *circuit* than a *device*, in that it has many internal feedback mechanisms. The tunnel diode is not the ultimately fast device, I think. The more fundamental electromagnetic control mechanism of an electron triode can be used to make a faster amplifying device, when the technology matures.

As we saw from the examples of a *pnpn* switch, a tunnel diode, and a negative-resistance device synthesized from triodes such as that shown in Figure 2.7, the boundary between *device* and *circuit* is not clear at all. This ambiguity may be strange from the traditional circuit-theoretical viewpoint, but that is to be expected, since an equivalent circuit contains a hierarchy that may even go down inside a device. The only elementary electron device is a triode: The triode cannot be decomposed any further. Any more complex device can be described as a *black box* that is synthesized from triodes. Black boxes can be described as circuits, have complex internal working mechanisms, and are ultimately slower than elementary triodes. Within a black box, the energy may be converted from one form to another and back, which may add to the delay: In an optocoupler device it may be converted to light energy, and in transferred-electron devices to the potential energy of electrons in the higher energy bands. Even a passive component, inductance, may be considered as a black box, in which the electrical energy is converted to magnetic energy.

The use of black boxes in circuit theory has an interesting implication. Later it will be convenient to consider that the black boxes are *outside* the circuit. Modeling a circuit by splitting it into the intrinsic part and the extrinsic part, or the *inside* and the *outside*, allows, above all, a description based on the assumption that the energy exchange between the two parts determines the circuit response. A circuit draws energy from the power supply and converts it to heat. This conversion is concomitant with the forward flow of time. If energy returns from the outside of the circuit and adds to the existing energy, it appears as if the time had flowed backward. This may also be expressed by saying that the energy resets the circuit's initial condition. Thus the circuit operation can be separated into two: forward and backward operation.

2.5 High-Frequency Circuit Model

In this section we study how to model the digital circuits of an IC chip. The observation in the last section, that the distinction between device and circuit is unclear, leads to the idea that the devices and the interconnects should be modeled together by a single equivalent circuit. The circuit-theoretical framework appropriate for this purpose is the *microstate* theory. In my last monograph [1] I introduced the concept of microstate, and of microstate sequence in digital circuits. In the microstate theory, semiconductor triodes in a digital circuit can be in the nonconducting, the triode (or short-circuit), or the active state (equivalent to a voltage-controlled current generator having infinite internal impedance). In this theory the active device's role is only to control the circuit connectivity. The state of the circuit is determined by the states of all the triodes of the circuit. In this model the role of the triodes is restricted, and the role of the interconnects is extended. Because of the built-in emphasis on the interconnects, the microstate model of a circuit is a convenient starting point of our high-speed theory. Microstates of scaled-down CMOS digital circuits are especially simple: They are described by a set of simultaneous linear differential equations of the node voltages V_i as

$$\frac{dV_i}{dt} = \sum_{j=1}^{N} A_{ij} V_j + B_i \qquad 1 \le i \le N \qquad (2.27)$$

where A_{ij} and B_i are constants. Every time the structure of the set of simultaneous equations changes, a microstate transition takes place. The sequence of transitions is described by the microstate sequence [1].

The solution of the above equation is in the form

$$V_i = \sum_j B_{ij} \exp(\alpha_j t) + C_i \qquad (2.28)$$

and this is the representation of a single microstate. The solution is the sum of exponential terms of the form $\exp(\alpha_j t)$, where

$$B_{ij} \exp(\alpha_j t) \qquad j = 0, 1, 2, \ldots \qquad (2.29)$$

is the contribution from the jth *mode* to the ith *node* voltage. Can this component be regarded as a more fundamental state of the circuit that might be called a *submicrostate*? A submicrostate is a mode of the circuit that has the specified triode connectivity set. Understanding the modes of a linear system is not always a simple task. In the following sections I discuss several methods to analyze the modes, by using the peculiarities of the equivalent circuit model of common digital circuits. If we observe the general mathematical structure, however, a microstate is a superposition of more than one submicrostate, and so the submicrostates have physical significance. To excite a single submicrostate is, however, impossible. At the time of the circuit's configurational change, a set of submicrostates are excited together. From this viewpoint submicrostates do not have much significance. By this reasoning a microstate is the most fundamental concept of the circuit theory, because it is the minimum *tangible* state. The relationship between microstates and submicrostates is very much like the relationship between baryons and quarks in elementary-particle physics. Submicrostates belong to a physically different world, where many exotic features, such as multiple time describing a single circuit phenomenon, become real, as we will see in the later chapters. To understand high-frequency effects such as inductance effects, however, we need to understand the submicrostates.

The mechanisms of circuit operation at high frequencies are still primarily the charging and discharging of capacitors through resistors. We wish to develop a circuit theory including inductances that reduces to the conventional microstate theory in the limit of zero inductance. The microstate representation of the circuit operation allows the inclusion of loop inductance. Each microstate is defined by a set of linear simultaneous differential equations. The solutions of these equations are written as sums of the exponential terms (sums of the time-constant processes) as

$$V_i(t) = \sum_j X_{ij} \exp(\alpha_j t) \qquad (2.30)$$

where $V_i(t)$ is the time-dependent node i voltage, and the term $X_{ij}\exp(\alpha_j t)$ of the sum is the contribution to it from the jth mode. If inductance is included, the general structure of the solution as the sum of time-constant processes does not change. The time dependence of each mode will change, however, and since the order of the secular equation increases, new modes emerge. How do we modify the RC-circuit theoretical framework to accommodate inductance? The fundamental requirement is that the node voltages including inductances should converge to the results of the RC circuit model in the limit as the inductances tend to zero, in a mathematically obvious way. To write down the solution of the LCR circuit equation is not the best way. If that is done, it is hard to understand how each mode is changed by the inductances. Our objective is achieved by introducing transformed times into the circuit theory. We modify each mode of the backbone RC circuit as

$$\exp(\alpha_j t) \rightarrow \exp[\alpha_j \theta_j(t)] \qquad (2.31a)$$

by introducing the transformed time $\theta_j(t)$: Each exponential term argument, $\alpha_j t$, is replaced by $\alpha_j \theta_j(t)$. The original time t used in the backbone circuit having no inductance is the *global* time, while the time that the mode j refers to in the LCR circuit is the *local* time for mode j.

This is not the only way to introduce local times. Under certain restrictive conditions the solution of the microstate equation can be written as

$$V_i(t) = \sum_j A_{ij} \exp[\alpha_j \theta_i(t)] \qquad (2.31b)$$

In this case $\theta_i(t)$ is the local time of node i. We note that the index of the local time can be either i, the node index, or j, the mode index. In any case, in the limit of zero inductance,

$$\theta_i(t), \ \theta_j(t) \rightarrow t \qquad (2.32)$$

Our objective is to construct the circuit theory using the many-time formalism. This idea is straightforward as long as the inductances are small, and the circuits do not go into the multiple overshoots, or ringing oscillations. How to extend the many-time formalism when ringing occurs will be discussed later. The many-time formalism has already been found useful if nonlinearities of the components are included in the theory [1], but its usefulness is not limited to that. The theory built along this line of thought has many new and interesting features.

2.6 Equivalent Circuit Model

A digital circuit in one microstate has the component connectivity shown by a microstate circuit diagram that is a simplified version of the complete circuit diagram. Simplification is carried out by substitutions: A device that is in the triode region is replaced by a connecting wire, a device in the nonconducting region is removed, and a device in the saturation region is retained as a controlled current generator. The capacitances associated with the devices are retained, even if the device vanishes from the diagram. An example of circuit simplification is shown in Figure 2.11(a) (a complete circuit diagram) and Figure 2.11(b) (a microstate circuit diagram). In Figure 2.11(a) the FETs MNA and MPB are in the triode region, and MPA and MNB are in the nonconducting region. They are removed in the microstate circuit diagram. To include the effects of the interconnect resistance and capacitance, lumped RC circuits are added to the diagram. These interconnect components retain their connectivity all the time, but their *effective* connectivity can be different for different modes excited in them. This is shown later in this chapter. On the submicrostate level the already restricted connectivity of the circuit diagram is restricted for the second time. This hierarchical simplification is a neat feature of the submicrostate theory.

The simplified microstate circuit diagram does not have a completely arbitrary component connectivity. A circuit node is always connected to one capacitance terminal, and the other terminal is connected to an AC ground that includes ground, V_{DD}, and a reference voltage source. This is consistent with the requirement of Section 1.9 that a node must have capacitance to ground if the node voltage relative to ground is measurable.

Only a small number of nodes are connected to ground (or V_{DD}) by a resistor (or by a current generator modeling a device). The resistor or de-

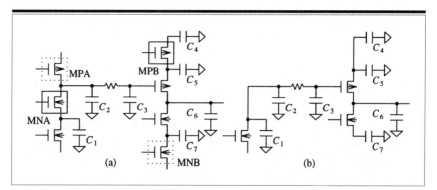

Figure 2.11 Complete Circuit Diagram and Microstate Circuit Diagram

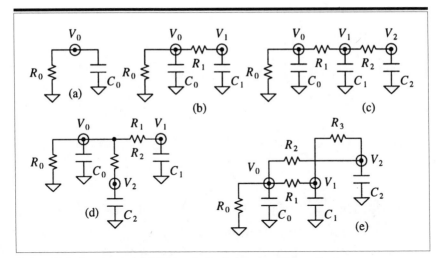

Figure 2.12 Fundamental *RC* Circuits Having Three or Fewer Nodes

vice discharges or precharges the connected nodes. Two physically meaningful nodes are connected by a resistor or by an FET. As long as we confine ourselves to circuit connectivity issues, a conducting FET and a resistor are interchangeable. In this chapter we replace an FET in the saturation region with constant gate voltage by a resistor, and an FET in the triode region by a short circuit.

The types of the circuits that satisfy the above restrictions are quite limited. Figure 2.12 shows all the circuits that have one node (a), two nodes (b), and three nodes (c)–(e) that have only one pulldown resistor. Since they are the most fundamental circuit structures that model digital integrated circuits, they are frequently used as the examples. We assume that the node voltages V_0, V_1, and V_2 are originally at V_{DD}, and we study how the circuit discharges. The central issue is how to understand the mechanisms of the modes of these circuits. This is not a simple task. Later we use both closed-form analysis and numerical analysis: We introduce the idea of using a Monte Carlo method in the process of understanding the modes. The first step toward this objective is to carry out the closed-form analysis. Rather than giving detailed algebra each time, I tabulate the relevant results for these fundamental circuits.

2.7 Analysis of the Simple Digital Circuit Models

The followings are the closed-form analysis results on the RC circuits shown in Figure 2.12(a)–(e).

- *Figure 2.12(a):*

$$\alpha = -1/R_0 C_0 \tag{2.33}$$

$$V_0(t) = A \exp(\alpha t) \quad \text{where} \quad A = V_{DD}$$

- *Figure 2.12(b):*

$$\alpha = (-\Gamma + \sqrt{\Delta})/2C_0 C_1, \quad \beta = (-\Gamma - \sqrt{\Delta})/2C_0 C_1 \tag{2.34}$$

where

$$\Gamma = C_1[(1/R_0) + (1/R_1)] + (C_0/R_1), \quad \Delta = \Gamma^2 - 4(C_0 C_1/R_0 R_1)$$
$$V_1(t) = A \exp(\alpha t) + B \exp(\beta t) \tag{2.35}$$
$$V_0(t) = C \exp(\alpha t) + D \exp(\beta t)$$
$$C = (1 + C_1 R_1 \alpha)A \quad D = (1 + C_1 R_1 \beta)B$$

and where

$$A = -\beta V_{DD}/(\alpha - \beta), \quad B = \alpha V_{DD}/(\alpha - \beta) \tag{2.36}$$

It is convenient to use a notation $T_{ij} = C_i R_j$ in the following complex equations.

- *Figure 2.12(c):* α, β, and γ are the solutions of the following secular equation:

$$S^3 + a_2 S^2 + a_1 S + a_0 = 0 \tag{2.37}$$

where

$$a_2 = (1/T_{00}) + (1/T_{11}) + (1/T_{22}) + (1/T_{01}) + (1/T_{12})$$
$$a_1 = (1/T_{00}T_{11}) + (1/T_{11}T_{22}) + (1/T_{22}T_{00}) \tag{2.38}$$
$$+ (1/T_{00}T_{12}) + (1/T_{11}T_{02}) + (1/T_{22}T_{01})$$
$$a_0 = 1/(T_{00}T_{11}T_{22})$$

This equation can be solved by Cardano's method as follows [25]. By setting

$$a = -(a_2^2 - 3a_1)/3, \quad b = (2a_2^3 - 9a_2 a_1 + 27a_0)/27 \tag{2.39}$$

and we define

$$u = \left[-\frac{b}{2} + \left(\frac{b^2}{4} + \frac{a^3}{27} \right)^{1/2} \right]^{1/3} \qquad v = -\frac{a}{3u} \qquad (2.40)$$

and we define the complex cube root of 1 as

$$\omega = (1/2) + j(\sqrt{3}/2) \qquad (2.41)$$

then we have

$$\alpha, \beta, \gamma = u + v \quad \omega u + \omega^2 v \quad \omega^2 u + \omega v \qquad (2.42)$$

They are all real, negative roots. It is convenient to assign their values such that $|\alpha| \leq |\beta| \leq |\gamma|$. Then

$$V_2(t) = A \exp(\alpha t) + B \exp(\beta t) + C \exp(\gamma t)$$

$$V_1(t) = D \exp(\alpha t) + E \exp(\beta t) + F \exp(\gamma t)$$

$$V_0(t) = G \exp(\alpha t) + H \exp(\beta t) + I \exp(\gamma t)$$

$$D = (1 + T_{22}\alpha)A$$

$$E = (1 + T_{22}\beta)B \qquad (2.43)$$

$$F = (1 + T_{22}\gamma)C$$

$$G = [(1 + T_{22}\alpha)[1 + (R_1/R_2) + T_{11}\alpha] - (R_1/R_2)]A$$

$$H = [(1 + T_{22}\beta)[1 + (R_1/R_2) + T_{11}\beta] - (R_1/R_2)]B$$

$$I = [(1 + T_{22}\gamma)[1 + (R_1/R_2) + T_{11}\gamma] - (R_1/R_2)]C$$

$$A = \frac{-\beta\gamma V_{DD}}{(\alpha - \beta)(\gamma - \alpha)} \qquad B = \frac{-\gamma\alpha V_{DD}}{(\alpha - \beta)(\beta - \gamma)} \qquad C = \frac{-\alpha\beta V_{DD}}{(\gamma - \alpha)(\beta - \gamma)}$$

- *Figure 2.12(d):* α, β, and γ are solutions of the following secular equation:

$$S^3 + a_2 S^2 + a_1 S + a_0 = 0 \qquad (2.44)$$

where

$$a_2 = (1/T_{00}) + (1/T_{11}) + (1/T_{22}) + (1/T_{01}) + (1/T_{02})$$

$$a_1 = (1/T_{00}T_{11}) + (1/T_{11}T_{22}) + (1/T_{22}T_{00}) + (1/T_{11}T_{02}) \qquad (2.45)$$
$$\quad + (1/T_{22}T_{01})$$

$$a_0 = 1/(T_{00}T_{11}T_{12})$$

with $|\alpha| \le |\beta| \le |\gamma|$. We have

$$V_2(t) = A \exp(\alpha t) + B \exp(\beta t) + C \exp(\gamma t)$$

$$V_1(t) = D \exp(\alpha t) + E \exp(\beta t) + F \exp(\gamma t)$$

$$V_0(t) = G \exp(\alpha t) + H \exp(\beta t) + I \exp(\gamma t)$$

$$G = a_{31}A = (1 + T_{22}\alpha)A$$

$$H = a_{32}B = (1 + T_{22}\beta)B$$

$$I = a_{33}C = (1 + T_{22}\gamma)C \tag{2.46}$$

$$D = a_{21}A = [[1 + (R_1/R_0) + (R_1/R_2) \\ + T_{01}\alpha](1 + T_{22}\alpha) - (R_1/R_2)]A$$

$$E = a_{22}B = [[1 + (R_1/R_0) + (R_1/R_2) \\ + T_{01}\beta](1 + T_{22}\beta) - (R_1/R_2)]B$$

$$F = a_{23}C = [[1 + (R_1/R_0) + (R_1/R_2) \\ + T_{01}\gamma](1 + T_{22}\gamma) - (R_1/R_2)]C$$

A, B, and C are determined by solving the simultaneous equations

$$\begin{bmatrix} 1 & 1 & 1 \\ a_{21} & a_{22} & a_{23} \\ a_{31} & a_{32} & a_{33} \end{bmatrix} \begin{bmatrix} A \\ B \\ C \end{bmatrix} = \begin{bmatrix} V_{DD} \\ V_{DD} \\ V_{DD} \end{bmatrix} \tag{2.47}$$

- *Figure 2.12(e):* α, β, and γ are solutions of the following secular equation:

$$S^3 + a_2S^2 + a_1S + a_0 = 0 \tag{2.48}$$

where

$$a_2 = (1/T_{00}) + (1/T_{11}) + (1/T_{22}) + (1/T_{01}) \\ + (1/T_{02}) + (1/T_{13}) + (1/T_{23})$$

$$a_1 = (1/T_{00}T_{11}) + (1/T_{00}T_{13}) + (1/T_{00}T_{22}) \\ + (1/T_{00}T_{23}) + (1/T_{11}T_{02}) + (1/T_{11}T_{03}) \\ + (1/T_{11}T_{22}) + (1/T_{11}T_{23}) + (1/T_{22}T_{01}) \\ + (1/T_{22}T_{03}) + (1/T_{22}T_{13}) + (1/T_{01}T_{23}) + (1/T_{02}T_{13}) \tag{2.49}$$

$$a_0 = (1/T_{00}T_{11}T_{22}) + (1/T_{00}T_{11}T_{23}) + (1/T_{00}T_{13}T_{22})$$

with $|\alpha| \le |\beta| \le |\gamma|$. We have

$$V_2(t) = A\,\exp(\alpha t) + B\,\exp(\beta t) + C\,\exp(\gamma t)$$

$$V_1(t) = D\,\exp(\alpha t) + E\,\exp(\beta t) + F\,\exp(\gamma t)$$

$$V_0(t) = G\,\exp(\alpha t) + H\,\exp(\beta t) + I\,\exp(\gamma t)$$

$$D = a_{21}A = [R_1 + R_2 + R_3(1 + T_{22}\alpha)]A/[R_1 + R_2 + R_3(1 + T_{11}\alpha)]$$

$$E = a_{22}B = [R_1 + R_2 + R_3(1 + T_{22}\beta)]B/[R_1 + R_2 + R_3(1 + T_{11}\beta)]$$

$$F = a_{23}C = [R_1 + R_2 + R_3(1 + T_{22}\gamma)]C/[R_1 + R_2 + R_3(1 + T_{11}\gamma)]$$

$$G = a_{31}A = \frac{(1 + T_{11}\alpha)[R_1 + R_2 + R_3(1 + T_{22}\alpha)] + R_1(T_{22} - T_{11})\alpha}{R_1 + R_2 + R_3(1 + T_{11}\alpha)}A$$

$$H = a_{32}B = \frac{(1 + T_{11}\beta)[R_1 + R_2 + R_3(1 + T_{22}\beta)] + R_1(T_{22} - T_{11})\beta}{R_1 + R_2 + R_3(1 + T_{11}\beta)}B$$

$$I = a_{33}C = \frac{(1 + T_{11}\gamma)[R_1 + R_2 + R_3(1 + T_{22}\gamma)] + R_1(T_{22} - T_{11})\gamma}{R_1 + R_2 + R_3(1 + T_{11}\gamma)}C$$

$$(2.50)$$

A, B, and C are determined by solving the simultaneous equations

$$\begin{bmatrix} 1 & 1 & 1 \\ a_{21} & a_{22} & a_{23} \\ a_{31} & a_{32} & a_{33} \end{bmatrix} \begin{bmatrix} A \\ B \\ C \end{bmatrix} = \begin{bmatrix} V_{DD} \\ V_{DD} \\ V_{DD} \end{bmatrix} \qquad (2.51)$$

We have analyzed one-, two-, and three-node circuits. To study more complex circuits, it is necessary to list all the circuits that have the specified number of nodes, and to classify them systematically. We need a simple scheme of circuit representation. In Figure 2.13(a) a node having capaci-

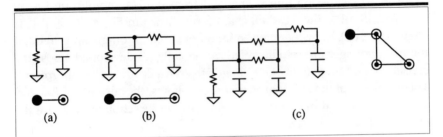

Figure 2.13 *RC* Circuit Connectivity Diagram

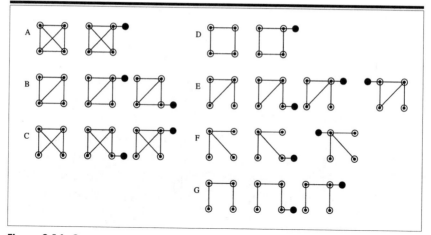

Figure 2.14 Connectivities of Four-Node *RC* Circuits

tance to ground is represented by an eye, and ground by a closed circle. The connecting line represents a resistance. Figure 2.13(b) and (c) show examples of representation of the two- and three-node circuits. In this representation, an *N*-node circuit has *N* eyes and one closed circle. All the circuits are generated by exhaustively listing the ways of connecting the *N* eyes by lines and connecting any one of the eyes to ground. In Figure 2.14 four-node *RC* circuits are classified using the symbolic representation. It is convenient to classify the circuits by the number of links connecting the eyes, that is, by the number of the resistors in the circuit. There are classes *A* (six resistors), *B* and *C* (five resistors), *D* and *E* (four resistors), and *F* and *G* (three resistors). In each class, there are a few varieties, according to where the ground connection is made.

Circuit analysis often assumes that there is an ideal ground. This assumption becomes invalid if inductance is included, as we saw in Section 1.10. Even if inductance is not included, ground of an IC can be less than ideal if the substrate is resistive. In this case the equivalent circuit of an *RC* transmission line is changed from the common-ground form [Figure 2.15(b)] to the *RC* dual-chain form shown in Figure 2.15(a). This circuit is described by the same equation as the common-ground circuit, if the node voltages are reinterpreted as the voltages developed across the capacitances, rather than the voltages referenced to a common ground. The equations of the *RC* dual chain are as follows (the currents of R_{1w} and of R_{1s} are equal in magnitude and opposite in direction, and so are the currents of R_{2w} and R_{2s}):

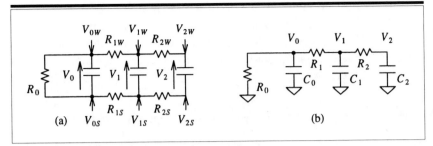

Figure 2.15 Simplification of the Equivalent Circuit

$$\frac{V_{1W} - V_{0W}}{R_{1W}} = \frac{V_{0S} - V_{1S}}{R_{1S}} \qquad \frac{V_{2W} - V_{1W}}{R_{2W}} = \frac{V_{1S} - V_{2S}}{R_{2S}}$$

$$C_0 \frac{d}{dt}(V_{0W} - V_{0S}) = \frac{V_{1W} - V_{0W}}{R_{1W}} - \frac{V_{0W} - V_{0S}}{R_0}$$

$$(2.52)$$

$$C_1 \frac{d}{dt}(V_{1W} - V_{1S}) = \frac{V_{2W} - V_{1W}}{R_{2W}} - \frac{V_{1W} - V_{0W}}{R_{1W}}$$

$$C_2 \frac{d}{dt}(V_{2W} - V_{2S}) = - \frac{V_{2W} - V_{1W}}{R_{2W}}$$

From the first two equations,

$$V_{1W} - V_{0W} = [R_{1W}/(R_{1W} + R_{1S})]$$
$$[(V_{1W} - V_{1S}) - (V_{0W} - V_{0S})]$$

$$(2.53)$$

$$V_{2W} - V_{1W} = [R_{2W}/(R_{2W} + R_{2S})]$$
$$[(V_{2W} - V_{2S}) - (V_{1W} - V_{1S})]$$

Then if we set $V_0 = V_{0W} - V_{0S}$, $V_1 = V_{1W} - V_{1S}$, $V_2 = V_{2W} - V_{2S}$, $R_1 = R_{1W} + R_{1S}$, and $R_2 = R_{2W} + R_{2S}$, we have

$$C_0 \frac{dV_0}{dt} = \frac{V_1 - V_0}{R_1} - \frac{V_0}{R_0}$$

$$(2.54)$$

$$C_1 \frac{dV_1}{dt} = \frac{V_2 - V_1}{R_2} - \frac{V_1 - V_0}{R_1} \qquad C_2 \frac{dV_2}{dt} = - \frac{V_2 - V_1}{R_2}$$

and they are the equations of the simplified circuit having the common ground, shown in Figure 2.15(b).

This simplification does not work if the RC transmission-line structure is more complex than this. The simplified equation never means the existence of a universal ground. Rather, it means that the voltage developed across a capacitance is the only information-carrying signal, as we shall see in Chapter 6.

2.8 Modes of a Uniform and Straight *RC* Chain

The discharge of a uniform and straight RC chain circuit shows many interesting details of the modes of the RC circuits used to model digital circuits. In the RC chain circuit of Figure 2.16 the node k voltage is V_k. The circuit equation is of the form

$$V_k = V_{k+1} + RC[(dV_{k+1}/dt) + (dV_{k+2}/dt) + \cdots + (dV_{N-1}/dt)] \qquad (2.55)$$

and if the mode depends on time like exp(St), the equations are converted to

$$V_k = V_{k+1} + \sum_{j=k+1}^{N-1} SV_j \qquad (2.56)$$

For *RC* chains of length 2 to 6 the secular equations that determine S are derived as

$$S^2 + 3S + 1 = 0 \qquad (N = 2)$$
$$S^3 + 5S^2 + 6S + 1 = 0 \qquad (N = 3)$$
$$S^4 + 7S^3 + 15S^2 + 10S + 1 = 0 \qquad (N = 4)(2.57)$$
$$S^5 + 9S^4 + 28S^3 + 35S^2 + 15S + 1 = 0 \qquad (N = 5)$$
$$S^6 + 11S^5 + 45S^4 + 84S^3 + 70S^2 + 21S + 1 = 0 \qquad (N = 6)$$

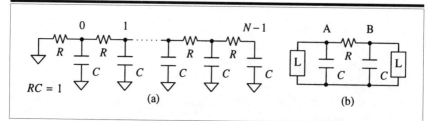

Figure 2.16 Uniform Straight *RC* Chain

Table 2.1 Roots of the Secular Equation

	Root					
N	1st	2nd	3rd	4th	5th	6th
1	−1.0	—	—	—	—	—
2	−0.3819	−2.6180	—	—	—	—
3	−0.1981	−1.5549	−3.2469	—	—	—
4	−0.1206	−1.0000	−2.3473	−3.5320	—	—
5	−0.0810	−0.6903	−1.7154	−2.8308	−3.6825	—
6	−0.0581	−0.5030	−1.2908	−2.2411	−3.1361	−3.7709

where we set $RC = 1$. The roots of the equations are listed in Table 2.1. If $RC \neq 1$, the roots are to be divided by RC.

The reciprocal of the absolute value of the root S is the time constant of the mode. Suppose that $N \to \infty$ (the RC chain is infinitely long). The impedance looking into the RC chain at the grounded end, Z, can be computed in closed form. If the leftmost link is removed, the impedance remains the same. Thus

$$Z = R + \frac{1}{(1/Z) + SC} \quad \text{or} \quad Z^2 - RZ - (R/SC) = 0 \quad (2.58)$$

Therefore

$$Z = (R/2)[1 + \sqrt{1 - (4/|S|CR)}] \quad (2.59)$$

This formula suggests that $|S| = 4/CR$ is the upper limit of the magnitude of S, so that for the time constant $CR/4$ is the lower limit. Table 1 shows that the roots of the secular equation have magnitude less than 4. Let us consider the physical meanings of this limit. In Figure 2.16(b) the time constant for equalizing the two nodes A and B, T_{EQ}, is the product of $C/2$ (we note that the two capacitors are connected in series) and R, so that $T_{EQ} = RC/2$. This is the upper limit of this time constant, since there are two circuits, L of Figure 2.16(b), loading the capacitor, and the effective values of the capacitance and the resistance are smaller than the values assumed. If the two circuits L are replaced by resistors R, a good estimate is $T_{EQ} = R \times C \times C/[C + C + C(R/R) + C(R/R)] = RC/4$. This approximation will be justified in the next section. In an RC circuit having the structure

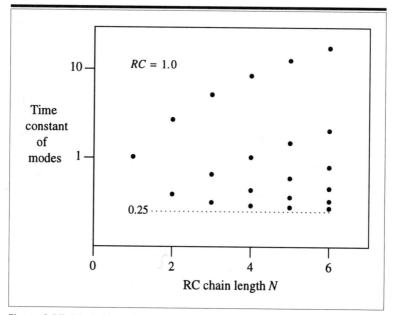

Figure 2.17 Mode Time-Constant Distribution Versus Length of the *RC* Chain

of Figure 2.16(a) it is impossible to make a time constant shorter than $RC/4$. Therefore the small time constants approach the limit. Figure 2.17 shows the time constants versus the chain length N. An N-link RC chain has N modes, and all the time constants of the modes up to $N = 6$ are plotted versus N. The small time constants approach the lower limit, $RC/4$.

The longest time constant is well separated from the second longest time constant, as is observed from Figure 2.17. The smallest solution of the secular equation (which gives the longest time constant) is approximated by neglecting the terms higher than second order in S in the secular equation. The time constant obtained by this approximation equals the numerical coefficient of the first-order term of the secular equation. This is the Elmore time constant, and its values in units of RC are

$$T_E = 3, 6, 10, \ldots, N(N + 1)/2, \ldots \qquad N \ge 2 \qquad (2.60)$$

Figure 2.18 shows the voltage profile for each of the N modes of the length-N chain, for $N = 2, 3, 4, 5,$ and 6. The profiles were determined by solving the set of circuit equations. The algebra is tedious but straightforward; only the results are shown in Figure 2.18. For the voltage profiles, the following rules apply. If the modes are arranged in the order of decreasing time constant, the longest-time-constant mode has $\frac{1}{4}$ standing wave in the chain, the second-longest-time-constant mode has about

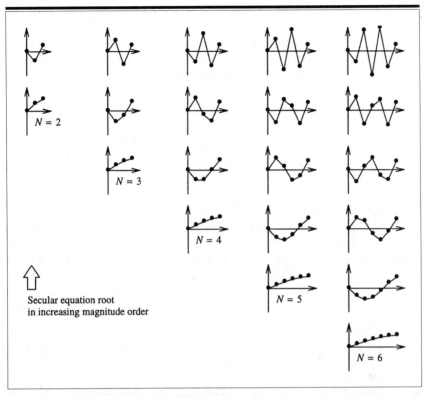

Figure 2.18 Voltage Profile of the *RC* Modes

$\frac{1}{2}$ standing wave in it, the third about 1 standing wave, the fourth about $1\frac{1}{2}$ standing waves, and so on. In the shortest-time-constant mode the node voltage polarity alternates. This is the mode whose time constant is closest to the lower limit. This mode's mechanism is indeed node voltage equalization of Figure 2.16(b).

Each mode has its characteristic voltage profile in the RC chain. The voltage profile and the time constant of the mode can be correlated as follows. We approximate an *N*-link RC chain by a continuous RC line. The voltage profile of the line is determined from the partial differential equation (Section 6.4)

$$\frac{\partial^2 V(x,t)}{\partial x^2} = RC \, \frac{\partial V(x,t)}{\partial t} \qquad (2.61)$$

This equation has a solution of the form

$$V(x,t) = \exp(St) \, P(x) \qquad (S < 0) \qquad (2.62)$$

where $P(x)$ satisfies the equation

$$P''(x) = -RC|S|P(x) \tag{2.63}$$

where the prime $'$ indicates differentiation with respect to x. The left end of the RC line is grounded. Now suppose that the right end is also grounded. The solution that is zero at $x = 0$ and $x = N$ is given by

$$V(x,t) = \text{const.} \times \exp(-|S|t) \sin(K_n x)$$

where

$$(2.64)$$

$$K_n = n\pi/N \text{ and } |S| = (n\pi)^2/RCN^2$$

The time constant is given by $1/|S|$. If $RC = 1$, $N = 6$, and $n = 1$ ($\frac{1}{2}$ standing wave in the RC chain), then $|S| = 0.2742$, and this does not agree with 0.5030 given in Table 1. In reality there is more than $\frac{1}{2}$ standing wave in the chain, as is observed from Figure 2.18 (the profile for $N = 6$, second from the bottom). If N is reduced accordingly to 4.3, then $|S| = 0.5337$, and this value agrees well with the exact solution. The third-longest time constant for the $N = 6$ chain is estimated at $S = -1.2588$, which agrees well with the exact solution $S = -1.2908$. This approximation breaks down if there are too many standing waves in the RC chain. Then the stage-to-stage voltage equalization mechanism of Figure 2.16(b) gives a better estimate.

2.9 Understanding the Modes of the Circuit—I

The closed-form results of the last two sections showed that an RC circuit node voltage is the sum of terms of the form $\exp(\alpha t)$. Each term is a *mode* of the circuit, which has the characteristic time constant $T_\alpha = -1/\alpha$. To understand the time constants is the first step toward understanding the mode. This is not easy, since many resistances and capacitances are involved. If some of them take extreme values, the closed-form time constants provide clear physical meanings. Let us consider a two-stage RC circuit.

The two-stage RC chain circuit of Figure 2.19(a) reduces to the circuit of Figure 2.19(b) in the limit as $R_1 \to 0$. In this limit

$$\lim_{R_1 \to 0} (-1/\alpha) = R_0(C_0 + C_1) \tag{2.65}$$

This is the time constant of the joint discharge of capacitances C_0 and C_1 through resistance R_0. In the other limit as $R_0 \to \infty$ we have

$$\lim_{R_0 \to \infty} (-1/\beta) = R_1 C_0 C_1/(C_0 + C_1) \tag{2.66}$$

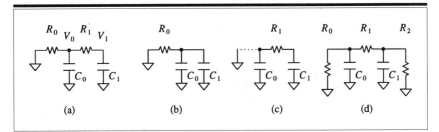

Figure 2.19 Discharge Mechanisms of Two-Node *RC* Chain

and this is the time constant of discharge of the two series-connected capacitors C_0 and C_1 through the resistance R_1, or the time constant of the capacitance voltage equalization. The two time constants are therefore related to the overall discharge process of the chain and to the node voltage equalization process, respectively.

Understanding the discharge mechanism in general is not always so easy as in this example. A Monte Carlo method is helpful [26]. We guess the functional dependence of the time constant on the values of the resistances and the capacitances, by a simple and physically meaningful formula. We compute the time constant accurately on one hand, and the value of the guessed formula on the other hand, for a large number of randomly generated resistance and capacitance values, and correlate the two sets of numbers. If a strong correlation is found, the guessed formula and the physical mechanism are correct. In the two-stage RC chain of Figure 2.19(a), the longer time constant is for the process of a joint discharge of C_0 and C_1 through R_0, as we saw before. We guess a formula

$$T_G = R_0(C_0 + C_1) \tag{2.67}$$

and we correlate the guessed and the accurate values of the time constants. The correlation for 1000 sets of randomly generated resistance and capacitance values in the range between 0 and 1 is shown in Figure 2.20(a). T_α is the accurate time constant, and T_G is the guessed time constant given above. Selection of random parameter values in the range is justified on the basis that a circuit that contains any extreme parameter values is not to be included, since such a circuit can be better modeled using a different circuit connectivity (either joining two nodes together, or disconnecting the circuit at an extremely high resistance). There is indeed a correlation in Figure 2.20(a), but there is significant upward scattering of the points. The guessed formula gives underestimates. The cause of the upward scatter is that R_1 delays discharge of C_1, thereby making $T_\alpha > T_G$. Indeed, the lower limit of T_α for a given T_G is approximately equal to T_G. The effect of R_1 is included by modifying the guessed function to

$$T_G = R_0(C_0 + C_1) + R_1 C_1 \qquad (2.68)$$

Then the correlation between the computed and the guessed time constants improves dramatically, as shown in Figure 2.20(b). Adding the term

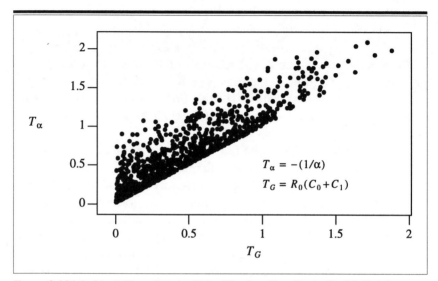

Figure 2.20(a)   Mode Time-Constant Identification (Two-Node Chain): First Guess

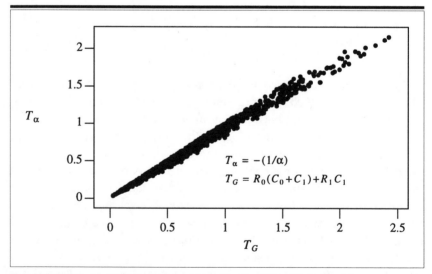

Figure 2.20(b)   Mode Time-Constant Identification (Two-Node Chain): Second Guess

R_1C_1 to the original time constant is an improvement to include the effects of C_1, which is on the far side of R_1. The guessed time constant is the Elmore time constant [27].

The second time constant of the two-stage RC chain circuit is that of the process of equalizing the voltages of the two nodes, if $R_0 \to \infty$. The guessed function is again

$$T_G = R_1 C_0 C_1 / (C_0 + C_1) \qquad (2.69)$$

and the correlation with the accurately computed time constant is as shown in Figure 2.21(a). There is significant downward scattering of the points, and this correlation is not sufficient to claim that the equalization of the node voltage is always the mechanism. From the circuit of Figure 2.19(a) the guessed time constant is an overestimate if R_0 is small, since the capacitance C_0 is effectively short-circuited to ground. By adding a term $(R_1/R_0)C_1$ to the denominator,

$$T_G = R_1 C_0 C_1 / [C_0 + C_1 + C_1(R_1/R_0)] \qquad (2.70)$$
$$= R_1 / [(1/C_1) + ((1 + (R_1/R_0))/C_0)]$$

we obtain a better guess function. This formula can be derived from the closed-form solution of the secular equation. Then the correlation improves dramatically, as shown in Figure 2.21(b). This observation shows that the second time constant is for the process of voltage equalization,

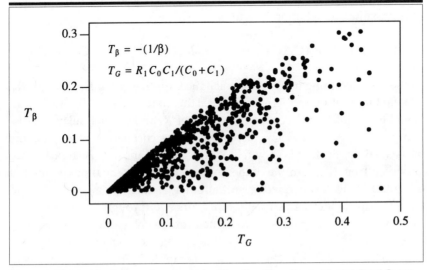

Figure 2.21(a) Mode Time-Constant Identification (Two-Node Chain): Third Guess

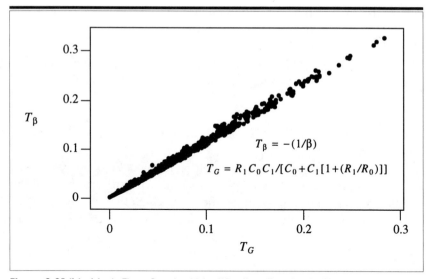

Figure 2.21(b) Mode Time-Constant Identification (Two-Node Chain): Fourth Guess

modified by including the effects of R_0, which tends to short-circuit C_0. The right-hand side of the formula shows that the capacitance C_0 is reduced to $C_0 R_0 / (R_0 + R_1)$ in computing the voltage equalization time constant. The mathematical procedure of adding $(R_1/R_0)C_1$ to the denominator is a standard way of including the effects of R_0 in parallel to C_0, which we used in Section 2.8.

Modes of the three-link straight RC chain are shown in Figure 2.22(a) as a second example. The chain has three modes. The longest time constant is approximated by Elmore's formula

$$T_E = R_0(C_0 + C_1 + C_2) + R_1(C_1 + C_2) + R_2 C_2 \quad (2.71)$$

In the corresponding mode the capacitances discharge jointly through R_0. The first term of Elmore's formula gives this time constant. The two additional terms take into account the effects of R_1 and R_2 that delay the discharge of $C_1 + C_2$ and C_2, respectively. The correlation of T_E and the accurate time constant is shown in Figure 2.23(a), and is very good. As we observed from the example of the uniform RC chain, the time constant of this mode is well separated from the other, smaller time constants, and therefore the mode has a strong *identity*. Since R_0 through C_2 are randomly generated in the range from 0 to 1, the average value of each component is 0.5. The Elmore time constant for this average component value is $0.5 \times 0.5 \times 3(3 + 1)/2 = 1.5$. The maximum computed (or guessed) time con-

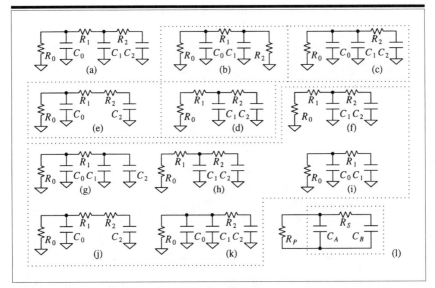

Figure 2.22 Analysis of the Mechanism of Modes of a Three-Stage Straight Chain

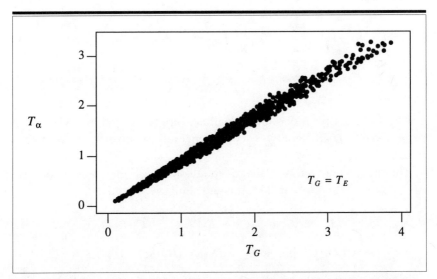

Figure 2.23(a) Mode Time-Constant Identification (Three-Node Chain): First Guess

stant for 1000 samples is about twice that value, but is less than the absolute maximum, 6. This is quite reasonable.

The shortest-time-constant mode of the chain is the reduction of the voltage difference between the nodes by transferring the charges among them, but exactly how this process takes place depends on the values of the components. We consider three candidate processes, shown by the equivalent circuits of Figure 2.22(b), (c), (d), and (e). Figure 2.22(b) shows the process of voltage equalization between nodes 0 and 1 through R_1, where R_0 and R_2 act as parallel loads to C_0 and C_2, respectively. The best estimate of the time constant of this process is to add to the denominator of the backbone time constant

$$T_B = \frac{R_1 C_0 C_1}{C_0 + C_1} \tag{2.72}$$

the two terms $C_1(R_1/R_0)$ and $C_0(R_1/R_2)$, as we did in the example of the two-link chain, to obtain

$$T_B = \frac{R_1 C_0 C_1}{C_0 + C_1 + C_1(R_1/R_0) + C_0(R_1/R_2)} \tag{2.73}$$

This is justified as follows. The secular equation of the circuit of Figure 2.22(b) is

$$C_0 C_1 S^2 + \left(\frac{C_0}{R_2} + \frac{C_0 + C_1}{R_1} + \frac{C_1}{R_0} \right) S$$

$$+ \frac{1}{R_0 R_1} + \frac{1}{R_0 R_2} + \frac{1}{R_1 R_2} = 0 \tag{2.74}$$

The largest negative root of this equation is approximated by dividing the negative of S^2 coefficient by S^3 coefficient, and the inverse of the magnitude of the largest root is T_B.

The time constant of voltage equalization of the circuits of Figure 2.22(c), (d), and (e) can be estimated similarly. We have

$$T_C = R_2(C_0 + C_1)C_2/[C_0 + C_1 + C_2 + C_2(R_2/R_0)]$$

$$T_D = R_2 C_1 C_2/[C_1 + C_2 + C_2(R_2/(R_0 + R_1)] \tag{2.75}$$

$$T_E = (R_1 + R_2)C_0 C_2/[C_0 + C_2 + C_2((R_1 + R_2)/R_0)]$$

As the guessed time constant T_G we take the smallest of T_B, T_C, T_D, and T_E. The correlation between the guessed and the computed time constants is good, as shown in Figure 2.23(d) [note that Figures 2.23(a)–(d) are arranged in decreasing order of the time constant]. The time constant of node voltage equalization should be centered at $0.5 \times 0.5/2 = 0.125$ (note

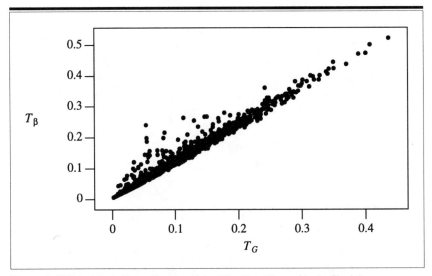

Figure 2.23(b)   Mode Time-Constant Identification (Three-Node Chain):
Second Guess

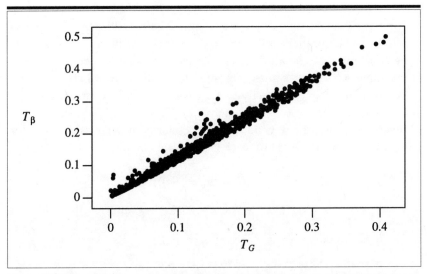

Figure 2.23(c)   Mode Time-Constant Identification (Three-Node Chain): Third Guess

that the two capacitors are in series). The maximum of 0.2 is reasonable.

The middle time constant is the hardest to guess. Since we chose the minimum of T_B through T_E as the shortest time constant, the maximum of the set is a reasonable guess. The correlation between the guessed and the

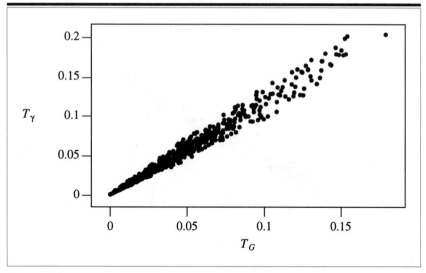

Figure 2.23(d) Mode Time-Constant Identification (Three-Node Chain): Fourth Guess

accurate time constants is shown in Figure 2.23(b). Although the correlation is reasonable, there are many data points that indicate the guessed time constant is too small. If the parameter values are generated at random, some parameter values can be quite small. Then it is better to use different component connectivity to guess the time constant, and that makes the assumed guess formula invalid. Figure 2.22(f)–(k) show a possible *degenerate RC* chain, valid for the peculiar parameter values. The formula for the time constant for each case is listed below:

$$T_F = R_2 C_1 C_2 / [C_1 + C_2 + C_2(R_2/(R_0 + R_1))] \qquad C_0 \approx 0$$

$$T_G = R_1 C_0(C_1 + C_2)/[C_0 + C_1 + C_2 + (C_1 + C_2)(R_1/R_0)] \qquad R_2 \approx 0$$

$$T_H = R_2 C_1 C_2 / [C_1 + C_2 + C_2(R_2/(R_0 + R_1))] \qquad R_0 \approx 0 \qquad (2.76)$$

$$T_I = R_1 C_0 C_1 / [C_0 + C_1 + C_1(R_1/R_0)] \qquad C_2 \approx 0$$

$$T_J = (R_1 + R_2)C_0 C_2 / [C_0 + C_2 + C_2((R_1 + R_2)/R_0)] \qquad C_1 \approx 0$$

$$T_K = R_2(C_0 + C_1)C_2 / [C_0 + C_1 + C_2 + C_2(R_2/R_0)] \qquad R_1 \approx 0$$

If the guessed time constants are replaced by T_F through T_K in the special cases, the correlation improves significantly, as shown in Figure 2.23(c).

2.10 Understanding the Modes of the Circuit—II

We continue to develop the mathematical techniques of understanding the circuit modes. The second way to understand the mechanism of the RC circuit mode is to study the polarities and the amplitudes of the node voltages. Let us consider the two-stage RC chain circuit of Figure 2.19(a) as an example. We have in Section 2.7

$$V_1(t) = A \exp(\alpha t) + B \exp(\beta t)$$

$$V_0(t) = C \exp(\alpha t) + D \exp(\beta t)$$

$$C = (1 + C_1 R_1 \alpha)A$$

$$D = (1 + C_1 R_1 \beta)B$$

(2.77)

The proportionality relationships $A \leftrightarrow C$ and $B \leftrightarrow D$ can be written as

$$C = (R_1/2C_0)(P + Q)A \qquad D = (R_1/2C_0)(P - Q)B \qquad (2.78)$$

where $P = (C_0/R_1) - (C_1/R_1) - (C_1/R_0)$ and $Q = \sqrt{P^2 + 4(C_0 C_1/R_1^2)}$. Then C and A have always the same sign, and D and B have always opposite signs. Then we may conclude that in the mode α, C_0 and C_1 are effectively connected in parallel, and in the mode β, C_0 and C_1 are effectively connected in series. Currents flow as shown by the dotted lines in Figure 2.24(a) and (b), respectively. In the circuit diagram of Figure 2.19(a) the connectivities of C_0 and C_1 are shown. On the level of the modes, their effective connectivity can be different from what we intuitively suppose from the circuit diagram. On the level of submicrostates, only some connections of the microstate circuit diagram are retained, as we observe from Figure 2.24(b) (R_0 is out of the loop). Starting from the complete circuit

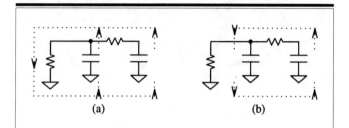

(a)   (b)

Figure 2.24 Current of Modes α and β

diagram, the connectivity decreases as the level goes down to the micro-state level and to the submicrostate level.

Digital circuits have a rather well-defined equivalent circuit structure, in that a capacitance exists between a node and ground. Thus by examining the voltages developed across the capacitances, it is possible to determine which capacitors are effectively in parallel, and which are in series. From this information it is possible to gain insights into the mode mechanisms. The Monte Carlo method can be applied to this problem as well. In the two-stage cascaded RC chain,

$$C/A = 1 + C_1 R_1 \alpha \qquad D/B = 1 + C_1 R_1 \beta \qquad (2.79)$$

We generate R_0, R_1, C_0, and C_1 randomly in the range between 0 and 1, and compute the ratios C/A and D/B by computer. Figure 2.25(a) and (b) show the results. There the horizontal axis is the index of 1000 randomly generated two-stage RC chains, and the vertical axis shows the ratios of the two node voltages. In Figure 2.25(a) the node 1 and node 0 voltages have same polarity, and the node 1 voltage is higher. Current is directed from node 1 to node 0 to ground, and the two capacitors discharge jointly. In Figure 2.25(b) the node 1 and node 0 voltages have the opposite sign, and the node 0 voltage can be the higher one. C_0 and C_1 are effectively connected in series.

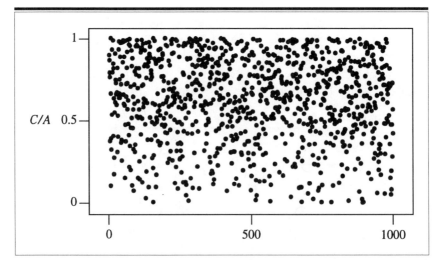

Figure 2.25(a) Mode Mechanism Identification of Two-Stage *RC* Chain

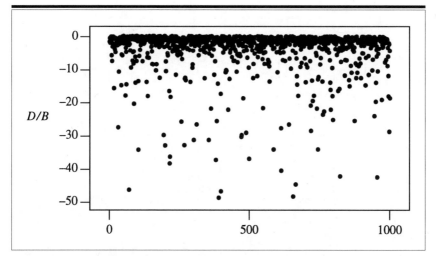

Figure 2.25(b) Mode Mechanism Identification of Two-Stage *RC* Chain

2.11 Non-monotonic *RC* Circuit Response

The circuits of Figure 2.12 have transient node waveforms that change in one direction only, decreasing. In this kind of circuit operation Elmore's formula gives the dominant mode (the longest) time constant accurately. Circuits having a nonmonotonic transient are also interesting, and we study the modes and the responses of an example, a twin-T notch filter circuit shown in Figure 2.26. If V_1 is a sinusoidal voltage source having the angular frequency ω, if the output node is open-circuited, and if

$$R_A = R_B = R, \qquad R_C = R/2$$
$$C_A = C_B = C \quad \text{and} \quad C_C = 2C \tag{2.80}$$

the twin-T circuit has a transmission zero at

$$\omega = \omega_0 \quad \text{where} \quad \omega = 1/CR \tag{2.81}$$

The parameter values can be chosen to build a frequency rejection or a frequency selection circuit.

If V_1 is a voltage source, the three node voltages V_R, V_C, and V_0 are all measurable, since they have capacitances to ground or to a voltage source. Using the voltages as the variables, the circuit equations are written as

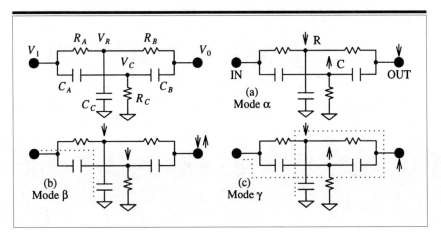

Figure 2.26 Modes of Twin-T Circuits

$$C_C \frac{dV_R}{dt} = \frac{V_1 - V_R}{R_A} + \frac{V_0 - V_R}{R_B}$$

$$C_A \frac{d}{dt}(V_1 - V_C) + C_B \frac{d}{dt}(V_0 - V_C) = \frac{V_C}{R_C} \qquad (2.82)$$

$$C_B \frac{d}{dt}(V_C - V_0) + \frac{V_R - V_0}{R_B} = 0$$

The derivatives dV_0/dt and dV_C/dt are solved for as

$$\frac{dV_0}{dt} = -\frac{(C_A + C_B)(V_0 - V_R)}{C_A C_B R_B} - \frac{V_C}{C_A R_C}$$

$$\frac{dV_C}{dt} = -\frac{C_B(V_0 - V_R)}{C_A C_B R_B} - \frac{V_C}{C_A R_C} \qquad (2.83)$$

If d/dt is replaced by S, we get

$$\left(C_C S + \frac{1}{R_A} + \frac{1}{R_B}\right) V_R - \frac{V_0}{R_B} = \frac{V_1}{R_A}$$

$$C_B S V_0 - \left[(C_A + C_B)S + \frac{1}{R_C}\right] V_C = -C_A S V_1 \qquad (2.84)$$

$$\frac{V_R}{R_B} - \left(C_B S + \frac{1}{R_B}\right) V_0 + C_B S V_C = 0$$

If V_R and V_C are eliminated,

$$V_0 = (N/\Delta)V_1 \tag{2.85}$$

where

$$N = (C_A R_A)(C_B R_B)(C_C R_C)S^3 + C_A C_B(R_A + R_B)R_C S^2$$
$$+ (C_A + C_B)R_C S + 1$$

$$\Delta = (C_A R_A)(C_B R_B)(C_C R_C)S^3 \tag{2.86}$$
$$+ [C_A C_B(R_A + R_B)R_C + R_A(R_B C_B)C_C$$
$$+ R_A(C_A + C_B)(R_C C_C)]S^2$$
$$+ [(C_A + C_B)R_C + C_B(R_A + R_B) + R_A C_C]S + 1$$

and where $\Delta = 0$ is the secular equation, which has roots α, β, and γ.

For a twin-T circuit to work as a notch filter, N must have a pair of pure imaginary roots, $S = \pm j\omega_0$ (sinusoid transmission vanishes at angular frequency ω_0). Then we have

$$\omega_0 = \frac{1}{\sqrt{C_A C_B(R_A + R_B)R_C}}$$

provided $\hspace{8cm}$ (2.87)

$$R_A R_B C_C = (C_A + C_B)(R_A + R_B)R_C$$

The three negative real roots of the secular equation, α, β, and γ, are chosen such that $|\alpha| < |\beta| < |\gamma|$. Each root defines a time constant:

$$T_\alpha = 1/|\alpha| \qquad T_\beta = 1/|\beta| \qquad T_\gamma = 1/|\gamma| \tag{2.88}$$

The solutions of the node voltages are written as

$$V_R = A \exp(\alpha t) + B \exp(\beta t) + C \exp(\gamma t)$$
$$V_0 = D \exp(\alpha t) + E \exp(\beta t) + F \exp(\gamma t) \tag{2.89}$$
$$V_C = G \exp(\alpha t) + H \exp(\beta t) + I \exp(\gamma t)$$

We then have

$$D/A = 1 + \frac{R_B}{R_A} + (R_B C_C)\alpha \qquad G/A = \frac{1}{R_A C_B \alpha} + \frac{C_C}{C_B} + (D/A)$$

$$E/B = 1 + \frac{R_B}{R_A} + (R_B C_C)\beta \qquad H/B = \frac{1}{R_A C_B \beta} + \frac{C_C}{C_B} + (E/B) \tag{2.90}$$

$$F/C = 1 + \frac{R_B}{R_A} + (R_B C_C)\gamma \qquad I/C = \frac{1}{R_A C_B \gamma} + \frac{C_C}{C_B} + (F/C)$$

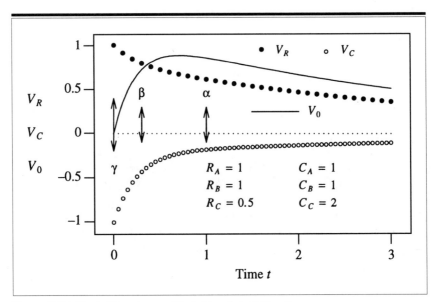

Figure 2.27 Node Waveforms of Switching Twin-T Circuit

Figure 2.27 shows plots of V_R, V_C, and V_0 when V_1 makes a unit negative step from 1 to 0 at $t = 0$. The voltage step is transmitted immediately to V_C and V_0, and the initial conditions are

$$V_R(+0) = 1 \qquad V_C(+0) = -1 \quad \text{and} \quad V_0(+0) = 0 \qquad (2.91)$$

The three modes α, β, and γ show up a long time, a medium time, and a short time after the voltage step, respectively, as shown in Figure 2.27. Let us consider what the ranges of the roots, α, β, and γ will be. Figure 2.28 shows the results of 200 trials, in which R_A, R_B, R_C, C_A, C_B, and C_C are randomly generated within the range from 0 to 1; α, β, and γ are computed; and the ratios β/α and γ/α are plotted. We observe that α, β, and γ are magnitudewise quite well separated, and the most frequent occurrences are in the range

$$3 < \beta/\alpha < 30 \qquad 10 < \gamma/\alpha < 300 \qquad (2.92)$$

This result shows that the three modes α, β, and γ must have distinct physical mechanisms. The mechanisms of the modes can be studied further by studying the amplitude ratio of the nodes.

Figure 2.29(a), (b), and (c) show plots of node voltage ratios V_0/V_R and V_C/V_R for the three modes α, β, and γ for randomly generated values of

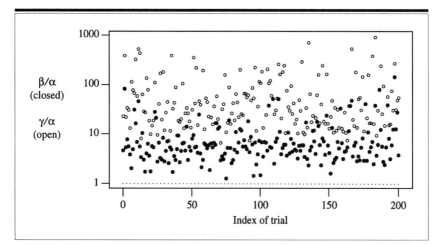

Figure 2.28 Separation of Time Constants of Twin-T Circuit

R_A through C_C. In mode α the output node is in phase with the node R, and C is out of phase with R. This is the variation observed in Figure 2.27 a long time after the input step (indicated by the double arrow α). In mode γ the output node and C are both out of phase with R. This is the variation observed in Figure 2.27 a short time after the input step (indicated by the double arrow γ). The two modes are indicated by D–D–U and D–U–U, respectively, where U and D (for up and down) are arranged in the order of nodes R–OUT–C. In this notation, if D and U are exchanged we get the same thing: D–D–U and U–U–D are synonymous, as are D–U–U and U–D–D. The combinations of the indicators are complete if D–D–D and D–U–D are added. These two combinations are for mode β.

As is observed from the random trials of Figure 2.29(b), the majority of the cases are D–D–D, having all the nodes in phase. There are cases of D–U–D, however, and therefore all the possible node polarity combinations exist. The polarities of the node voltage changes are indicated by the arrows in Figure 2.26(a)–(c).

Let us refer to Figure 2.26(a). In mode α the nodes R and OUT are in phase. If the left terminal of C_B is grounded, the delay time T_α is estimated from Elmore's time constant T_E as

$$T_E = R_A(C_C + C_B) + R_B C_B \qquad (2.93)$$

In the real twin-T circuit the left terminal of C_B is not grounded. Since the node C is out of phase, T_E is an underestimate of T_α (note that current from

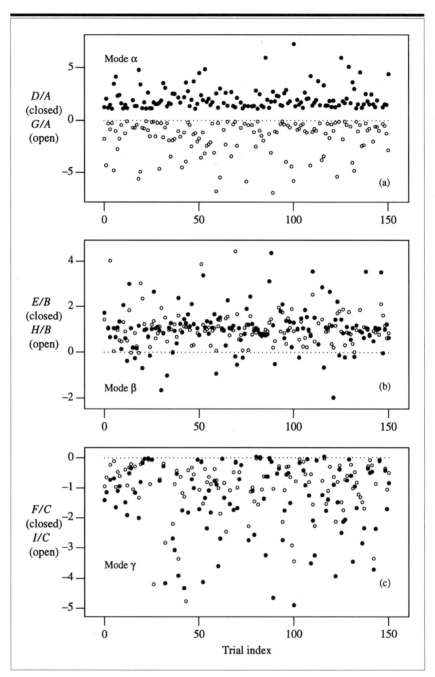

Figure 2.29 Node Voltage Amplitude Ratios of the Three Modes of a Twin T

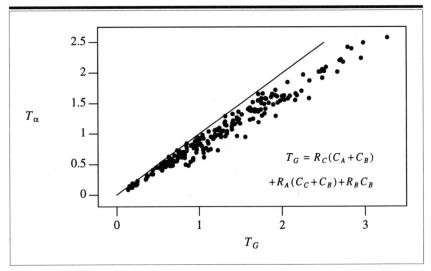

Figure 2.30 Correlation of Time Constants (Mode α)

C to OUT delays the discharge). The time constant of the discharge of $C_A + C_B$ by R_C is added to T_E, and

$$T_G = R_C(C_A + C_B) + R_A(C_C + C_B) + R_B C_B \qquad (2.94)$$

is a reasonable guess. Figure 2.30 shows that there is indeed a good correlation between the accurately computed T_α and T_G. We note that the form of T_G is the coefficient of S in the cubic expression for Δ. Since the three roots are well separated in magnitude, the coefficient gives a good estimate of the longest time constant.

In the majority of the cases the nodes R, OUT, and C are in phase in mode β. Then R_B and C_B are ineffective. Then the middle time constant is estimated by

$$T_G = R_A C_C \qquad (2.95)$$

This is quite remarkable, since four out of six independent parameter values do not matter. The correlation between T_β and T_G in Figure 2.31(a) is quite significant, considering the drastic simplification. We note that the middle time constant of the three-node RC chain was very hard to explain by a simple mechanism.

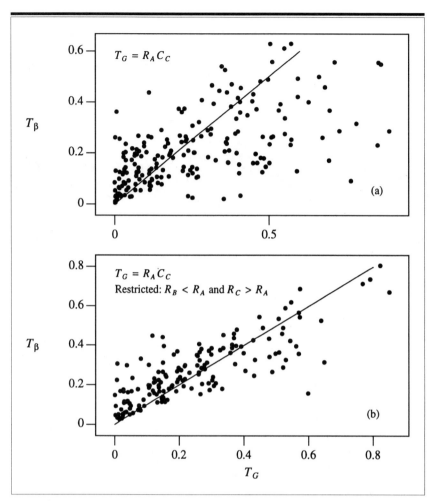

Figure 2.31 Correlation of Time Constants (Mode β)

If the three nodes are forced to be in phase by the condition

$$R_B < R_A \quad \text{and} \quad R_C > R_A \qquad (2.96)$$

and selecting the samples, the correlation improves quite significantly, as shown in Figure 2.31(b).

Mode γ has the shortest time constant. T_γ ranges from 0 to about 0.1. Since the average of the randomly generated parameter values is 0.5, the small time constant can be generated by connecting all the capacitances in the circuit in series, and then connecting them to a single resistor. From Figure 2.26 the resistor must be R_B. We then have

$$T_G = \frac{R_B \cdot C_A C_B C_C}{C_A C_B + C_B C_C + C_C C_A} \tag{2.97}$$

This connection gives the required polarities of the node voltage (charge of C_C is transferred to C_A and C_B through R_B). The correlation between T_y and T_G shown in Figure 2.32(a) is quite significant, but it can be improved by restricting the values of R_A and R_C so that

$$R_A > R_B \quad \text{and} \quad R_C > R_B \tag{2.98}$$

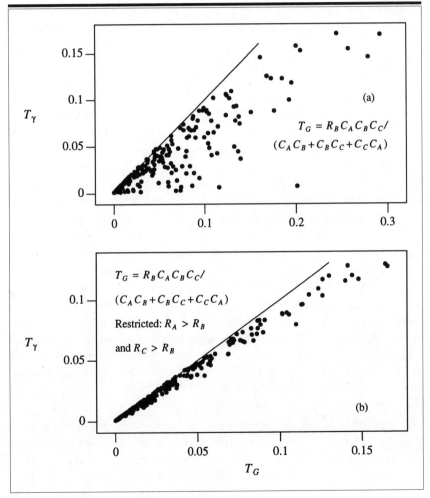

Figure 2.32 Correlation of Time Constants (Mode)

to improve the accuracy of the model. The correlation indeed improves, as shown in Figure 2.32(b). We are able to explain the three modes of the unloaded twin-T circuit quite clearly.

2.12 Mechanisms of Long-Lasting Metastable States in CMOS D-Latches

In this chapter we have studied methods of understanding the modes of RC circuits. The modes of the RC circuits that model digital circuits are energy-dissipating: their electrical activity decays with time. There is an exception, however, and that is the modes in bistable data-storage circuits. The electrical activity of a bistable circuit grows with time, and then, limited by the circuit's nonlinearity, settles at a final microstate that represents the logic level. Such an active mode presents curious, and practically important, phenomena of metastability [28].

We consider a system sending a digital signal (for example, a human finger), and a system receiving the signal and processing it (for example, the keyboard of a PC). The sending and the receiving systems have different *clocks* (time references). The receiving system must realign the signal with its own clock. Digital signal realignment is carried out by *capturing* the asynchronous input data into a *D-latch*, which is timed by the clock signal of the receiving system. Since the sending and the receiving clock are different in frequency and in phase, the following unfortunate situation may happen: At the time when the D-latch input is closed, the latch's internal circuitry has not acquired enough information from the input. Then the latch takes a long time to settle, and the final digital state of the latch depends on unpredictable external perturbation, such as noise.

In recent years, metastable states have become significant even in synchronous data-processing systems (those where the entire system is clocked by a single clock). A large digital crosspoint switch system that serves 1000 customers requires hardware that fills up several bays. Interconnections between the bays are twisted wire transmission lines that are several to 10 feet long. Since the data rate is several hundred megabits per second, several bits of data are held within the transmission-line length. Thus the timing relationship between the sent and the received signals becomes uncertain. To prevent this problem, the transmission-line lengths must be precisely adjusted, and must be maintained against environmental variations. This is a very expensive system assembly procedure.

Metastable states are unstable states that last unusually long times. Many similar effects exist in nature. A metastable state is sensitive to unpredictable external perturbation, or noise. The effects of noise on metastable states of MOS D-latches were studied by Veendrick [29]. He concluded that noise has *statistically* no effect on metastable states, since equal numbers of metastable states are created and destroyed by noise. He presented convincing theoretical arguments and experimental evidence, and the conclusion has been accepted for some time. There is no doubt that it is valid in many realistic circumstances. This theory provided qualitatively correct probability distributions of metastable states. What I wish to propose in this section is to modify his model and his analysis in the limit of very long-lasting metastable states. This modification provides a clear idea of the role of the noise in metastable states. To achieve this objective I must separate the effects of noise on the data storage circuits into two distinct phases: the effects before the latch closure (the preeffect) and after the latch closure (the aftereffect). What has been studied by Veendrick is the preeffect. I wish to concentrate on the aftereffect. During the two phases, noise has different effects on the latch circuit.

Before going into the details of the new model, it is necessary to estimate the amplitude of noise in CMOS digital circuit nodes. I assume that the latch is in a CMOS VLSI chip. Noise in a digital IC node may be caused by electrostatic induction, substrate voltage fluctuation, electromagnetic induction, thermal noise, and generation-recombination noise of the MOS devices. To determine the ranges of the noise voltages, let us consider two extreme cases: 1) the electrostatic induction, and 2) the thermal noise of the FET. The thermal noise voltage developed at a digital circuit node is estimated by a geometric mean of the thermal voltage ($kT/q = 25$ mV at room temperature) and the voltage developed across the gate capacitance C_G of the FET by a single electronic charge:

$$V(\text{thermal noise}) = \sqrt{(kT/q) \cdot (q/C_G)} = 1 \text{ mV} \qquad (2.99)$$

assuming $C_G = 10^{-14}$ F [1, 30, 31]. This formula was derived using the van der Ziel-Jordan-Shoji noise formula for MOSFETs, assuming the MOSFET is in saturation. The noise bandwidth was estimated from the -3-dB point of the load capacitance ($\approx 2C_G$) and the Early-effect drain resistance.

The amplitude of the electrostatically induced noise is estimated from the capacitive voltage division ratio. The capacitive division ratio is approximately the ratio of the thickness of the dielectric layer underneath the metal level to the distance between the two interacting nodes. In scaled-down CMOS there are many switching nodes within about 30 μm. If the thickness of the insulating oxide is 1 μm, the voltage swing of the target node will be of the order of V_{DD} (1 μm)/(30 μm) = 150 mV. Since only differential induced voltage is effective in perturbing a metastable state, the effect can be smaller than this. The range of induced noise is estimated to be in the range 10–100 mV. We may conclude that conventional digital circuit nodes have amplitude of noise from various sources, of the order of 1- to 100-mV range.

A D-latch in a metastable state is characterized by the difference between the DATA and $\overline{\text{DATA}}$ node voltages. If the difference at the beginning ($t = 0$), $v_D(0)$, is not zero, then the difference grows exponentially with a time constant t_0 whose value is determined by the latch circuit parameters. The latch settles within a time about $t_0 \log[V_{DD}/v_D(0)]$. If there is noise of amplitude V_N, is it reasonable to consider $v_D(0)$ much smaller than V_N? At first sight it seems the longest metastable state lasts only about $t_0 \log(V_{DD}/V_N)$. Is this a right conclusion?·

To answer to this question, it is necessary to formulate the metastability problem as follows. Figure 2.33 shows the equivalent circuit of a CMOS D-latch in a metastable state. The current control mechanism of both NFETs and PFETs of the CMOS inverters is represented by a single current generator having transconductance g_m. The load resistance R is the *effective* load resistance, due to the Early effect of the MOSFETs. The capacitance C is the cumulative node capacitance (contributions from the drain islands, from the gates, and from the wiring). The current generators I_0 and I_1 represent noise induced on the nodes N_0 and N_1, respectively. The circuit equations satisfied by the node voltages V_0 and V_1 are

$$C \frac{dV_0}{dt} = -g_m V_1 - \frac{V_0}{R} + I_0(t)$$

$$C \frac{dV_1}{dt} = -g_m V_0 - \frac{V_1}{R} + I_1(t)$$

(2.100)

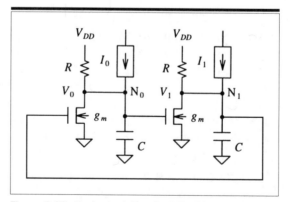

Figure 2.33   Equivalent Circuit of CMOS D-Latch in Metastable State

If we define the differential node voltage v_D by $v_D = V_0 - V_1$, then v_D satisfies the equation

$$\frac{dv_D}{dt} - \frac{1}{t_0}v_D = \frac{1}{C}[I_0(t) - I_1(t)] = \frac{1}{C}I_D(t) \qquad (2.101)$$

where $I_D(t) = I_0(t) - I_1(t)$ is the *differential* noise current. Equation (2.101) is solved as

$$v_D(t) = \exp(t/t_0)\left(v_D(0) + \frac{1}{C}\int_0^t I_D(\theta)\,\exp(-\theta/t_0)\,d\theta\right)$$

where $\qquad\qquad\qquad\qquad\qquad\qquad\qquad\qquad\qquad$ (2.102)

$$t_0 = \frac{CR}{g_m R - 1}$$

Equation (2.102) is normalized to

$$v_D(t_0 \cdot \tau) = \left(\frac{t_0 I_0}{C}\right)\exp(\tau)\left[\int_0^\tau i_D(\zeta)\,\exp(-\zeta)\,d\zeta + \frac{C}{t_0 I_0}v_D(0)\right] \qquad (2.103)$$

where $\tau = t/t_0$ is the normalized time, $i_D(\zeta) = I_D(t)/I_0$ is the normalized differential noise current, and I_0 is the maximum noise current to the node. The parameter $(t_0 I_0/C)$ is the noise voltage developed across the node capacitance during the characteristic time t_0. This is the parameter that is equivalent to the thermal noise and the induced noise voltages estimated before. Since that much differential node voltage fluctuation is inevitable in a metastable state, the latch may be considered to be in a metastable state as long as

$$|v_D| < (t_0 I_0/C) \qquad (2.104)$$

is satisfied. In this study we define the *duration* of a metastable state in this way. This definition is different from that used by Veendrick, and the difference is crucial in the later discussions.

Let us consider the simplest case of a special initial condition $V_D(0) = 0$:

$$v_D(t_0 \cdot \tau) = \left(\frac{t_0 I_0}{C}\right)\int_0^\tau i_D(\zeta)\,\exp(\tau-\zeta)\,d\zeta \qquad (2.105)$$

Suppose that i_D is a random function that takes values between -1 and

+1 with equal probability. v_D is then bounded within the range given by

$$-\frac{t_0 I_0}{C} \exp(\tau) < v_D < \frac{t_0 I_0}{C} \exp(\tau) \qquad (2.106)$$

The probability that a metastable state lasts a time $t_0 \tau$ is determined as follows. We compute Equation (2.105) for all the possible functions i_D in the range of the integral from 0 to τ. The probability of the metastable state lasting longer than $t_0 \tau$ is the number of functions that keep the integral of Equation (2.105) within the range -1 to $+1$, divided by the total number of functions. The practical way of computing the probability is to use the Monte Carlo method [26].

We compute Equation (2.105) subject to the following simplifications. The function $i_D(\zeta)$ takes a random value uniformly distributed within the range $-1 \le i_D \le +1$, and the value changes every time ζ increases by $\Delta\zeta$. It is convenient to assume that $1/\Delta\zeta$ is an integer, and to assume that τ takes integer values as well. Figure 2.34 shows a function $i_D(\zeta)$ schematically. There is a reason to choose the discrete step $\Delta\zeta$ at $\frac{1}{2}$. In Equation (2.105), t_0 is the time constant of the increase of the differential node voltage, which is about the same as the delay time of the inverter of Figure 2.33. Induced noise has a pulse width typically half the delay time, since it originates from similar gates on the same chip. The delay time is about the same as the reciprocal of the frequency where the small-signal gain of the amplifier of Figure 2.33 begins to fall with increasing frequency. It is reasonable to assume that one period t_0 contains two changes of value of the random variable.

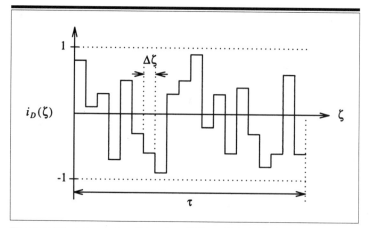

Figure 2.34 Schematic of Random Differential-Noise-Current Waveform

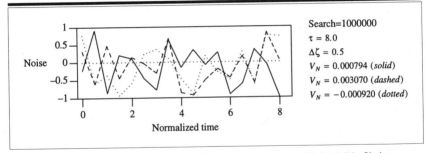

Figure 2.35 Random Search for Noise Waveforms Creating Metastable State

Figure 2.35 shows the results of random search for the function i_D that gives the minimum of the integral

$$V_N = \int_0^\tau i_D(\zeta)\, \exp(\tau-\zeta)\, d\zeta \qquad (2.107)$$

for 1 million random samples of functions, assuming $\tau = 8$ and $\Delta\zeta = 0.5$. The results for three independent searches are shown. The value of $|V_N|$ is less than 1, which satisfies the criterion of metastability defined before. The functions i_D are quite featureless to casual observation, but the noise waveforms are such that the differential node voltage established earlier is later canceled repeatedly.

Figure 2.36 shows the results of a similar random search for i_D that gives the maximum value of $|V_N|$. The results show that $i_D(\zeta)$ has values close to either -1 or $+1$ for small ζ, and the level is maintained for some time. Then the initial noise effect grows with time by the exponential factor, and the effect dominates during the later times. The point I wish to make is that among the variety of the noise waveforms, certain ones prevent growth of the differential node voltage. If such a special noise waveform arrives, the waveform actively maintains the metastable state. The probability of a very long-lasting metastable state is determined by how many such special waveforms exist. In my opinion, that probability depends essentially on the noise. This is different from the conclusion by Veendrick, and the difference originates from inclusion of the noise aftereffect.

The Monte Carlo method of computation of the probability appears straightforward, but there is one mathematical issue. The probability is a ratio of the number of functions, within the specified range, whose weighted integral is less than unity to the number of all possible functions. Both *numbers* are infinite, and they are of a high order of infinity. According to set theory, the cardinal of the set of all the functions in the range, $\aleph_2$, is higher than $\aleph_1$, the cardinal of continuum. In physics, probabilities

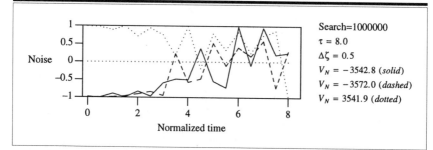

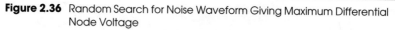

Figure 2.36   Random Search for Noise Waveform Giving Maximum Differential
Node Voltage

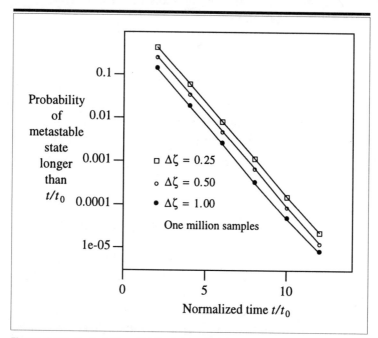

Figure 2.37   Probability of Metastable State

that are a ratio of the numbers both belonging to $\aleph_1$ are commonly computed by Monte Carlo method, but this case involves numbers belonging to $\aleph_2$. Use of the Monte Carlo method may require justification in such a case.

Figure 2.37 shows the probability of a metastable state versus time of duration. The best approximation of the numerical results is given by

$$P = f(\Delta\zeta)\,\exp(-\zeta) \quad \text{where} \quad f(\Delta\zeta) = \frac{4.14}{\sqrt{\Delta\zeta} + 3.19\,\Delta\zeta} \quad (2.108)$$

The probability is proportional to $\exp(-t/t_0)$. The functional dependence is the same as that of Veendrick's theory, but its physical interpretation is quite different. $f(\Delta\zeta)$ is a factor that reflects the *dominant* frequency of the noise, since a random voltage level is created at every multiple of $\Delta\zeta \cdot t_0$. The probability is a decreasing function of $\Delta\zeta$, since the earlier noise voltage levels have a greater effect on the differential voltage growth. The theory presented here is complementary to Veendrick's theory. If significant noise aftereffect exists, the mechanism of metastable states is explained only by this theory. The two theories give the same *functional* dependence of the probability of metastability on its duration. The dominant factor, $\exp(-t/t_0)$, occurs in both probability formulas. The coefficients of that factor are, however, different.

This theory gives significant insight into the electrical phenomena in a data storage circuit. It is unrealistic to assume that the circuit captures the processed signal with high precision. Rather, in the metastability limit the latch digitizes noise and stores a wrong Boolean value. The storage circuit works only if there is a significant timing margin to capture processed data against the noise.

CHAPTER THREE

Ground and Voltage Sources

3.1 Introduction

In this chapter we discuss the properties of voltage sources that supply DC power to high-speed circuits. The power supply of a high-speed circuit is either a voltage regulator using a negative-feedback stabilization circuit, or a chemical battery, both bypassed by a capacitor. At low speeds the arrangement has worked well, especially since high-capacitance electrolytic capacitors have become available and integrated integrated circuit voltage regulators have become well developed. Little attention was paid to its electrical properties, except for the obvious power rating, voltage regulation, overload capability, and efficiency, until the advent of high-speed VLSI technology. At the high speeds, not the power supply itself, but how it is connected to the circuit, becomes crucial. Traditionally the power supply was often connected just by a pair of twisted wires, and the entire system was expected to work as soon as the switch was turned on. For low-speed circuits, this is not an unrealistic expectation, but how the power supply connection will respond to high-speed transients is not obvious. This is as fundamental a problem as how to connect a voltmeter to measure voltage, which we studied in Section 1.10. The problem can be resolved by starting from the basic principles of physics.

One terminal of the power supply is connected to the *ground* of the circuit. The potential of the ground is regarded as the global reference voltage, and it must stay constant all the time. The ground is the most basic concept of electronics [32]. Often its properties are assumed to be well understood and well agreed upon. At high signal frequencies, however, where the period of the signal is comparable to the propagation delay time of the electromagnetic wave across the ground conductor, the concept of ground is no longer clear. A ground has a certain minimum size, which is about the size of a circuit in which signals can be transmitted by a single wire without restriction. Any node that cannot be contracted below the circuit's size is not allowed, if inductance is included in the model. Ground is in principle a concept of *RC*-only circuits, and as such, it is a classical concept. Because of that, it is useful to consider the reason why a special circuit node of an RC circuit is singled out and is called a ground.

1. In the early days of electronics, the only available amplifying device was a vacuum tube, in which electrons were the current carriers. What was conventional in vacuum tube circuits has been carried over to semiconductor circuits, since vacuum tubes, *npn* BJTs, and NFETs are all very similar. If vacuum tubes are used as voltage amplifiers, the cathode is connected to the negative terminal of the power supply, and the circuit that applies a DC bias to the grid and that brings the signal in (referenced to the negative power supply voltage level) must have one connection to the negative terminal as well. This is true for every amplifier stage. The negative power bus is the single node to which the largest number of connections are made. This special node consisted of a thick metal bus, or the backplane metal itself. The *chassis* of a vacuum tube circuit was almost always ground. This node was connected to the earth to avoid electrical shocks to the human body, which was almost always grounded. Thus the node was called *ground*.

2. The signal voltages the circuit produced were often measured relative to the negative power supply voltage level. This was the easiest way to understand the circuit operation, since then, in most circuits of conventional design, the voltages were read as positive numbers. The ground serves as the zero voltage *reference* of the circuit.

3. If more than one circuit in separate packages exchange signals, signals are voltages relative to the negative power supply voltage level. If all the circuits can be constructed such that the signal is referred to the ground, a single wire is able to transport a signal from one circuit to the other, if all the grounds are connected.

4. In practice an electronic circuit can never be constructed ideally. Two signal nodes and two current loops are never completely isolated. If two signal wires are routed side by side, the two wires couple capacitively. The capacitive coupling can be reduced if a constant-voltage conductor is placed nearby, or between the two wires. Availability of a constant-ground-potential conductor at any location of the circuit improves the circuit reliability.

These observations on shielding, packaging, and voltage reference suggest that the ground must have the same size as the circuit, and this is the fundamental reason why the concept cannot be carried over to a circuit model including inductance. If the physical meanings of ground are reconsidered on including inductances, the notion loses much of its sense.

From this point of view, the ground plays two distinct roles in an electronic system. The first is as one of a pair of wires for transporting a signal, thereby saving interconnects. The second is to enclose and isolate the electronic subsystem, to make it less susceptible to outside influence. We reiterate that both capabilities require a ground conductor having about the same size as the electronic circuit. Thus the ground as a single node ceases to exist at high speeds.

3.2 Communication between Circuits

Electronic circuits send and receive information. Information is carried by some form of energy. In conventional semiconductor circuits, the energy is the electrostatic energy of charge stored in capacitors, and its measure is the terminal voltage V. The voltage developed across a pair of terminals of a capacitor is the variable representing a signal. The voltage is measurable, according to the definition of Section 1.10. The energy and the charge in a capacitor C charged to V volts are given, respectively, by

$$\text{energy} = CV^2/2 \qquad \text{charge} = CV \qquad (3.1)$$

The charge and the energy must be transported from one location of the circuit (source) to the other (destination). Interconnects and ground provide the path for energy and charge.

To move energy and charge, a medium of transportation is required. The medium can be conducting wires, nonconducting space, or a mixture of both. If the distance does not exceed the size of the enclosure of the electronic system (such as a building), the choice depends on the electrical environment of the two circuit locations. The requirements for the transport medium are as follows:

1. On connecting the two circuits through the medium, neither circuit should be destroyed.

2. The information should not be altered during the transportation.

3. Many independent transportation *channels* should be available.

Requirement 1 determines whether or not conducting wires can be used. On connecting the two circuits, no excessive DC or long-lasting surge current should flow. For this requiremernt to be satisfied, one circuit should not be charged as a whole relative to the rest, either permanently or temporarily: The connecting wires should not be a path of the charge neutralization. If this condition is not satisfied, the sending and the receiving circuits have their own voltage references, but they have no common reference voltage. The potential difference between the two circuits may be high and potentially destructive. The two circuits are able to communicate if the signal is sent and received as electromagnetic wave through a non-conducting medium, as with a radio transmitter and receiver. The medium need not be free space: It can be optical fibers. There are other ways of bridging high voltage differences, but electromagnetic waves are the only way to maintain the highest possible transmission speed. If the reference voltage difference is not excessively high, capacitive coupling, which allows transmission of transient or AC signals, can be used.

If the sending and the receiving circuits have small, but nonzero difference in the reference voltage, and if the difference is time-dependent, balanced signal transmission can be used. The signal is carried by a pair of wires from one circuit to the other. At the receiving end of the pair of the wires there is a capacitor, whose charge represents the transmitted signal at the destination. The name "balanced transmission" indicates that the two wires are equivalent. A small variation of the reference voltages of the sending and the receiving circuits does not matter for signal transmission, but it must be kept small enough to guarantee that the circuits are not damaged by the connection.

If the reference voltages of the two circuits are nearly the same, and if they are maintained close enough at all times, the two circuits can be connected together without restriction. One terminal of the capacitor at the receiving end is connected to the common reference voltage, and the other terminal to a single wire from the signal source. This scheme saves one wire, and this is obviously convenient. The most complex and dense circuits can be assembled only in this way. This scheme depends, however, on the quality of the common reference voltage level, or ground. As is shown in Figure 3.1, the ground conducts many signal currents that carry the signals to the destinations simultaneously. We note that the signals are mixed together. The cost of saving one wire per interconnect is that the signals are not entirely pure at the destination.

In a large system, all three signal transmission methods are used. Communication between computers is often through optical fibers. Commu-

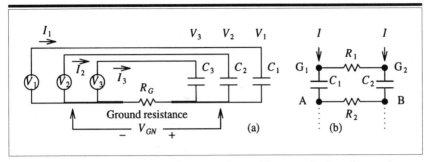

Figure 3.1 Noise Induction in the Common-Ground Communication Channels

nication between processors within a computer often uses twisted pairs of wires. Within a single processor board and within a single VLSI chip, a single wire carries each signal.

From this observation, the ground (common voltage reference) has different grades. One-wire signal transmission is feasible only if the best-grade ground is available. We are able to classify the circuits into groups: Within a small group a well-defined ground voltage level exists, and single-wire communication is feasible. Between the groups a vaguely agreed-upon voltage reference exists, but the precision is not high enough to carry even digital signals. The precision is, however, high enough to maintain the circuit's integrity: On connecting circuits of the different groups, one circuit does not destroy the other. Balanced transmission is used in this case. Between the largest groups of circuits, there is no voltage reference at all. Direct electrical connection may result in destruction of one by the other. In this case transmission by electromagnetic waves is used. Ground has clear meanings only within the smallest groups of electronic circuits.

Electronic circuits use energy and current for signal processing, and after processing is over, the waste energy and current must be thrown away. Waste energy is mostly converted to heat, and is dumped to the package. The waste current is dumped to the ground. Dumping current creates an IC environmental problem: noise. Figure 3.1 shows a common-ground, multiple-channel communication medium. The ground is not ideal, and it includes series resistance R_G. The ground resistance generates ground noise voltage V_{GN} given by

$$V_{GN} = R_G \sum_{n=1}^{N} I_n \qquad (3.2)$$

where N is the number of channels and I_n is the current in channel n. If each switching transient lasts τ_n and the logic-voltage swing is V_{LOGIC}, $I_n \approx C_n V_{\text{LOGIC}}/\tau_n$. If the averages of C_n and τ_n are C and τ, respectively, then

$$V_{\text{GN}} = f \cdot \sqrt{N} \, [(R_G C)/\tau] V_{\text{LOGIC}} = f \cdot \sqrt{N} \, (R_G/R_D) V_{\text{LOGIC}} \quad (3.3)$$

where f is a numerical factor of the order of unity, $\sqrt{N}$ is the factor that reflects the sum of the noise voltages generated at random, and τ can be approximated by the gate's delay time. If the delay time is expressed as the product of the driver impedance R_D and load C, the expression on the right-hand side is derived. If $V_{\text{GN}}/V_{\text{LOGIC}} < \frac{1}{2}$ is the criterion of the upper limit on the ground noise,

$$R_{\text{GN}} \leq R_D/(2f\sqrt{N}) \quad \text{or} \quad N \leq (1/4)(R_D/fR_{\text{GN}})^2$$

The ground symbols used in *RC* circuit diagrams require only that the voltages of the grounded nodes be constant (often set at 0), and information about the current flow within the ground conductor is not specified. The restriction simplifies low-speed circuit analysis, but it does not provide enough information at high speeds. If the current within the ground is not known, the inductance of the current loop cannot be determined. Free use of ground symbols makes sense only if the inductances are negligible. Circuit diagrams at high frequencies must be drawn to include information on the ground current. Drawing an appropriate equivalent circuit diagram demands understanding the circuit's operation, or its *mode*.

This issue has significance even at low frequencies, especially if high precision is required. A resistive ground is often constructed in hierarchy, using bypass capacitors. Suppose that current I is injected into a local ground node G_1 of Figure 3.1(b) and the same I is drained out from another local ground node G_2. Two ground nodes G_1 and G_2 are connected to the high-level ground nodes A and B via bypass capacitors C_1 and C_2. The high-level ground nodes A and B are connected together by a ground resistor R_2 that is smaller than R_1. The ground system can be represented by the equivalent circuit shown in Figure 3.1(b). In response to current I, voltage V is developed within the ground. The voltage is given by

$$V = Z_G I \quad (3.4)$$

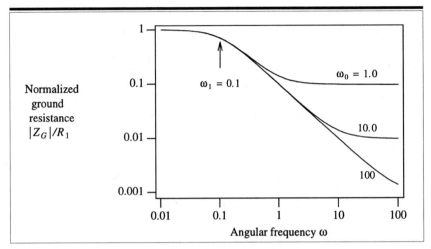

Figure 3.2 Ground Resistance versus Angular Frequency

where Z_G is the ground impedance. We are interested in the magnitude of the voltage. We have

$$|Z_G| = R_1\sqrt{[1 + (\omega/\omega_0)^2]/[1 + (\omega/\omega_1)^2]} \qquad (3.5)$$

where ω is the angular frequency of the current I, $\omega_0 = 1/R_2C$, $\omega_1 = 1/(R_1 + R_2)C$, and $C = C_1C_2/(C_1 + C_2)$. Figure 3.2 shows normalized ground resistance $|Z_G|/R_1$ versus angular frequency ω using $\omega_1 = 0.1$ and $\omega_0 = 1, 10, 100$.

Figure 3.3 shows a ground model including inductance. Wires A and B fabricated on the common ground G have the loop inductances L_A and L_B, respectively, and the terminating capacitances C_A and C_B, respectively, which develop the signal voltage. The common ground return has resis-

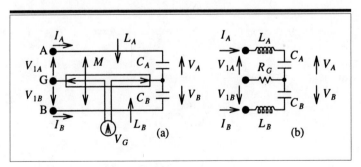

Figure 3.3 Ground as a Shared Interconnect

tance R_G. The voltage developed to R_G can be measured by connecting a voltmeter as shown in Figure 3.3(a). L_A and L_B are magnetically coupled by mutual inductance M. In this configuration L_A and L_B are never closely coupled, so that $\Delta = L_A L_B - M^2 > 0$. Suppose that the input voltages to A and B make a step-function change at $t = 0$ from 0 to V_{1A} and V_{1B}, respectively. We solve for V_A and V_B subject to the initial condition

$$V_A(+0) = 0 \qquad V_B(+0) = 0 \qquad I_A(+0) = 0 \qquad I_B(+0) = 0 \quad (3.6)$$

We have the circuit equation

$$V_{1A} = V_A + R_G(I_A + I_B) + L_A \frac{dI_A}{dt} + M \frac{dI_B}{dt} \qquad I_A = C_A \frac{dV_A}{dt} \tag{3.7}$$

$$V_{1B} = V_B + R_G(I_A + I_B) + M \frac{dI_A}{dt} + L_B \frac{dI_B}{dt} \qquad I_B = C_B \frac{dV_B}{dt}$$

Let n be an integer, and write

$$V_A(t) = \alpha_n t^n + \alpha_{n+1} t^{n+1} + \cdots$$

$$V_B(t) = \beta_n t^n + \beta_{n+1} t^{n+1} + \cdots$$

where V_A and V_B have the same n, since the circuit is structurally symmetrical. By substituting V_A and V_B in the infinite series expansion into the circuit equation and by setting the lowest-order term to cancel the constants V_{1A} and V_{1B}, we have $n = 2$ and a set of simultaneous equations

$$L_A C_A \alpha_2 + M C_B \beta_2 = (V_{1A}/2) \qquad M C_A \alpha_2 + L_B C_B \beta_2 = (V_{1B}/2) \quad (3.8)$$

which is solved using $\Delta = L_A L_B - M^2$ (> 0) to obtain

$$\alpha_2 = \frac{1}{2C_A \Delta}(L_B V_{1A} - M V_{1B}) \qquad \beta_2 = \frac{1}{2C_B \Delta}(-M V_{1A} + L_A V_{1B}) \quad (3.9)$$

and similarly

$$\alpha_3 = -\frac{R_G(L_B - M)}{6C_A \Delta^2}[(L_B - M)V_{1A} + (L_A - M)V_{1B}]$$

$$\tag{3.10}$$

$$\beta_3 = -\frac{R_G(L_A - M)}{6C_B \Delta^2}[(L_B - M)V_{1A} + (L_A - M)V_{1B}]$$

From the expressions for V_A and V_B as power series in t, V_A is a function of V_{1A} alone if $M = 0$ and $R_G = 0$. The same conclusion holds for V_B. The mutual inductance and the common ground resistance are the two coupling mechanisms between the two lines. Inclusion of self-inductances alone does not couple the two lines. This is because self-inductance is an attribute of a single loop. Nonzero R_G couples signals, since the resistance is locatable on the common ground.

3.3 Ground as an Isolation Wall

In the early days of electronics, it was found that circuits operated stably if the circuit's negative power supply terminal was connected to the earth. The earth is a conducting sphere, and its potential in the equilibrium state is uniform everywhere. Suppose that the earth, having radius $a = 6378$ km, has charge Q_E, uniformly distributed over the surface. If the earth were the only body in the universe, its potential ϕ_E would be given by

$$\phi_E = \frac{Q_E}{4\pi\epsilon_0 a} + \phi(\infty) \tag{3.11}$$

where $\phi_E(\infty)$ is the potential at the outer edge of the universe, which is assumed a constant. The charge Q_E is not independent of time. If a spaceship charged to $-\Delta Q_E$ is sent out, Q_E increases by ΔQ_E. The potential ϕ_E increases by $\Delta\phi_E$. The ratio $\Delta Q_E / \Delta\phi_E$ is the *capacitance* C_E of the earth, given by

$$C_E = \Delta Q_E / \Delta\phi_E = 4\pi\epsilon_0 a = 709.6 \ \mu\text{F} \tag{3.12}$$

The earth is the largest accessible reference body, but not with regard to its electrical capacitance. To change the earth's potential relative to the edge of the universe, charge must be added or removed from the earth, so that *extraterrestial* charge transfer is required. If the earth is to be used as the ground, one must assume the existence of a larger ground that surrounds it, the edge of the universe, to which the earth ground voltage is referred. Grounds have a hierarchy, as shown in Figure 3.4. We consider the response of the ground to charge transfer, such as sending a charged spacecraft out. Suppose that a charge transfer Q takes place between grounds G_1 and G_2. How much the difference of the voltages of the two grounds is affected depends on the capacitance C_1. The stability of the potential between G_1 and G_2 increases with C_1. The circuit whose ground

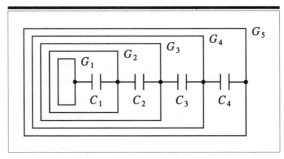

Figure 3.4 Hierarchy of Grounds

is G_1 is completely surrounded by G_2 and is completely isolated from the world outside G_2. Similarly how stable the ground G_2 is depends on the capacitance C_2 to the next ground G_3. A ground is similar to an inertial coordinate reference in mechanics. As it was pointed out by Brillouin [33], if the inertial coordinate frame is a global reference, it must have an infinite mass. Similarly, what characterizes the stability of a ground is the capacitance it has to the next ground. This issue is significant if more than one electronic system exchanging signals share a high-level ground.

Electrical activity of a circuit on a single ground does not change the potential of the ground as a whole, but transfer of charge from one part of the circuit to the other part creates a dipolar field. The time-dependent dipolar field emits radiation to the surrounding space if the frequency is high. A digital circuit operates over a frequency spectrum from 0 Hz (or DC) to a certain maximum frequency f_{max}. If the wavelength given by

$$\lambda_{min} = c/f_{max} \qquad (3.13)$$

is comparable to the size of the circuit, it works effectively as a resonant antenna, and electromagnetic energy is radiated. This constitutes environmental contamination, which may cause malfunction somewhere else. The electromagnetic wave is contained only by the next higher-level ground, which encloses the lower-level circuits completely. This containment prevents malfunction of the circuits outside the enclosure, but not those inside. The electromagnetic wave energy may accumulate in a resonant mode, and significant oscillation of the electric and the magnetic fields may result. To prevent the resonant mode, the higher-level ground should have blackbody inside walls that absorb the electromagnetic wave. At the frequencies where the radiation is significant, resistive loss is essential to accomplish the isolation effects expected of a ground. This situation exists in waveguide circuits.

The electromagnetic energy dumped at the infinity by a diverging electromagnetic wave has a thermodynamic implication. The universe can be closed and finite and still appear infinitely large if the outer boundary is made of a blackbody. In a hierarchical ground system the energy that returns from the next-level enclosure has the same problem and the same solution. If the frequency is low and the radiation loss is negligible, the energy dumped to the ground is consumed by the resistive loss in the ground. To study ground characteristics, the loss effect must be considered.

A ground stretches over the entire dimension of the circuit, and the next-level ground encloses the entire circuit. Electrostatics shows that every point in an empty space enclosed completely by a conducting material wall has the same potential at equilibrium. An electrical signal from the outside is *shielded* by the enclosure, and is not detectable inside. If the space inside the enclosure is occupied by an operating electronic circuit, its activities should not be detectable from the outside of the enclosure. The conducting enclosure works as a wall that separates the inside from the outside. The wall is impenetrable by a low-frequency signal, but it is penetrable by a high-frequency signal. In Figure 3.5 the electronic circuit is above a conducting wall. The circuit has two nodes, A and B. The circuit drives the nodes; node A has positive charge, and node B has negative charge. The positive charge at node A induces negative charge α on the wall of the conducting enclosure, and the negative charge at node B a positive charge β. To create the charges α and β on the conducting wall, current must flow through it as shown by the arrow. The current decays after some time, as the electrical energy is dissipated as heat or radiation from the enclosure. As the steady state is reached, the current vanishes and the charge distributions α and β together with A and B cancel precisely the electric field outside the enclosure. It takes some time to establish the state,

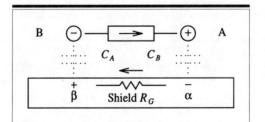

Figure 3.5 Electrostatic Shield

however. The arrangement of Figure 3.5 can be modeled by an equivalent circuit consisting of a series connection of C_A, C_B, and the ground resistance R_G. The current decays with time constant $R_G C_A C_B / (C_A + C_B)$. Before this current becomes negligible, the internal electrical activities are observable from the outside. After the steady state is reached, the enclosure isolates the inside from the outside. The state before (no dipole moment inside) and after (dipole moment inside) are indistinguishable by observation from the outside.

To prevent the external magnetic field from inducing noise voltage in the circuit, the circuit must be enclosed in a magnetic shield. The magnetic shield of Figure 3.6(a) is a box made from high-magnetic-permeability material. If a magnetic field is applied from the outside of the box, the magnetic flux at the location of the box is conducted by the wall material, and only small fraction of the field creeps inside the box. If magnetic field is created by a coil inside the box, the magnetic flux is conducted by the wall of the box, and it does not leak out significantly. A coil, shown in Figure 3.6(b), has the capability of magnetically shielding the object inside the loop. If the external magnetic field $\vec{H}$ increases, a current in the direction shown by the arrows is induced in the conducting loop, and it maintains the magnetic flux linked by the loop, by compensating the increment. The induced current decays if the loop has series resistance. The magnetic shielding effect is significant only within time L/R, where L is the loop inductance and R is the loop series resistance. Magnetic shielding is effective only at high frequency, that is, above a certain lower limit. If a superconductor is used, the lower frequency limit is zero (DC).

A conducting enclosure has resonant modes, which affect the shielding. Shielding by a conducting body surrounding a circuit is effective only up to the frequency that equals the light velocity divided by the dimension of the body. At higher frequency the simple electric shield mechanism ceases to work. At such frequencies there are many modes of electromagnetic

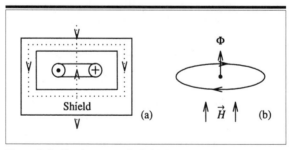

Figure 3.6 Electromagnetic Shield

oscillation of the body, and the oscillation can be excited either by the internal circuit or by the external field. At a mode frequency the inside and the outside are able to communicate, and the isolation effects are lost. At higher frequencies there are more modes. Excitation of mode oscillation means that the surface of a conducting body can no longer be considered a constant-potential surface.

3.4 Impedance of a Power Supply

The ground-and-power terminal pair delivers electrical energy from the power supply to a chip. Impedance of the power supply as seen from the chip's bonding pad has a complex frequency dependence. We get an insight into it by studying the equivalent circuit shown in Figure 3.7. The following analysis is a small-signal circuit response analysis: The lower-case variables are deviations from the DC operating points. The DC power source is almost always regulated by negative-feedback control. The regulator section comprises a current control device (controlled current generator having transconductance G_m), a differential amplifier, and a one-stage RC filter circuit that models the dominant pole of the differential amplifier. The differential amplifier generates the amplified regulation error voltage v_3 given by

$$v_3 = A(v_{\text{REF}} - v_1) \tag{3.14}$$

where A is the gain of the amplifier. At DC, the RC filter circuit does not matter. Current $G_m(v_2 - v_1)$ is generated by the controlled current generator, and that compensates for increased or decreased load current i_0. Then at DC,

$$v_0 = \frac{A}{1 + A} v_{\text{REF}} - \frac{i_0}{G_m(1 + A)} \tag{3.15}$$

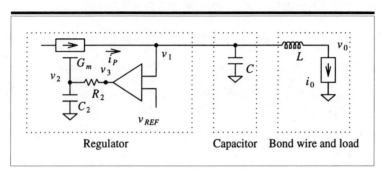

Figure 3.7 Small-Signal Equivalent Circuit of Power Source

If A is large, v_0 follows v_{REF} closely, and the voltage variation due to i_0 is reduced by a factor of $1/(1 + A)$. If circuit delays are included, the node voltages satisfy the equations

$$C \frac{dv_1}{dt} = i_p - i_0 \qquad C_2 \frac{dv_2}{dt} = \frac{v_3 - v_2}{R_2}$$

$$v_3 = A(v_{REF} - v_1) \qquad i_p = G_m(v_2 - v_1) \qquad v_1 - v_0 = L \frac{di_0}{dt} \qquad (3.16)$$

By replacing the differential operator d/dt with $j\omega$ and by eliminating the variables other than v_0 and i_0, we obtain

$$v_0 = -\frac{1}{\Delta}[(1 - \omega^2 LC + j\omega LG_m)(1 + j\omega C_2 R_2)$$

$$+ j\omega LAG_m]i_0 + \frac{G_m A}{\Delta} v_{REF} \qquad (3.17)$$

where

$$\Delta = G_m[A + (1 + j\omega C_2 R_2)[1 + j\omega(C/G_m)]] \qquad (3.18)$$

If $L \to 0$,

$$v_0 \to \frac{-(1 + j\omega C_2 R_2)i_0 + AG_m v_{REF}}{AG_m + (1 + j\omega C_2 R_2)(G_m + j\omega C)} \qquad (3.19)$$

and if $\omega \to 0$, we get Equation (3.15) back again. If $\omega \to \infty$ then $v_0 \to -j\omega L i_0$, and if $i_0 = 0$ and $\omega \to 0$ then $v_0 \to [A/(1 + A)]v_{REF}$. These limits have the obvious physical meanings.

Figure 3.8 shows the plots of the magnitude of the internal impedance looking into the power supply from the chip, defined by

$$|Z| = \sqrt{Re\ (Z)^2 + Im\ (Z)^2}$$

where

$$Z = v_0/i_0 \qquad (3.20)$$

with

$$v_{REF} = 0$$

versus angular frequency ω. For $\omega \to 0$, $|Z| \to 1/[G_m(1 + A)]$, and if $A =$

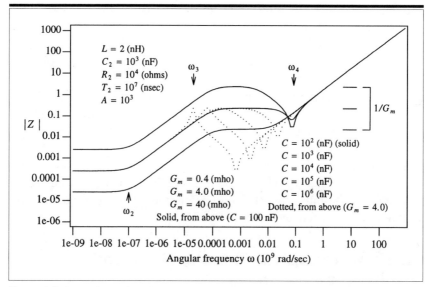

Figure 3.8 Impedance Looking into a Power Supply

10^3, we have $|Z| \approx 2.5 \times 10^{-3}$, 2.5×10^{-4}, and $2.5 \times 10^{-5} \, \Omega$ for $G_m = 0.4$, 4.0, and 40.0 S respectively. This is the DC *regulation* of the power supply.

The impedance of the power supply begins to increase at the angular frequency where the gain of the feedback loop begins to decrease with increasing frequency. The angular frequency is given by

$$\omega_2 = 1/R_2 C_2 = 10^{-7} \text{ (Grad/sec)} = 100 \text{ (rad/sec)}$$

and the impedance settles at the higher level, where the output impedance of the source follower current control device determines the power supply impedance. If the capacitance C is high, an impedance maximum appears at angular frequency ω_3 given by

$$\omega_3 = \left[\frac{2A(C_2 R_2)(C/G_m) - (C_2 R_2)^2 - (C/G_m)^2}{2(C_2 R_2)^2 (C/G_m)^2} \right]^{1/2}$$

$$\rightarrow \left[\frac{A}{(C_2 R_2)(C/G_m)} \right]^{1/2} \quad (A \rightarrow \infty) \tag{3.21}$$

This is the angular frequency where the phase margin of the negative-feedback loop approaches zero due to the additional G_m-C phase shift at the regulator output. At still higher frequencies the effects of the series inductance L becomes significant. L and C resonate at the series resonant frequency ω_4 given by

$$\omega_4 = 1/\sqrt{LC} \qquad (3.22)$$

At this angular frequency the internal impedance of the power supply has a minimum due to the series resonance. For $\omega > \omega_4$ the internal impedance of the power supply increases linearly with frequency due to the inductance. The impedance increase is limited, however, by the characteristic impedance of the power interconnect as an LC transmission line, which was modeled here using a single L. At this high frequency a more precise modeling of the interconnect than a single inductance is required. This problem will be discussed in the next section.

3.5 Power Supply Connection in Hybrid Integrated Circuits

The response of a power supply at high frequencies depends how it is connected to the chip. If a chip is connected by a bonding wire as shown in Figure 3.9(a) and (b) (cross section), the bonding wire works effectively as a series inductance at moderately high frequencies. In many high-level chip assembly media, such as hybrid integrated circuits (HICs), the bonding wire is connected to one of the power supply *planes* of the assembly medium as shown in the cross section, Figure 3.9(b). How does the metal plane work? We need to compute the impedance looking into the power-plane pairs in closed form and to study its frequency dependence. If the parallel power planes in the HIC extend to infinity in all directions, the characteristic of the pair of HIC power ground planes observed from the terminals A and B is approximated by a nonuniform LC ladder circuit.

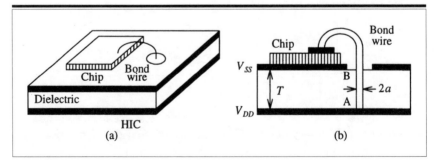

Figure 3.9 HIC Power-bus Connection

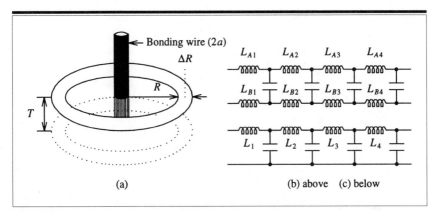

Figure 3.10 Inductance and Capacitance Computation

Figure 3.10(a) shows an expanded view of the HIC's port. As the diameter of the concentric ring in Figure 3.10(a) increases, the parallel capacitance increases and the series inductance decreases. The structure can be represented by the nonuniform LC transmission-line equivalent circuits of Figure 3.10(a) and (b). The two inductances L_{Ai} and L_{Bi} are not independent. We have

$$L_i = L_{Ai} + L_{Bi} \tag{3.23}$$

and L_i can be determined unambiguously, but L_{Ai} and L_{Bi} cannot, as we observed in Section 1.11.

Figure 3.9(b) shows the cross section of the bonding wire connection. The HIC has metal planes V_{SS} and V_{DD}, which are isolated by an insulating layer having dielectric constant ϵ and magnetic permeability μ (practically $\mu = \mu_0 = 4\pi/10^9$ H/cm) and thickness T. If the HIC package is large, we may assume that the boundaries are far away from the bonding wire. The electromagnetic field has axial symmetry with respect to the axis of the bonding wire. We consider concentric rings whose inner and outer radii are R and $R + \Delta R$, as shown in Figure 3.10(a). The capacitance of the pair of concentric rings (in the V_{SS} and V_{DD} planes) is given by

$$C = 2\pi R \, \Delta R \, \epsilon/T \tag{3.24}$$

The inductance of the concentric ring for radial current can be determined by computing the magnetic energy stored in the space between the conductors using the formula

$$\text{energy} = LI^2/2 \tag{3.25}$$

Alternatively, if we note that the electromagnetic wave that propagates in the space between the two conducting planes is a TEM (transverse electromagnetic) mode wave, we have velocity $= 1/\sqrt{\epsilon\mu_0}$. From this relationship the inductance can be determined as

$$L = T\,\Delta R\,\mu_0/2\pi R \qquad (3.26)$$

This is the inductance of the pair of rings (V_{DD} and V_{SS}) whose current path is closed by a pair of cylindrical walls at the inside (radius R) and the outside (radius $R + \Delta R$). Then the concentric structure can be modeled by a nonuniform transmission line. The inductances $L_1, L_2, \ldots$ and capacitances $C_1, C_2, \ldots$ are given by

$$L_n = L/n \qquad C_n = nC \qquad (3.27)$$

where L and C are the inductance and capacitance of a concentric ring that has internal and external radii a and $2a$, respectively, where a is radius of the bonding wire, from which the nonuniform transmission line starts outward. We need to compute the driving-point impedance of the nonuniform transmission line. Unfortunately, it is impossible to solve this problem in a simple closed form. We have, however, the closely similar problem of an exponentially scaled transmission line whose parameters are

$$L_n = (1 + \rho)^{n-1}L \qquad C_n = (1 + \rho)^{-n+1}C \qquad (3.28)$$

where ρ is a scale factor that can be positive or negative. This problem can be solved in a closed form. The driving-point impedance of the exponentially scaled transmission line Z satisfies an implicit equation (remove one section at the driving end and recompute the line impedance using the scaling of Z)

$$Z = j\omega L + \cfrac{1}{j\omega C + \cfrac{1}{(1 + \rho)Z}} \qquad (3.29)$$

which is solved as

$$Z = \frac{1 - \omega^2 LC - \dfrac{1}{(1 + \rho)} \pm \sqrt{\left[1 - \omega^2 LC - \dfrac{1}{(1 + \rho)}\right]^2 - \dfrac{4\omega^2 LC}{(1 + \rho)}}}{2j\omega C}$$

$$(3.30)$$

We subdivide L and C into m sections,

$$L \to L/m \qquad C \to C/m \qquad \rho \to \rho/m \qquad (3.31)$$

substitute into Equation (3.30), and take the limit of $m \to \infty$ to obtain the driving-point impedance of a continuous exponentially scaled transmission line as

$$Z = \frac{\rho \pm \sqrt{\rho^2 - 4\omega^2 LC}}{2j\omega C} \qquad (3.32)$$

The $\pm$ sign must be determined so that the impedance has the proper frequency limits. In the limit as $\omega \to \infty$, the $+$ sign gives

$$Z = \sqrt{L/C} + (\rho/2j\omega C) + o(\omega^{-2}) \qquad (3.33)$$

Therefore for $\omega > \omega_0 = |\rho|/2\sqrt{LC}$ we have

$$\frac{Z}{\sqrt{L/C}} = -\frac{1}{\Omega}\frac{\rho}{|\rho|}j + \sqrt{1 - (1/\Omega^2)} \qquad (3.34)$$

where $\Omega = \omega/\omega_0$. For $\omega \to 0$, we have for the $+$ sign

$$Z \approx j\omega L/|\rho| \quad (\rho < 0) \qquad Z \approx |\rho|/j\omega C \quad (\rho > 0) \qquad (3.35)$$

This is because if $\rho < 0$ the transmission line has infinite total parallel capacitance and finite total series inductance $L/|\rho|$. If $\rho > 0$, it has infinite series inductance and finite parallel capacitance $C/|\rho|$. Here we note that

$$1 + (1 + \rho) + (1 + \rho)^2 + \cdots = 1/\rho \qquad (3.36)$$

We have

$$\frac{Z}{\sqrt{L/C}} = -j\frac{(\rho/|\rho|) + \sqrt{1 - \Omega^2}}{\Omega} \qquad (3.37)$$

We consider the dimensions of L, C, and ρ. Let the inductance of a concentric ring having the inner radius a and the outer radius b be written as $L(a,b)$. Then

$$L = L(a, 2a)/a \quad \text{and} \quad \rho = [[L(2a, 3a)/L(a, 2a)] - 1]/a \qquad (3.38)$$

so that the units are

$$L: \text{ H/cm}; \quad C: \text{ F/cm}; \quad \rho: \text{ 1/cm}. \qquad (3.39)$$

One section of the discrete exponentially scaled line used to derive Equation (3.30) was 1 cm long. Figure 3.11 shows an example of the real and the imaginary parts of the driving-point impedance versus normalized angular frequency.

There is a paradox in this problem. If radius of the bonding wire a becomes small, L increases and C decreases, and the characteristic impedance of the nonuniform transmission line at the feeding port becomes

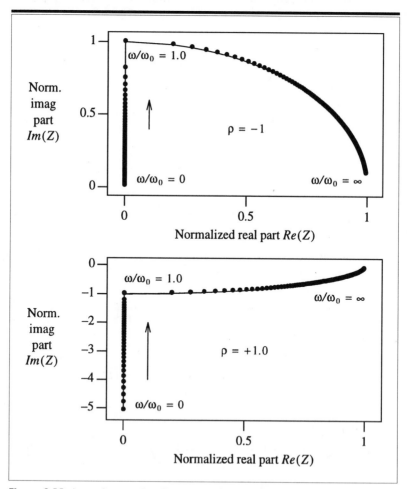

Figure 3.11 Impedance of an Exponentially Scaled Transmission Line

quite high. This should be observable as an HIC power-bus noise problem, but it is not. That is because the impedance has a frequency dependence. The nonuniform transmission line presents high impedance to high frequencies, but at low frequencies it presents a small inductive impedance that tends to zero as the frequency decreases. This paradox shows that wavelength of the signal launched on the nonuniform transmission line is crucial. If wavelength is much longer than the nonuniform region, the high local characteristic impedance never shows up. The similar effect exists in optics. Antireflection coating on an optical lens is effective for light waves, but useless for preventing reflection of microwaves from the lens.

Nonuniform transmission lines have many interesting characteristics. Properties of an exponentially scaled RC transmission line were studied in detail in my previous works [1, 7]. Here we study the properties of a scaled LC transmission line. In the discretized equivalent circuit of Figure 3.12(a), L_i and C_i depends on the location index i. In the limit as very fine discretization, we use the coordinate x instead of the index i, and L and C depend on x as $L(x)$ and $C(x)$. We have a pair of partial differential equations satisfied by the voltage $V(x,t)$ and current $I(x,t)$:

$$\frac{\partial V(x,t)}{\partial x} = -L(x)\frac{\partial I(x,t)}{\partial t} \qquad \frac{\partial I(x,t)}{\partial x} = -C(x)\frac{\partial V(x,t)}{\partial t} \qquad (3.40)$$

By eliminating $I(x,t)$ between the two equations we obtain

$$\frac{\partial^2 V(x,t)}{\partial x^2} - \beta\frac{\partial V(x,t)}{\partial x} = \gamma\frac{\partial^2 V(x,t)}{\partial t^2}$$

$$\beta = \frac{L'(x)}{L(x)} \quad \gamma = L(x)C(x)$$

$$(3.41)$$

Let us consider the simplest case, where $L(x)C(x) = LC = \gamma$ (constant) and $L'(x)/L(x) = \beta$ (constant). We have $L(x) = L(0)\exp(\beta x)$ and $C(x) = C(0)$

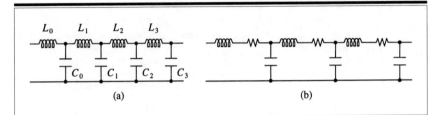

(a) (b)

Figure 3.12 Scaled *LC* Transmission Line

$\exp(-\beta x)$. The LC transmission line is exponentially scaled, and the two coefficients of the modified wave equation are both constants.

It is interesting to compare this equation with the equation of a lossy LC transmission line shown in Figure 3.12(b). By the same derivation we obtain

$$LC \frac{\partial^2 V(x,\,t)}{\partial t^2} + RC \frac{\partial V(x,\,t)}{\partial t} = \frac{\partial^2 V(x,\,t)}{\partial x^2} \tag{3.42}$$

This equation has the same structure as for the exponentially scaled transmission line, except that the time and the space variables are exchanged. From this symmetry we may say the following: The LCR transmission line-equation describes a wave that grows or attenuates with t. The exponentially scaled LC transmission-line equation describes a wave that grows or attenuates with x. The wave amplitude increases as it travels to the right (we assume $\beta > 0$). This is the case of increasing transmission-line impedance to the right. The *local* transmission-line impedance is $\sqrt{L(x)/C(x)}$. The local wave velocity $1/\sqrt{L(x)C(x)}$ remains constant. The wave amplitude increases by the stepup mechanism of a transmission line that is effectively a stepup transformer.

The potential profile in the exponentially scaled LC transmission line can be determined by numerically solving Equation (3.40) on a computer. We assume

$$L_k = L\rho^k \qquad C_k = C/\rho^k \tag{3.43}$$

Figures 3.13(a) and (b) show the cases for $\rho > 1$ and $\rho < 1$, respectively. If $\rho > 1$, the inductance increases and the capacitance decreases to the right.

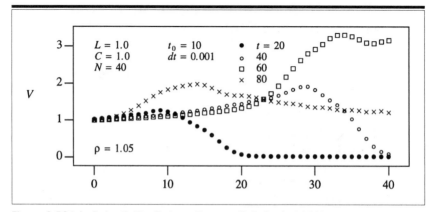

Figure 3.13(a)   Potential Profile in an Exponentially Scaled *LC* Line

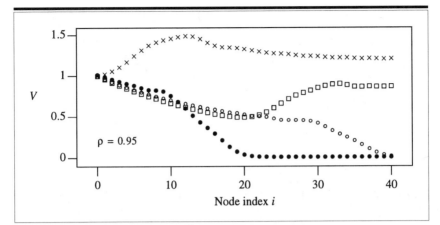

Figure 3.13(b) Potential Profile in an Exponentially Scaled *LC* Chain

At any point on the scaled LC transmission line, the current from the left is higher than required to charge the right-side line section of the wavefront to the same voltage. Then the voltage amplitude increases as the wavefront moves to the right (closed and open circles). Since the right end of the transmission line is open-circuited, the voltage amplitude increases (squares) upon reflection. The current from the right side is not enough to charge the left side to the same potential, since the capacitance on the left is higher than on the right. Then the voltage of the reflected wave decreases as it travels back to the left ($\times$). If $\rho < 1$, the inductance decreases and the capacitance increases to the right. The wave amplitude decreases as it propagates to the right (closed and open circles), and it increases as the reflected wave travels back.

The effect of exponentially scaling an LC transmission line is different from what we observe in an exponentially scaled RC transmission line. The variation of the wave amplitude can be understood completely from the voltage stepup-stepdown mechanism due to the *local* parameter variation. In the scaled RC transmission line an essentially new effect, the drift of the potential profile of the line, appears [1, 7]. The amplitude of the wavefront does not change in the scaled RC transmission line.

3.6 Interconnect Response Analysis Methods

A pair of resistive wires is the simplest interconnect. The pair of wires can be modeled by a distributed *LCR* transmission line. In this section we give analysis methods for the line, useful both for closed-form and for numerical analyses. The method is based on the *ABCD* matrix formalism for linear four-terminal circuits [12, 34].

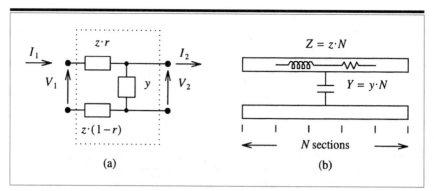

Figure 3.14 Analysis of Integrated Circuit Interconnects

Figure 3.14(a) shows a four-terminal circuit, which may be considered as a black box shown by the dotted enclosure. The ABCD matrix of the black box relates the input variables, voltage V_1, and current I_1 to the output variables V_2 and I_2 as

$$\begin{bmatrix} V_1 \\ I_1 \end{bmatrix} = \begin{bmatrix} A & B \\ C & D \end{bmatrix} \begin{bmatrix} V_2 \\ I_2 \end{bmatrix} \tag{3.44}$$

We note that the polarity of I_2 is opposite to the conventional definition of the impedance or the admittance matrix. This is for convenience of cascading many stages of the boxes. The reciprocity of a linear passive circuit is given by

$$AD - BC = 1 \tag{3.45}$$

so that the determinant of the ABCD matrix is unity.

If two four-terminal circuits are cascaded as shown in Figure 3.15, then V_1, I_1, V_2, I_2, V_3, and I_3 are related by

$$\begin{bmatrix} V_1 \\ I_1 \end{bmatrix} = \begin{bmatrix} A_1 & B_1 \\ C_1 & D_1 \end{bmatrix} \begin{bmatrix} V_2 \\ I_2 \end{bmatrix} \qquad \begin{bmatrix} V_2 \\ I_2 \end{bmatrix} = \begin{bmatrix} A_2 & B_2 \\ C_2 & D_2 \end{bmatrix} \begin{bmatrix} V_3 \\ I_3 \end{bmatrix} \tag{3.46}$$

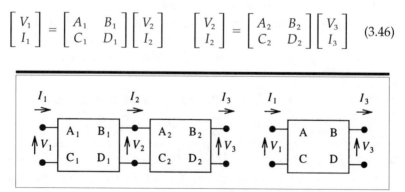

Figure 3.15 Cascading Two Stages of Four-Terminal Circuits

From these we have

$$\begin{bmatrix} V_1 \\ I_1 \end{bmatrix} = \begin{bmatrix} A & B \\ C & D \end{bmatrix} \begin{bmatrix} V_3 \\ I_3 \end{bmatrix}$$

where (3.47)

$$\begin{bmatrix} A & B \\ C & D \end{bmatrix} = \begin{bmatrix} A_1 & B_1 \\ C_1 & D_1 \end{bmatrix} \begin{bmatrix} A_2 & B_2 \\ C_2 & D_2 \end{bmatrix}$$

The ABCD matrix of the circuit built by cascading the two stages of the circuits having $A_1 B_1 C_1 D_1$ and $A_2 B_2 C_2 D_2$ matrices is the product of the two matrices.

Suppose that N stages of the same four-terminal unit circuits are cascaded. The ABCD matrix of the entire circuit Δ_N is given by

$$\Delta_N = \Delta^N$$

where (3.48)

$$\Delta = \begin{bmatrix} A_0 & B_0 \\ C_0 & D_0 \end{bmatrix}$$

here Δ is the ABCD matrix of the unit circuit. Δ is generally not a diagonal matrix, and therefore its Nth power is not easy to compute. If a transformation matrix T and a diagonal matrix Γ are found such that

$$\Delta = T^{-1} \Gamma T$$

where

$$T = \begin{bmatrix} t_{11} & t_{12} \\ t_{21} & t_{22} \end{bmatrix}$$ (3.49)

and

$$\Gamma = \begin{bmatrix} \gamma_0 & 0 \\ 0 & \gamma_1 \end{bmatrix}$$

we have

$$\Delta_N = \Delta^N = T^{-1} \Gamma^N T \quad \text{where} \quad \Gamma^N = \begin{bmatrix} \gamma_0^N & 0 \\ 0 & \gamma_1^N \end{bmatrix} \quad (3.50)$$

The matrices Γ and T are found using the standard matrix diagonalization method [35]. γ_0 and γ_1 are the eigenvalues of the unit circuit ABCD matrix, given by

$$\gamma_0 = [A_0 + D_0 + \sqrt{(A_0 + D_0)^2 - 4}]/2$$

$$\gamma_1 = [A_0 + D_0 - \sqrt{(A_0 + D_0)^2 - 4}]/2$$

(3.51)

where the reciprocity $A_0 D_0 - B_0 C_0 = 1$ has been used. All the components of matrix T are given in terms of the two arbitrarily selected components t_{11} and t_{22} as

$$\frac{t_{12}}{t_{11}} = -\frac{A_0 - \gamma_0}{C_0} \qquad \frac{t_{21}}{t_{22}} = -\frac{D_0 - \gamma_1}{B_0} \qquad (3.52)$$

The ABCD matrix of the unit circuit of Figure 3.14(a) is given by

$$A_0 = 1 + yz \qquad B_0 = z \qquad C_0 = y \qquad D_0 = 1 \qquad (3.53)$$

where y and z are given by

$$y = Y/N \qquad z = Z/N \qquad (3.54)$$

in which Y and Z are the total parallel admittance and the total series impedance, respectively, of the line, and N is the number of units of the LCR transmission line. We have

$$\gamma_0(N) = (1/2)[2 + \chi + \sqrt{\chi^2 + 4\chi}]$$

$$\gamma_1(N) = (1/2)[2 + \chi - \sqrt{\chi^2 + 4\chi}]$$

(3.55)

where $\chi = yz = YZ/N^2$. If Equation (3.55) is substituted into Equation (3.50) and the limit as $N \to \infty$ is taken,

$$\gamma_{0\infty} = \lim_{N \to \infty} \gamma_0(N)^N = \exp(\sqrt{YZ}) \qquad \gamma_{1\infty} = \lim_{N \to \infty} \gamma_1(N)^N = \exp(-\sqrt{YZ})$$

$$\lim_{N \to \infty} [t_{12}(N)/t_{11}(N)] = \sqrt{Z/Y} \qquad \lim_{N \to \infty} [t_{21}(N)/t_{22}(N)] = -\sqrt{Y/Z}$$

(3.56)

and therefore the ABCD matrix of the continuous transmission line is given by

$$\begin{bmatrix} A_\infty & B_\infty \\ C_\infty & D_\infty \end{bmatrix} = \begin{bmatrix} t_{22}(\infty)/M & -t_{12}(\infty)/M \\ -t_{21}(\infty)/M & t_{11}(\infty)/M \end{bmatrix} \tag{3.57}$$

$$\times \begin{bmatrix} \gamma_{0\infty} & 0 \\ 0 & \gamma_{1\infty} \end{bmatrix} \begin{bmatrix} t_{11}(\infty) & t_{12}(\infty) \\ t_{21}(\infty) & t_{22}(\infty) \end{bmatrix}$$

where $M = t_{11}t_{22} - t_{12}t_{21} = t_{11}t_{22} - (\sqrt{Z/Y}\,t_{11})(-\sqrt{Y/Z}\,t_{22}) = 2t_{11}t_{22}$. Thus we have

$$\begin{bmatrix} A_\infty & B_\infty \\ C_\infty & D_\infty \end{bmatrix} = \begin{bmatrix} \cosh\sqrt{YZ} & \sqrt{Z/Y}\,\sinh\sqrt{YZ} \\ \sqrt{Y/Z}\,\sinh\sqrt{YZ} & \cosh\sqrt{YZ} \end{bmatrix} \tag{3.58}$$

$$\rightarrow \begin{bmatrix} 1 + (YZ/2) & Z \\ Y & 1 + (YZ/2) \end{bmatrix} \quad (\omega \rightarrow 0)$$

The ABCD matrix was written using the line's total series impedance Z and the total parallel admittance Y. In the LCR transmission line shown in Figure 3.14(b),

$$Y = j\omega C \qquad Z = R + j\omega L \tag{3.59}$$

The right-hand side of Equation (3.58) is, in the limit as $\omega \rightarrow 0$, using $Y \rightarrow 0$,

$$\begin{bmatrix} A_\infty & B_\infty \\ C_\infty & D_\infty \end{bmatrix} \rightarrow \begin{bmatrix} 1 + (j\omega C/2)(R + j\omega L) & R + j\omega L \\ j\omega C & 1 + (j\omega C/2)(R + j\omega L) \end{bmatrix} \tag{3.60}$$

In the shorted-end line, the driving-point impedance in the limit as $\omega \rightarrow 0$ is given by

$$Z_S(0) = \lim_{\omega \rightarrow 0} \frac{V_1(\omega)}{I_1(\omega)} = R \tag{3.61}$$

and in the open-end line, the driving-point impedance has the limit

$$Z_O(\omega) \rightarrow 1/j\omega C \tag{3.62}$$

In the limit as $\omega \rightarrow 0$ the shorted-end transmission line acts as a pure resistance R and the open-end transmission line as a pure capacitance C. The limits are both reasonable. In the limit of $\omega \rightarrow \infty$ we note that

$$\sqrt{YZ} = j\omega\sqrt{LC}\sqrt{1 - (jR/\omega L)} \rightarrow j\omega\sqrt{LC}$$
$$+ (1/2)(R/\sqrt{L/C}) \tag{3.63}$$

We consider the high-frequency limit ($\omega \to \infty$) of an infinitely long transmission line ($L, C, R \to \infty$, and the ratios of the component values remain unchanged). Both the real and the imaginary part of $\sqrt{YZ}$ diverge. From Equation (3.58) we have

$$Z_S(\omega) = \lim_{\omega, L, C, R \to \infty} [V_1(\omega)/I_1(\omega)] = \sqrt{L/C} \qquad (3.64)$$

The transmission line looks as if it were a constant characteristic resistance $\sqrt{L/C}$. Equation (3.58) has all the expected limits.

Components of the ABCD matrix depend on the angular frequency ω as follows. Figure 3.16(a) shows the trajectories of the point [Re(A_∞), Im(A_∞)] when the angular frequency ω is increased from 0 to ∞. The trajectories of D_∞ are identical. The parameter values are $L = 1$ and $C = 1$, and the three values of R chosen are given in the figure. The closed and open circles and the crosses show A_∞ at every increment $\Delta\omega = 0.1$, starting from 0. All the trajectories start from (1, 0) at $\omega = 0$, and they make a spiral motion, approaching the *limit cycle* shown by the closed ellipse, as ω increases. That every trajectory ends up with a limit cycle is proven as follows. In the limit as $\omega \to \infty$ we have

$$A_\infty = \cosh\sqrt{YZ} = X_A \cos \omega\sqrt{LC} + jY_A \sin \omega\sqrt{LC}$$
$$= \mathrm{Re}(A_\infty) + j\,\mathrm{Im}(A_\infty) \qquad (3.65)$$

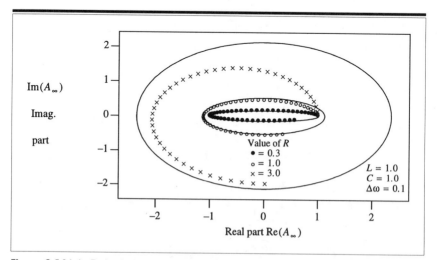

Figure 3.16(a) Trajectories and Limit Cycles of the Matrix Element A_∞

where

$$X_A = \cosh\left(\frac{1}{2}\frac{R}{\sqrt{L/C}}\right)$$

and   (3.66)

$$Y_A = \sinh\left(\frac{1}{2}\frac{R}{\sqrt{L/C}}\right)$$

We have

$$\frac{Y_A}{X_A} = \frac{\exp(R/\sqrt{L/C}) - 1}{\exp(R/\sqrt{L/C}) + 1} \qquad (3.67)$$

and

$$[\mathrm{Re}(A_\infty)/X_A]^2 + [\mathrm{Im}(A_\infty/Y_A)]^2 = 1 \qquad (3.68)$$

This is the equation of an ellipse in the $\mathrm{Re}(A_\infty) - \mathrm{Im}(A_\infty)$ coordinate plane. The axes of the ellipse are $2X_A$ and $2Y_A$. As ω increases, the trajectory turns counterclockwise and approaches the ellipse from the inside. In the limit as $\omega \to 0$

$$A_\infty = \cosh\sqrt{YZ} \to X_1 + jY_1 \qquad (3.69)$$

where

$$X_1 = 1 - (1/2)\omega^2 LC \qquad Y_1 = (1/2)\omega CR \qquad (3.70)$$

and therefore the trajectory in the impedance plane is given by

$$Y_1 = \frac{R}{\sqrt{2L/C}}\sqrt{1 - X_1} \qquad (3.71)$$

In the limit as $\omega \to \infty$, B_∞ and C_∞ are given by

$$B_\infty = X_B \cos(\omega\sqrt{LC}) + jY_B \sin(\omega\sqrt{LC}) \qquad (3.72)$$

$$C_\infty = X_C \cos(\omega\sqrt{LC}) + jY_C \sin(\omega\sqrt{LC})$$

where

$$X_B = \sqrt{L/C} \sinh\left(\frac{1}{2}\frac{R}{\sqrt{L/C}}\right) \qquad Y_B = \sqrt{L/C} \cosh\left(\frac{1}{2}\frac{R}{\sqrt{L/C}}\right)$$

$$X_C = \sqrt{C/L} \sinh\left(\frac{1}{2}\frac{R}{\sqrt{L/C}}\right) \qquad Y_C = \sqrt{C/L} \cosh\left(\frac{1}{2}\frac{R}{\sqrt{L/C}}\right)$$

$$(3.73)$$

and

$$\frac{Y_B}{X_B} = \frac{Y_C}{X_C} = \frac{\exp(R/\sqrt{L/C}) + 1}{\exp(R/\sqrt{L/C}) - 1} > 1 \qquad (3.74)$$

The trajectories of B_∞ and C_∞ are shown in Figures 3.16(b) and 3.16(c), respectively.

The *ABCD* matrix components are the fundamental parameters characterizing an *LCR* transmission-line segment. As an application, we study the driving-point impedance of the segment. The resistance of the segment whose remote end is short-circuited, R_{IS}, and the capacitance of the segment whose remote end is open-circuited, C_{IO}, have values R and C in the limit as $\omega \to 0$. For $\omega \neq 0$ they are given, respectively, by

$$\frac{R_{IS}(\omega)}{R} = \mathrm{Re}\left[\frac{B_\infty}{D_\infty R}\right] \qquad \frac{C_{IO}(\omega)}{\omega C} = \mathrm{Im}\left[\frac{C_\infty}{A_\infty \omega C}\right] \qquad (3.75)$$

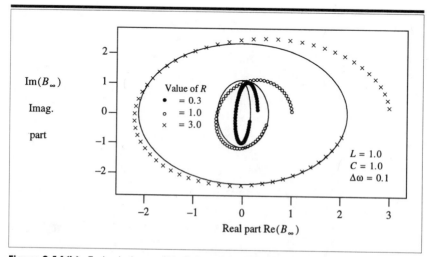

Figure 3.16(b) Trajectories and Limit Cycles of the Matrix Element B_∞

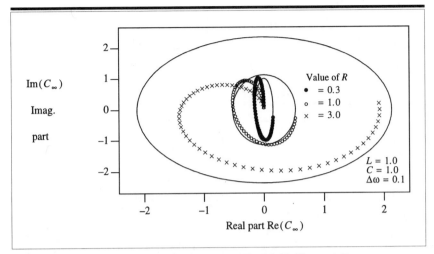

Figure 3.16(c) Trajectories and Limit Cycles of the Matrix Element C_∞

The normalized resistance and capacitance are plotted versus angular frequency in Figure 3.17(a) to (d). In the four graphs, the solid curves show the effective resistance of the shorted-end line, and the dotted curves show the effective capacitance of the open-ended line. The results are interpreted as follows. If R is large, the normalized series resistance approaches $\sqrt{L/C}/R$ in the limit as $\omega \to \infty$, while making an oscillatory variation with increasing ω. The limit as $\omega \to \infty$ can be explained as follows: Only the short section of the line close to the driving point, which has inductance and capacitance L_S and C_S, respectively, satisfying $\omega^2 L_S C_S \approx 1$, is effective in determining the driving-point impedance. The rest of the transmission line is too far away to influence the driving point. The transmission line presents its characteristic impedance to the driver, and as the section becomes short, the effects of R vanishes. There is an effective length of the transmission line that actually loads the driver, and it depends on the transition time, or the frequency. The normalized effective capacitance begins to decrease at about the angular frequency ω determined from $\omega^2 LC \approx 1$, and approaches zero as ω increases. The driver feels the capacitance of the short section only, and the length of the section decreases with increasing frequency.

Figure 3.17(a) and (b) show that *resonance* peaks of the real part of the impedance of the shorted transmission line appear with period of $\Delta\omega = \pi$. This is because the round trip of the signal introduces a delay $2\sqrt{LC}$, which corresponds to angular frequency $\Delta\omega = 2\pi/2\sqrt{LC} = \pi$, since $L = C = 1$. They are parallel resonances.

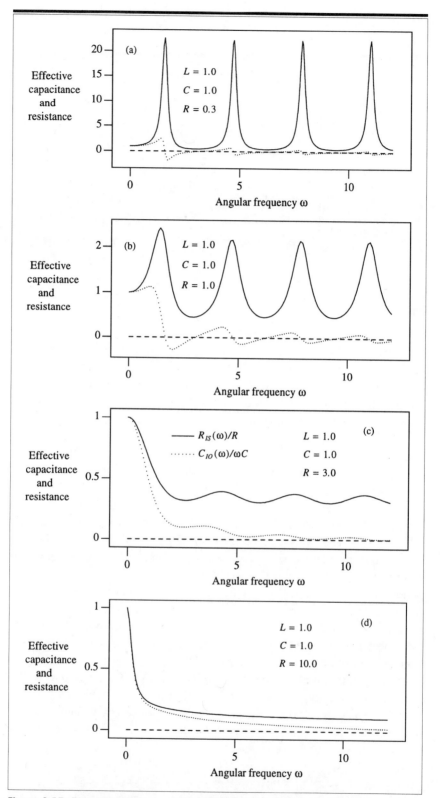

Figure 3.17 Effective Resistance of Shorted-End and Effective Capacitance of Open-End Transmission Line

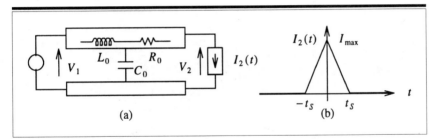

Figure 3.18 Loading a Resistive Wire with a Current Generator

The results of this section can be used to determine the high-speed power supply response. A low-impedance power supply is connected to the load by a pair of wires, as shown in Figure 3.18(a). The wires can be modeled by a continuous LCR transmission line. Suppose that the load is a current generator $I_2(t)$. Let the regulated input power supply voltage V_1 be zero for simplicity. Then

$$0 = A(\omega)V_2(\omega) + B(\omega)I_2(\omega) \tag{3.76}$$

$$I_1(\omega) = C(\omega)V_2(\omega) + D(\omega)I_2(\omega)$$

From the first equation we have

$$V_2(\omega) = -Z_O(\omega)I_2(\omega)$$

where $\hspace{8cm}$ (3.77)

$$Z_O(\omega) = B(\omega)/A(\omega)$$

Here $Z_O(\omega)$ is the power supply output impedance contributed by the connecting wire, and $V_2(\omega)$ and $I_2(\omega)$ are the Fourier-transformed output voltage and the load current, respectively, defined by [36]

$$V_2(t) = -\frac{1}{\sqrt{2\pi}} \int_{-\infty}^{\infty} \exp(j\omega t)\, Z_O(\omega)I_2(\omega)\, d\omega \tag{3.78}$$

$$I_2(\omega) = \frac{1}{\sqrt{2\pi}} \int_{-\infty}^{\infty} \exp(-j\omega t)\, I_2(t)\, dt$$

We have

$$Z_0(\omega) = \sqrt{Z/Y}\ \sinh(\sqrt{YZ})/\cosh(\sqrt{YZ}) = Z[N(X)/D(X)] \quad (3.79)$$

$$N(X) = 1 + (X/6) + (X^2/120) + (X^3/5040)$$
$$+ (X^4/362880) + (X^5/39916800) + \cdots$$

$$D(X) = 1 + (X/2) + (X^2/24) + (X^3/720)$$
$$+ (X^4/40320) + (X^5/3628800) + \cdots$$

where

$$X = YZ = -\omega^2 L_0 C_0 + j\omega C_0 R_0 \qquad (3.80)$$

$$Y = j\omega C_0 \quad \text{and} \quad Z = R_0 + j\omega L_0$$

The power-series expansion of the numerator and the denominator of $Z_0(\omega)$ allows easy numerical computation of the complex integral. We assume the load current waveform shown in Figure 3.18(b):

$$I_2(t) = 0 \quad (t \le -t_S) = I_{max}(t + t_S)/t_S \quad (-t_S < t \le 0) \qquad (3.81)$$
$$= I_{max}(t_S - t)/t_S \quad (0 < t \le t_S) = 0 \quad (t > t_S)$$

We then have

$$I_2(\omega) = \sqrt{2/\pi}\, t_S I_{max}[1 - \cos(\omega t_S)]/(\omega t_S)^2 \qquad (3.82)$$

The Fourier inversion integral can be computed accurately enough if the terms up to order $(YZ)^5$ are retained in Equation (3.79). Figures 3.19(a) and (b) show the results.

In Figure 3.19(a) the series resistance R_0 was changed while keeping L_0 and C_0 constant. As the load begins to draw current at $t = -t_S = -1$, V_2 becomes negative. At the beginning $V_2(t)$ is approximately proportional to the elapsed time $\Delta t = t + t_S$ (we note that $t < 0$). The proportionality constant is determined as follows. Let the total length of the line be 1, and the fraction of the length of the section the signal has reached within Δt be k. The charge stored in the capacitance of the section equals the charge the current generator consumed. We have

$$\Delta t = k\sqrt{L_0 C_0} \quad \text{and} \quad -kC_0 V_2(t)/2$$
$$= (I_{max}/t_S)(\Delta t)^2 \times (1/2) \qquad (3.83)$$

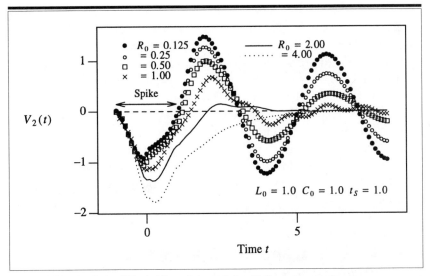

Figure 3.19(a) Output Node Waveforms: Higher Voltage across R_0

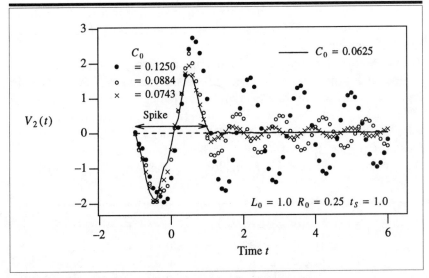

Figure 3.19(b) Output Node Waveforms: Higher Voltage across L_0

The second relation assumes that the voltage profile near the right end of the line is linearly dependent on the distance from the right end. By eliminating k we find

$$V_2(t) \approx -\sqrt{L_0/C_0}\, I_{max}(\Delta t / t_S) \qquad (3.84)$$

The relationship is independent of R_0. It agrees with the accurately computed voltage waveforms. The initial negative voltage spike generates ringing oscillation if R_0 is small. If R_0 is large, the oscillation damps, and V_2 returns to zero quickly [crosses in Figure 3.19(a) for $R_0 = 1.0$ and for the higher R_0]. The parameter values for Figure 3.19(a) reflect the case where the resistance develops more voltage than the inductance: The initial output voltage waveform is close to a replica of the current waveform. The parameter values for Figure 3.19(b) reflects the case where the inductance develops more voltage than the resistance: Then the initial output voltage waveform is close to a replica of the derivative of the current waveform. The two regimes are distinguished by the relative magnitude of t_s and $\sqrt{L_0 C_0}$.

3.7 Grounds of Strongly Connected Circuit Blocks

If two circuit blocks are connected by a large number of wires, they are called *strongly connected*. Each circuit has a ground, to which its own signal voltages are referred. The two grounds are connected together by a dedicated ground wire as well, to maintain their average DC voltages equal. The connection does not guarantee that the two ground voltages are equal all the time. The ground voltage difference fluctuates around zero. In Figure 3.20 two circuit blocks A and B exchange signals. The ground of each block is shown by the thick line. The two grounds are connected by a wire $\overline{CDEF}$. The pulldown driver MN1 conducts, and it sends a signal to block B. Current I flows through the driver output interconnect. How does the current return to block A from block B and close the loop to satisfy the

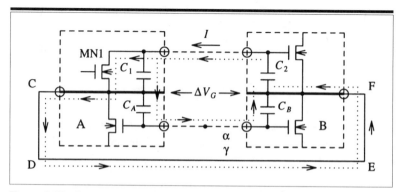

Figure 3.20   Current Flow Paths between Integrated Circuits

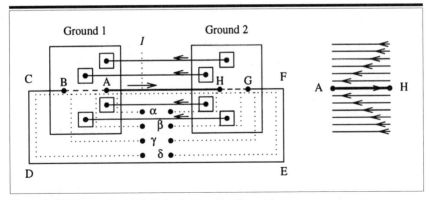

Figure 3.21 Hierarchy of Multiply Connected Grounds

current conservation law? The ground connection $\overline{CDEF}$ does carry some current, but not all. The current closes the loop through many other signal paths that exist between the two blocks, such as path α. Any circuit node has a parasitic capacitance to the ground. C_1, C_2, C_A, and C_B of Figure 3.20 show some of the parasitic capacitances through which the high-speed currents close loops. If there are many connection wires, and if the ground connection $\overline{CDEF}$ is not significantly shorter or significantly more conductive than the other wires, the current carried by all the nonground interconnect wires can be quite significant, or it can even be dominant.

Figure 3.21 shows two circuit blocks connected by many interconnect wires. Their DC ground levels are maintained by the wire $\overline{CDEF}$. A small square having a closed circle at the center indicates a gate that sends or receives signal. Let us consider the inductance of a new interconnect $\overline{AH}$ added to the circuit. The inductance of the loop including the section $\overline{AH}$ appears to be the inductance of the large closed loop *ABCDEFGHA*, but it is actually smaller than that. If voltmeter connection loops having successively larger sizes, α, β, γ, . . . are used to measure the voltage, the voltage reading increases from connection α to connection β, but beyond that the larger loops γ, . . . may give the same readings. All the currents I return by the many interconnects that are enclosed by the loop γ. Since the loop β is smaller than the loop δ (this loop approximates the ground connection loop ABCDEFGHA), its inductance is accordingly smaller. As we observe from this example, all the interconnect wires except for the one we singled out, $\overline{AH}$, may be considered as ground return paths. They are all in parallel, and they are at various distances from $\overline{AH}$.

The inductance of the loop including $\overline{AH}$ is determined by taking the return current distribution into consideration. From this ground connection model we arrive at two interesting conclusions: (1) If current is distributed symmetrically to $\overline{AH}$, the inductance of the loop including $\overline{AH}$ is small, since the magnetic field is confined within a narrow area around $\overline{AH}$. (2) Inductances of two nearby interconnect segments are about the same if the interconnects are more or less uniformly placed (as in IC bonding wires) and if the segments are not too close to the edge of the circuit, where the return current paths are scarce.

Examples of the strongly connected grounds of this type are as follows.

1. The ground of a VLSI chip and the ground of the package.

2. The ground of the VLSI package and the ground of the PC board.

3. The grounds of two heavily communicating VLSI chips, such as the CPU and the memory management chips. In an extreme case, the ground of the PC board in between the chips may become rather irrelevant.

4. The ground of the two PC boards connected by backplane interconnects.

3.8 Transients in an Input/Output Buffer Frame—I

The circuits on a chip are powered by the external power supply, through the power distribution circuits of the chip's I/O buffer frame and the package. Circuits inside the silicon chip see the I/O buffer frame as their immediate power supply. The intended power supply is behind the I/O buffer frame. We need to understand how the I/O frame distributes the power.

Figure 3.22 shows the chip-package interconnects schematically: the top view on the left, and three cross sections at α, β, and γ on the right. The chip's ground (GND) and V_{DD} pads are connected to the package's GND and V_{DD} conductor layers, respectively, by bonding wires. Input and output signals to the chip are conducted to the package lead wires by bonding wires as well. The chip is fabricated using an $n+$ substrate twin-tub CMOS [37]. The substrate is connected to V_{DD}. The regions in the Figure 3.22 cross sections connected to V_{DD} are shadowed by the oblique lines. When we study the chip's ground-bus noise, the chip's ground busses are the dynamic nodes. The $n+$ substrate is the reference ground of the backbone circuit of the chip's ground-bus system. For the transient response analysis we do not distinguish between ground and V_{DD}, since they are always strongly connected. The chip substrate is connected to the package V_{DD} by

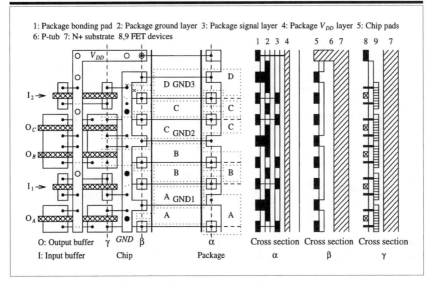

Figure 3.22 Schematic of an Output Buffer Frame Structure

so many wires that the substrate can be used as the reference ground for transient analysis.

Figure 3.22 must be converted to an equivalent circuit. Three output buffers O_A, O_B, and O_C and two input buffers I_1 and I_2 are shown. O_A is in the high-impedance (tri-) state, O_B switches and dumps the discharge current to the ground bus, and O_C is in the quiescent conducting state (the load capacitor has discharged some time before). These are the three possible states of an output buffer. The I/O buffer frame consists of current loops A, B, C, ... They are the discharging current paths of the output drivers A, B, C, . . . , respectively. They are multiply connected loops, since the discharge current of a single driver flows through many return paths. Of the chip's ground connections, to a 1st order approximation, the GND1 bond wire is a segment of loop A, the GND2 bond wire is the common segment of loops B and C, and the GND3 bond wire is a common segment of loops C and D. In the package, the current from the chip ground pad finds connections to the many signal lines back to the chip, through capacitances of the package itself and of the external circuit board beyond the package. Back at the chip, the current finds connection paths to the reference ground (the substrate) through the device, wiring, pad, or input protection-circuit parasitic capacitances. Since there are a huge number of signal lines and device components, and their total capacitance is large, the current path is well closed from the starting chip ground pad to the

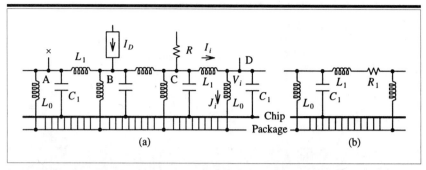

Figure 3.23 Simplified Equivalent Circuit of the I/O Frame Ground bus of a VLSI Chip

package and back to the chip substrate. Thus the capacitance that closes the loop is the capacitance between the chip ground bus pad (dynamic nodes) and the substrate, lumped to the node where the chip's ground connection is made, that is, C_1 of the equivalent circuit of Figure 3.23.

A major component of C_1 is the capacitance of the on-chip ground bus, the bonding pad, and the device parasitics. The closed-loop inductance L_0 and the capacitance C_1 work jointly as a harmonic oscillator at each location of the ground connection. The harmonic oscillators at two neighboring locations are connected by the chip ground bus. The go and the return current paths, consisting of the on-chip ground bus and the $n+$ substrate, are closed to form a loop by the lumped capacitance C_1 at the location of the ground-wire connection. This loop has inductance L_1. Thus we obtain the equivalent circuit of Figure 3.23(a). The thick line shows the chip substrate as the reference ground, and the thin line shows the package ground. The two grounds are strongly connected, as shown by the many parallel vertical lines in Figure 3.23.

In the equivalent circuit of the ground bus the driver at location A is in tri-state. The node A has no connection (the device capacitance is lumped to C_1 there). The driver of node B switches, and the current is dumped to node B. The device is represented by a current generator I_D. Node C is connected to a conducting driver, and the driver internal resistance is modeled by a resistor R (the device is in the triode region). The other end of R is connected to the ground reference. The voltages V_i ($i =$ A, B, C, D, . . .) of Figure 3.23 characterize the ground-bus noise. In the equivalent circuit, only those voltages developed across capacitances like C_1 have valid physical meanings. The equivalent circuit for the I/O buffer frame includes loop D, which is located at an input buffer. The noise voltage is developed between the two terminals of C_1 at that location, and this volt-

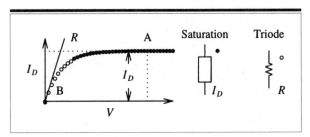

Figure 3.24 Driver FET Characteristics

age appears in series with the input signal voltage. It is not possible, however, to claim that this voltage is the noise voltage to the input buffer. This point will be discussed after the analysis of the ground-bus noise is completed (Section 3.11). The ground-bus noise analysis can be extended to include the resistance of the on-chip ground bus. In Figure 3.23(b) a section that includes the locatable on-chip ground-bus resistance R_1 is shown.

The output driver FET is a nonlinear device. Its current-voltage characteristic is shown schematically in Figure 3.24. At the beginning of the discharge process the gate of the FET is pulled up, and the FET is turned on. The FET sustains the load capacitor voltage, and it is at the bias point A of Figure 3.24. It is in the *saturation* region. The FET is equivalent to a current generator. As the discharge continues, the voltage sustained by the FET decreases, and the bias point approaches B, where the FET moves into the triode region. In the later phase of the discharge, the FET is equivalent to a linear resistor. Thus the equivalent circuit of the output buffer frame uses a current generator I_D and resistor R for the respective discharge phases.

The FET current I_D establishes current in L_0, thereby transferring the energy stored in the output capacitor to the inductor. The inductor and the capacitor make a resonant oscillator, which is excited into oscillation by the current surge. In the later phase of the discharge the linear resistor R of the FET absorbs the energy stored in the resonant circuit. The entire process of the ground bus-noise is separated into two distinct phases. This problem was discussed in detail in my preceding book [1].

3.9 Transients in an Output Buffer Frame—II

Noise generated on the ground bus affects the circuit's integrity. Understanding the *power-bus* noise is hard, however, since many noise sources interact through the complicated ground-bus circuits that include inductances. The complexity of the ground-bus noise is aggravated in that an output driver can be in any of the three states:

1. discharging the output capacitance through the chip's ground bus,

2. conducting in a quiescent state,

3. nonconducting.

There are 3^N different states of the N-driver output buffer frame. We study a few illustrative cases to gain information, and we use physical insight into the circuit model to derive general conclusions. For simplicity, we consider the later phase of the ground-bus noise process first, where the response of the output buffer frame determines the noise.

Figure 3.25 shows an equivalent circuit that includes the on-chip ground-bus resistance for an N-member output buffer frame (ground bus). The series resistances of the on-chip ground bus are the only components that dissipate the energy. The drivers are all in the nonconducting state except the mth one, which discharges the external capacitance through the ground bus. After the short discharge phase, the driver FET goes into the nonconducting state as well. In this case the initial process of the discharge is equivalent to a deposition of charge on the mth node, which raises the node voltage to unit potential. We solve the set of circuit equations

$$L_1 \frac{dI_i}{dt} + R_1 I_i = V_{i-1} - V_i$$

$$L_0 \frac{dJ_i}{dt} = V_i \tag{3.85}$$

$$C_1 \frac{dV_i}{dt} = I_i - I_{i+1} - J_i$$

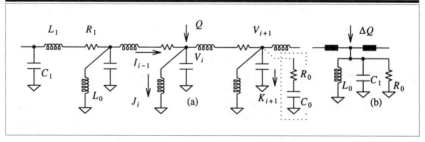

Figure 3.25 Equivalent Circuit of an Output Buffer Frame

subject to the initial conditions

$$I_i = J_i = 0 \qquad V_i(i \neq m) = 0 \qquad V_m = 1.0 \qquad (3.86)$$

The numerical analysis results for a seven-node output buffer frame whose center node is excited is shown in Figure 3.26. If inductance, capacitance, resistance, and time are measured in nanohenries, nanofarads, ohms, and nanoseconds, the parameter values reflect a realistic case. At the beginning (time $0 \leq t \leq 0.3$) the response depends on how node m is excited (not shown). After $t = 0.3$ the waveform becomes a slowly decaying sinusoidal wave. The phase and amplitude of the sinusoid at different nodes are all approximately the same. In the equivalent circuit of Figure 3.25 each mode has a lossless LC harmonic oscillator, whose period of oscillation is given by

$$T_P = 2\pi \sqrt{L_0 C_1} = 0.4 \qquad (3.87)$$

and this is the period of the sinusoidal oscillation. Since all nodes are approximately synchronous, only small current flows through L_1 and R_1, and the only resistance in the circuit, R_1, is ineffective in dissipating the energy. This is the reason why the sinusoid wave decays very slowly. The circuit ends up in a quasistable oscillation. This is not surprising, if we remember the thermodynamic principle of minimum entropy production [38]. A thermodynamic system arrives at a steady state that generates the minimum heat. If the circuit has a mode that does not dissipate energy, all the modes that dissipate energy decay after some time, and only the nondissipative mode survives forever.

In a linear circuit, the amplitude of a mode is determined by the initial condition. There is no transfer of energy from one mode to the other. Then, after some time, only the mode having the longest (or ideally, infinite) time constant remains alive, and that is the *steady state* of the circuit. This is what the principle of minimum entropy production dictates. The simulation of Figure 3.26 identified the nondissipative mode of the output buffer frame circuit. Minimum entropy production can be used to determine the steady-state DC electric field in a nonuniform body. A steady-state DC potential distribution maintains itself as long as the field is supported by the DC bias. This is the state whose time constant is infinitely long. If the principle of the minimum entropy production is interpreted this way, that often mysterious thermodynamic principle becomes easy to understand.

In this circuit another thermodynamic principle controls the circuit operation. The harmonic oscillators at all the nodes carry the same energy.

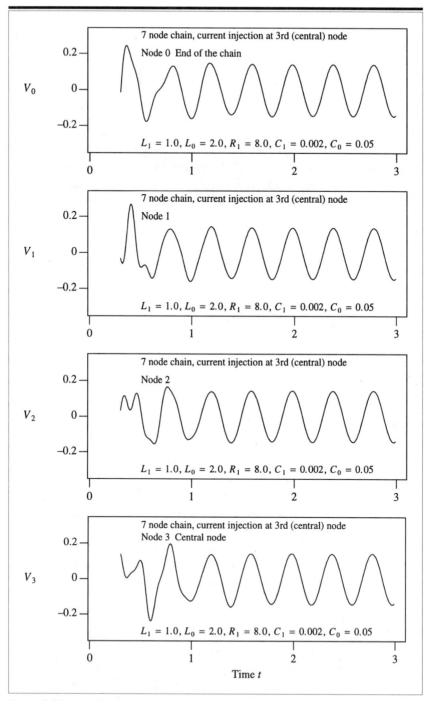

Figure 3.26 Node Voltage Waveforms, Unloaded

This is the energy equipartition law. The law demands that energy be distributed statistically equally to all the harmonic oscillators of the system [19]. This is consistent with the dictates of minimum entropy production. That principle provides a stronger constraint on the steady state than the equipartition law, however, since it specifies the phases of the oscillators as well. If every other oscillator of the output buffer frame were in out-of-phase oscillation, the resistor R_1 would dissipate energy quite effectively. The operation of the output buffer frame is subject to both basic principles of thermodynamics.

To attenuate the long-lasting sinusoidal oscillation, each harmonic oscillator must have an effective resistive loading. A conducting driver device at each node adds a significant resistance loading. With reference to Figure 3.25, the third circuit equation is modified to

$$C_1 \frac{dV_i}{dt} = I_i - I_{i+1} - J_i - K_i \qquad \frac{dK_i}{dt} + \frac{K_i}{R_0 C_0} = \frac{1}{R_0} \frac{dV_i}{dt} \quad (3.88)$$

where K_i is the current in the series combination of R_0 (the device resistance) and C_0 (the driver load capacitance). The same seven-node output buffer frame was simulated as shown in Figure 3.27. In this example, all the nodes are loaded by conducting drivers. The harmonic oscillation decays rapidly.

With reference to Figure 3.25, if each of the N nodes is connected to the neighbor nodes via inductance L_1 and if each node has capacitance C_1, the electrical signal takes time $N\sqrt{L_1 C_1}$ to travel from one end to the other. This time is the measure of equalization of energy among the harmonic oscillators. If the time is less than the decay time constant of the harmonic oscillation, all the harmonic oscillations decay together. We have $N\sqrt{L_1 C_1} = 0.27$ in the examples, and this is much less than the decay time constant of oscillation (about 1 in Figure 3.27). If the energy equalization mechanism is working strongly, which node is loaded by a conducting driver does not matter for the overall decay of the oscillation. In Figure 3.28, only the first, the center, and the last node of a seven-member output buffer frame are loaded by conducting drivers. The rest of the nodes have no loading. In spite of the loading difference, the oscillation amplitudes at different nodes are all about the same: the output buffer frame loses energy as a whole, and the instantaneous energy distribution is about uniform. This is dictated by the equipartition law. As is observed from these examples, the electrical phenomena in an output buffer frame can be understood from the energy balance among harmonic oscillators.

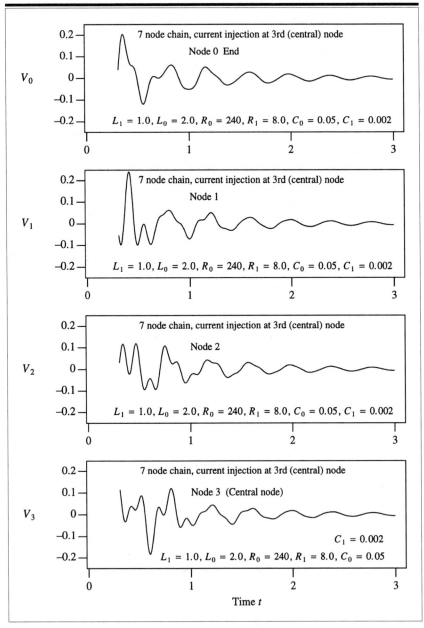

Figure 3.27 Node Voltage Waveforms, All Nodes Loaded

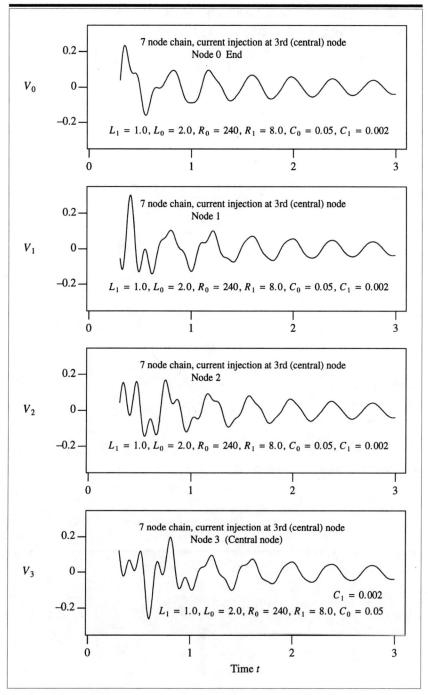

Figure 3.28 Node Voltage Waveforms, Nodes 1, 2, 4, 5 Unloaded

3.10 Response of the Output Buffer Frame Immediately after Discharge

Within the short time following the output driver discharge, an electrical signal has no time to travel from the discharging node to the other nodes. The rest of the nodes of the output buffer frame are still in their quiescent states, and only the discharging-node voltage is affected. Then the equivalent circuit of the frame is simplified as shown in Figure 3.29. In this equivalent circuit the two immediate neighbor nodes are in the quiescent states: Their voltages are at ground potential. The impedance looking into the discharging node, Z, is given by

$$\frac{1}{Z} = \frac{1}{Z_R + jZ_I} = j\omega C_0 + \frac{1}{j\omega L_0} + \frac{2}{R_1 + j\omega L_1} \qquad (3.89)$$

where

$$Z_R = 2\omega^2 L_0^2 R_1 / \Delta$$

$$Z_I = j\omega L_0 [R_1^2 (1 - \omega^2 L_0 C_0) + \omega^2 L_1 [L_1 (1 - \omega^2 L_0 C_0) + 2L_0]] / \Delta$$

$$(3.90)$$

in which

$$\Delta = R_1^2 (1 - \omega^2 L_0 C_0)^2 + \omega^2 [L_1 (1 - \omega^2 L_0 C_0) + 2L_0]^2 \quad (3.91)$$

The driving-point impedance of the output buffer frame, on taking two sections on both sides, is given by

$$\frac{1}{Z} = j\omega C_0 + \frac{1}{j\omega L_0} + \frac{2}{Z_{SR} + jZ_{SI}} \qquad (3.92)$$

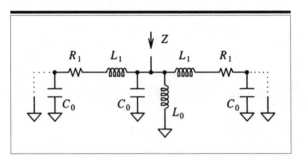

Figure 3.29 Driving-Point Impedance of an Output Buffer Frame

where

$$Z_{SR} = R_1 + \frac{1}{\Delta} [-R_1(1 - \omega^2 L_0 C_0)\omega^2 L_0 L_1$$
$$+ (L_0 + L_1 - \omega^2 L_0 L_1 C_0)\omega^2 L_0 R_1]$$

$$Z_{SI} = \omega L_1 + \frac{1}{\Delta}[R_1^2(1 - \omega^2 L_0 C_0)\omega L_0 + (L_0 + L_1 - \omega^2 L_0 L_1 C_0)\omega^3 L_0 L_1]$$

(3.93)

in which

$$\Delta = R_1^2(1 - \omega^2 L_0 C_0)^2 + (L_0 + L_1 - \omega^2 L_0 L_1 C_0)^2\omega^2 \quad (3.94)$$

The real and the imaginary parts of the impedances are plotted in Figure 3.30(a) and (b). Using the impedance, the driving-point response to the injected discharge current can be computed by using the Fourier transform. The injected current $I(t)$ is transformed as [36]

$$I(t) = \frac{1}{\sqrt{2\pi}} \int_{-\infty}^{\infty} \exp(j\omega t)\, I(\omega)\, d\omega$$

where

(3.95)

$$I(\omega) = \frac{1}{\sqrt{2\pi}} \int_{-\infty}^{\infty} \exp(-j\omega t)\, I(t)\, dt$$

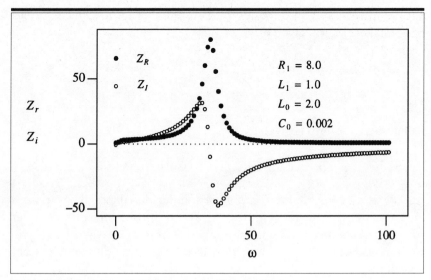

Figure 3.30(a) Driving-Point Impedance (One Section)

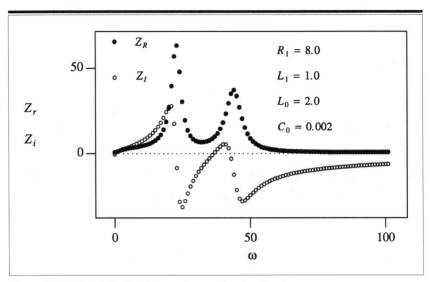

Figure 3.30(b) Driving-Point Impedance (Two Sections)

and the voltage $V(t)$ developed at the node is given by

$$V(t) = \frac{1}{\sqrt{2\pi}} \int_{-\infty}^{\infty} \exp(j\omega t)\ Z(\omega)I(\omega)\ d\omega \qquad (3.96)$$

The current delivered to the output buffer frame is the optimal waveform (the minimum ground-bus noise subject to the specified delay time). The optimum waveform for an inductive output buffer frame is the isosceles triangular waveform given by [1]

$$I(t) = 0 \quad (t < -t_s) \quad (I_{max}/t_s)(t + t_s) \quad (-t_s < t < 0) \qquad (3.97)$$
$$= (I_{max}/t_s)(t_s - t) \quad (0 < t < t_s) \quad 0 \quad (t > t_s)$$

Its Fourier transform is given by

$$I(\omega) = \sqrt{2/\pi}\, t_s I_{max}[(1 - \cos(\omega t_s))/(\omega t_s)^2] \qquad (3.98)$$

By substituting the Fourier component and the impedance, the responses are computed as shown in Figures 3.31(a) and (b). Since the output buffer frame presents predominantly an inductive impedance, the node waveform appears like the first derivative of the optimum discharge current waveform.

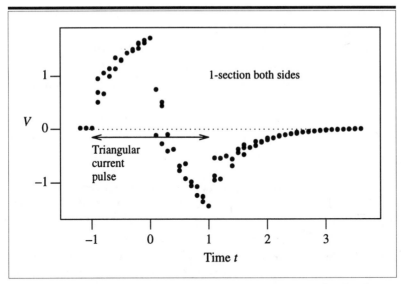

Figure 3.31(a)   Driving-Point Response of Output Buffer Frame: One Section

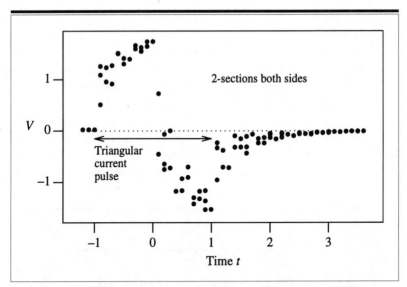

Figure 3.31(b)   Driving-Point Response of Output Buffer Frame: Two Sections

3.11 Interpretation of I/O-Buffer-Frame Noise Simulation

A circuit including naked inductances, such as an I/O buffer frame, is peculiar in many respects. Analysis results based on a simple equivalent circuit must be interpreted carefully. By an I/O-buffer-frame analysis, two design issues must be settled: (1) is the chip's integrity maintained under realistic operating conditions, and (2) is the noise induced in the input circuit tolerable? Let us consider a CMOS chip.

Figure 3.32 shows a schematic diagram of two CMOS chips communicating via PC board traces. The ground-bus bonding wire $\overline{fga}$ is shown as the inductance L, which is the inductance of many current loops sharing segment $\overline{fga}$, such as the loop $\overline{ahijklmnfga}$. The return paths of the loop cross the chip boundary at many locations, one of which is shown by $\overline{jk}$. In chip I the output driver pulldown device is shown with a layout diagram. At locations m and n, the ground bus is connected to the source of the output driver pulldown device. The device is fabricated on a p-tub, shown enclosed by the dotted boundary. The p-tub is also connected to the ground bus at location x. Is the connection effective? A p-tub is a thin sheet of p-type semiconductor, whose sheet resistance is quite high, typically of the order of several kilohms per square. Because of the high sheet resistance and the associated capacitance to the n-type substrate and to many n+ islands diffused on it, the p-tub is unable to change voltage rapidly—not within the short time scale typical of ground-bus noise. To study the effects of the n+ source-drain region of the output driver device fabricated on a p-tub, we may assume, as the first approximation, that the p-tub is always at the average ground voltage set by the ground connection, being maintained at it by many parasitic capacitances. This simplification has a practical consequence. If the equivalent circuit of Figure 3.23 or 3.25 is used to analyze the noise generated by the output buffer discharge, the voltage developed across C_1 may be interpreted as the voltage developed across the pn junction between n+ and p-tub. If this voltage becomes less than $-V_{BE}$ (where V_{BE} is the turn-on voltage of a pn junction)

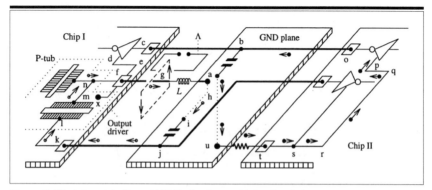

Figure 3.32 Interpretation of I/O-Buffer-Frame Noise Analysis

for a significant length of time, minority carriers are injected into the
p-tub, and there is a risk of latchup.

Evaluation of the noise induced at the input buffer requires examina-
tion of the I/O-buffer-frame equivalent circuit. Suppose that an I/O buffer
frame like that shown in Figure 3.32 is modeled by the equivalent circuit
of Figure 3.33. The output driver discharges the load capacitance C_0, and
the discharge current generates a noise voltage V_N across the inductance
L_0 representing the ground connection. Then it appears that the noise at
the input buffer that shares the ground connection as one of the input
signal paths is V_N. Is this interpretation correct? If L_0 were an isolated inductance, it would be. If L_0 is a naked inductance, the interpretation is
incomplete, and is often misleading. This crucial point is as follows. According to this interpretation, if the input and the output buffers do not
share the same ground connection, the input buffer gets no noise. This is
not true. Suppose that a conductor loop Λ is placed to the side of the
ground connection bonding wire $\overline{fga}$ as shown in the upper middle part
of Figure 3.32. The voltage measured across the small gap between the two
closed circles equals the time derivative of the magnetic flux created by
the current of the bonding wire $\overline{fga}$ that links the loop. This induced voltage is never zero. A naked inductance of the bonding wire in a crowded
place is likely to have a nearby loop that is as large as the ground current
loop and is strongly coupled to the ground current loop. Then the loop
may generate as much noise as V_N. Yet the loop does not share the ground

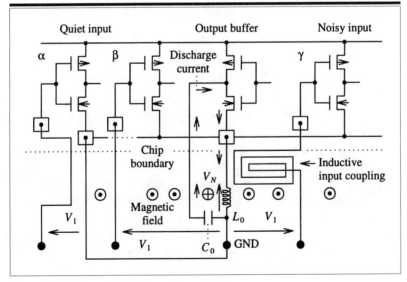

Figure 3.33 Power-Bus Noise and Input Noise

connection. This observation clearly contradicts the simple interpretation given to Figure 3.33. The reason for this contradiction is that L_0 is not the inductance of the bonding wire, but the inductance of the ground current loop. The input noise is determined by how much magnetic flux generated by the ground current loop links the loop made by the input signal path. To determine the noise induced at the input circuit, it is necessary to know the actual ground current distribution and the structure of the circuit that carries the signal from the outside to the input buffer. The physical mechanism that generates the noise voltage in the input circuit is not the *sharing* of the naked inductance L_0, but is the mutual induction between the ground and the input current loops. An unrealistic but informative example is shown in Figure 3.33: The input connection to buffer γ couples to the ground loop in a stepup transformer configuration. Then the induced noise can be higher than V_N. Input buffer α of Figure 3.33 has its own ground connection and an input signal wire closely placed on the side of the ground connection. The buffer gets small noise. This arrangement is similar to a differential signal transmission, used in the ECL logic circuit.

According to this observation, V_N is not a useful measure of input noise. V_N provides, however, an upper-limit estimate of the induced input noise, subject to the following assumptions. Suppose that the inductance of the ground connection loop is L_0, that of the input circuit loop is L_1, and the mutual inductance between them is M. We have $L_0 L_1 > M^2$, and the two loops make a transformer. In Section 3.7 we observed the following: If two circuits are strongly connected by many wires of similar structure, as with bonding wires of an IC in a package, any wire will have about the same inductance. If we may assume that the ground connection and the input connection loops have the same structure and dimension, then $L_0 = L_1 = L$. If the primary and the secondary loops are closely coupled, we have $M = L$. Then in the limit of close coupling the two loops make a unity-ratio transformer, and the primary and the secondary voltages are the same. The noise voltage V_N is induced in the secondary, or input, circuit, subject to this condition. Since M is generally less than L because of the leakage magnetic flux, the induced voltage never exceeds V_N. Thus V_N is the most pessimistic estimate of the noise.

Bonding-wire resistance creates input noise as well, as schematically shown in Figure 3.32 (right side). The chip II ground-bus voltage varies as the output buffer discharges. The resistive powerbus noise affects all the input drivers supplied by the groundbus equally, independent of the distance between the input and the output buffers. The resistive and inductive components of power-bus noise have quite different mechanisms.

Peculiarities of the inductive noise affect power-bus noise measurement as well. If a CRT oscilloscope is used to measure a noise waveform, the probe tip and the ground connection must be smaller than the typical size of the structure on the chip, such as a bonding pad. In practice an accurate noise measurement is quite involved and difficult.

CHAPTER FOUR

Digital Circuit Theory Including Inductance

4.1 Introduction

Integrated circuit technology has made progress with feature size reduction. Device linear dimensions have been reduced by more than an order of magnitude during the last 15 years. As we observed in Section 2.2, a device can be separated into the triode mechanism generating the DC current-voltage characteristic, and the essential and nonessential capacitance parasitics. Because of the scaledown of the dimension, the effects that make the high-speed current-voltage characteristic of the stripped-off device different from its DC current-voltage characteristic, such as the transit-time effects, are reduced as well. High-speed device operation can still be described using the DC current-voltage characteristic. A triode can be described in terms of microstates, as being in one of three operational regions: the N (nonconducting), the S (saturation), or the T (triode) region. Since the device has only a small dimension in the direction of the current flow, and since it has the essential drain-source capacitance parasitic, the voltage developed between the drain and the source nodes is measurable. The gate-to-source voltage is measurable as well, since the input capacitance is essential for the device operation. These capacitance parasitics are moved to the interconnect part of the circuit diagram. If the device is an element that closes a current loop, the device dimension is so small that the loop inductance is determined by the interconnect alone.

  With this simplification, the device and the interconnect combination can be analyzed by representing the device as an equivalent resistor or as a collapsible current generator. If the device is represented as a resistor, the circuit analysis is essentially simplified to linear LCR circuit analysis, and if it is modeled as a collapsible current generator, only three different analyses for the three device microstates are required to cover all the possible cases. The theory of this chapter is developed along this line of thought.

The conventional circuit model is a description of the charge and discharge of a capacitance through a resistance. Circuit performance evaluation using the model became a routine circuit design practice. Modeling parasitic inductances, including determination of their accurate values, is difficult. Capacitance and inductance are different in that the electrostatic energy of capacitance is confined within the well-defined space between the capacitor electrodes, but the magnetic energy of inductance cannot be confined; it exists essentially outside the current path. Thus the theory must be constructed so that the qualitative difference caused by inductance is clearly observable. Therefore, if inductance is added to the circuit, the circuit analysis acquires an essentially new feature. The inductance originates from devices and components connected into a loop. The loop may or may not be observable from the circuit diagram, and it carries no identification in the backbone *RC* circuit diagram. Obviously the circuit diagram must be redrawn. To explain the LCR circuit operation in terms of the backbone RC circuit model and to highlight the new features, we need to introduce more than one local time to describe the circuit. The local time is defined for each *mode* of the backbone RC circuit. A mode is a substructure of a microstate, and therefore it is convenient to call modes submicrostates. In the LCR circuit model, the formerly unidentified object, an inductance, exchanges energy with the outside of the circuit. The role of the inductance cannot be called strictly passive when it retrieves energy stored outside the circuit. The pseudoactive nature of the inductance complicates the microstate theory. The complication is on the level of the submicrostates of the circuit, and consists in qualitative features extracted from the behavior of the submicrostates: the direction of flow of the local times.

Effects of parasitic inductance begin to show up at a clock frequency of a few megahertz in special circuits such as ground-bus connections, and a few hundred megahertz in most digital circuits. The practical application of the theory of this chapter assumes that the inductance is small. The theory is useful as the first-order correction to the conventional resistance-capacitance circuit theory. In this limit a general theory can be constructed using the well-developed perturbation theory for differential equations, created by H. Poincaré [39], and later developed into the general *boundary-layer method* [40]. The mathematical formulation uses a time-axis transformation, which leads to the concept of the transformed time, so that there is more than one local time in the theory. Poincaré's idea of time-axis transformation is historically famous, because it was a seminal contribution to the theory of relativity. His idea was the basis on which I developed the theory of this chapter. If inductance is dominant, as a natural extension of the theory, the direction of flow of the local time becomes the issue.

4.2 Inductance Effects

In this section I introduce the general framework of a circuit theory including inductance, using the simplest example. Figure 4.1(a) shows a CMOS inverter, whose input node was grounded and output node was at V_{DD} before the input was driven up to V_{DD} at $t = 0$. The input transition took place within a negligibly short time. The PFET turns off instantly, and the NFET turns on and its drain-source (two-terminal) characteristic can be approximated by a resistor R. The subsequent switching process is described by the simple equivalent circuit of Figure 4.1(b). The output node voltage $V(t)$ decreases with time as

$$V(t) = V_{DD} \exp(-t/RC) \tag{4.1}$$

The two ground symbols in both Figure 4.1(a) and (b) indicate that the two nodes are connected by ground. If inductance is included, the circuit diagram must be redrawn as shown in Figure 4.1(c) and (d). The output circuit loop has naked inductance L. The inductance L is defined only for the complete loop consisting of the conducting NFET, L, and C, and as for voltages, only those developed across *small* devices (FET and C) have physical meaning. $V_0(t)$ is the voltage developed across the capacitance C, and it has a clear physical meaning. It satisfies the equation

$$LC \frac{d^2V_0(t)}{dt^2} + RC \frac{dV_0(t)}{dt} + V_0(t) = 0 \tag{4.2}$$

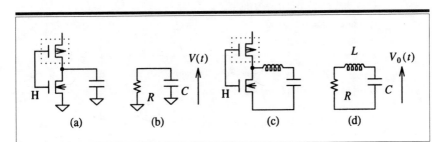

Figure 4.1 Inductance Effects in CMOS Circuits

which is solved by assuming the form $V_0(t) \approx \exp(St)$, where the *complex* frequency S is determined from

$$LCS^2 + RCS + 1 = 0$$

or (4.3)

$$S = [-(RC) \pm \sqrt{(RC)^2 - 4(LC)}]/2LC$$

In the RC-only circuits of Chapter 2 the complex frequency was always a real and negative number. In LCR circuits S is generally a complex number. We write

$$S = S_R + jS_I \qquad (4.4)$$

where S_R and S_I are the real and the imaginary parts of the complex frequency, respectively. If $L < L_C = R^2C/4$, then $S_I = 0$ and the two solutions are both real and negative. If $L > L_C$, the two solutions are complex numbers. Then

$$S_R = -R/2L$$

and (4.5)

$$S_I = \sqrt{(1/RC)^2 - [S_R + (1/RC)]^2}$$

If L is increased from a small value, the track of S on a complex frequency plane is shown in Figure 4.2. The trajectories start at X and Z for small L, and go through the critical damping point C at $L = L_C$. Here S is a pair of

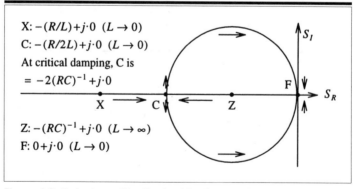

Figure 4.2 Trajectory of the Roots of the Secular Equation

complex conjugate numbers in the underdamping region if $L > L_C$. In the limit as $L \to \infty$ the pair S tends to 0, indicated by F. The node voltage is written as

$$V_0(t) = A \exp(\alpha t) + B \exp(\beta t) \qquad \text{(overdamping)} \qquad (4.6)$$

where

$$A = \frac{1 + \chi}{2\chi} V_{DD}$$

$$B = -\frac{1 - \chi}{2\chi} V_{DD}$$

$$\alpha = -\frac{1}{2T_{LR}}(1 - \chi) \qquad\qquad (4.7)$$

$$\beta = -\frac{1}{2T_{LR}}(1 + \chi)$$

in which χ, T_{LR}, and T_{RC} are defined by

$$\chi = \sqrt{1 - 4(T_{LR}/T_{RC})} \qquad T_{LR} = L/R \qquad T_{RC} = RC \quad (4.8)$$

This solution is valid if χ is real and not zero. If $\chi = 0$ (that is, $T_{LR}/T_{RC} = 0.25$), we take the limit as $\chi \to 0$ in the solution. We obtain

$$V_0(t) = V_{DD}[1 + (t/2T_{LR})]\exp(-t/2T_{LR}) \qquad \text{(critical)} \quad (4.9)$$

If χ is imaginary, then A, B, α, and β are complex numbers. Then Equation (4.6) must be reinterpreted by taking the real part only. We have, after some algebra,

$$V_0(t) = V_{DD} \exp(-t/2T_{LR}) \, [\cos(\chi_A t/2T_{LR})$$
$$+ (1/\chi_A) \sin(\chi_A t/2T_{LR})] \qquad \text{(underdamping)} \quad (4.10)$$

where $\chi_A = \sqrt{4T_{LR}/T_{RC} - 1}$.

An example of the node waveforms is shown in Figure 4.3. The dotted curve is the discharge waveform of the *RC*-only *backbone* circuit, to which L is added to obtain the solid curves. At $L = 0.25$ the circuit is in the critical damping state, a state exactly between the overdamping and the under-

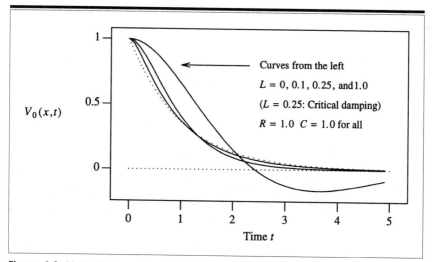

Figure 4.3 Node Voltage Waveforms of *LCR* Circuit

damping (oscillatory) states. Inductance increases the delay at the beginning of the transient, but later the transient speeds up, and goes ahead of the transient of the *RC*-only backbone circuit. To describe this response, the introduction of local time is most convenient. Figure 4.4 shows the waveforms of the *RC*-only backbone circuit (dotted) and of the critically damped *LCR* circuit (solid). We consider that the solid curve is obtained by shifting the dotted curve horizontally by a time Δ. At time $t = 0.45$ the

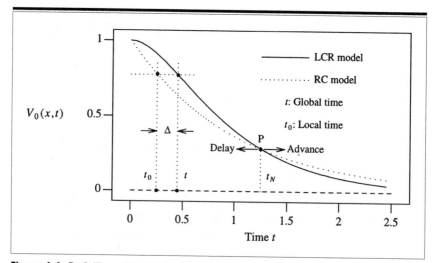

Figure 4.4 Definition of the Global and the Local Time

node voltage of the LCR circuit is $0.775\,V_{DD}$. The same node voltage of the RC circuit was reached at $t_0 = 0.25$. Then we consider that the effective time of the LCR circuit, which we call a local time, is 0.25, and the local time is delayed by $\Delta = 0.45 - 0.25 = 0.2$ relative to the time of the RC-only backbone circuit, which we call the global time. The LCR circuit is an RC circuit whose time variable is transformed. In this case t and t_0 are related by

$$t - t_0 = \Delta(t) \tag{4.11}$$

$$V_{DD}\,\exp(-t_0/CR) = A\,\exp(\alpha t) + B\,\exp(\beta t)$$

Δ can be expressed as a function of either the global time t or the local time t_0. Since the left side of the equation is simpler than the right side and it is easy to solve for the local time t_0, it is convenient to express Δ as a function of t. The time t is called "global" because it is the time used in the backbone RC circuit, to which inductances are added. The circuit is first modeled using an RC-only equivalent circuit that has only one time, the global time. When the inductances are added, the single global time is not adequate to describe the entire circuit: We introduce many *local* times. The local time of the example is delayed at the beginning, but it catches up with the global time at point P of Figure 4.4, and after that the local time is ahead of the global time. A typical relationship between local time and global time is shown in Figure 4.5.

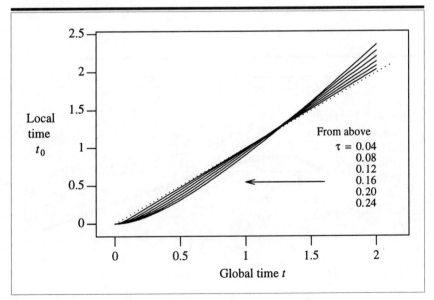

Figure 4.5 Local Time Versus Global Time

Let us consider the case of small L. If the small parameter $\tau = T_{LR}/T_{RC} = L/(R^2C)$ is introduced, we have

$$A = (1 + \tau + 3\tau^2 + 10\tau^3 + \cdots)V_{DD}$$

$$B = -\tau(1 + 3\tau + 10\tau^2 + 35\tau^3 + \cdots)V_{DD}$$

$$\alpha = -(1 + \tau + 2\tau^2 + 5\tau^3 + \cdots)/T_{RC}$$

$$\beta = -(1 - \tau - \tau^2 - 2\tau^3 + \cdots)/T_{LR}$$

(4.12)

In the vicinity of $t = 0$ we have

$$V_0(t) \rightarrow V_{DD}[1 - (1/2)(t^2/T_{LR}T_{RC})]$$ (4.13)

The RC and LCR circuit responses are plotted in Figure 4.6 as dotted and solid curves, respectively. The time shift Δ is plotted versus the global time t as filled circles. In the limit as $t \rightarrow 0$

$$\Delta \rightarrow t - (1/2)(t^2/T_{LR})$$ (4.14)

and in the limit as $t \rightarrow \infty$

$$\Delta \rightarrow T_{RC}\tau[1 + (5/2)\tau + (19/3)\tau^2 + \cdots]$$
$$- \tau(1 + 2\tau + 5\tau^2 + \cdots)t$$
(4.15)

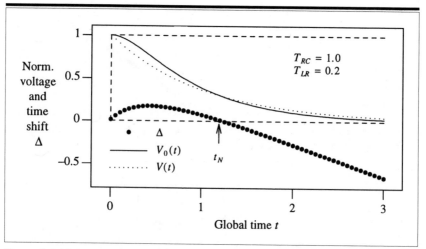

Figure 4.6 Voltage and Time Shift

At $t = t_N$, $\Delta = 0$, so the two curves cross. The time t_N is given by

$$t_N = [1 + (1/2)\tau + (1/3)\tau^2 + \cdots\ ]T_{RC} \qquad (4.16)$$

The physical meanings of these results are as follows: If positive voltage is applied to an inductance, it takes some time to build up the current. Once the current is established, however, it continues for some time even if the terminal voltage becomes zero, or even negative. In this example it takes time t_N to establish the inductance current. The circuit response $V_0(t)$ is delayed to $V(t)$ in the period $0 \le t \le t_N$. The capacitor discharge current of the RC-only circuit [Figure 4.1(b)] decreases as $V(t)$ decreases. In the LCR circuit [Figure 4.1(d)] the inductance maintains the high discharge current, and discharge of the capacitor continues at a high rate. This is the reason why $\Delta < 0$ after the inductance current is established. In the limit as $\tau \to 0$,

$$t_0 = [1 + (Z/R)^2]t - T_{LR} \quad (t \to \infty)$$

where $\qquad\qquad\qquad\qquad\qquad\qquad\qquad\qquad\qquad (4.17)$

$$Z = \sqrt{L/C}$$

here Z is the characteristic impedance of L and C. The relationship between t_0 and t are as follows: The t_0 time scale is stretched by a constant factor $1 + (Z/R)^2$ (>1), and t_0 is delayed from t by $T_{LR} = L/R$, to the first order of the small parameter τ. The new theory converges to the RC-only theory in the limit of zero inductance.

We solve the same problem by a different method, to clarify the mathematics of the time-axis transformation. We assume that the LCR transient is a simple modification of the RC transient by allowing the time axis transformation. We introduce a *local* time θ, which is a function of the global time t, by

$$V_0(t) = V_{DD}\ \exp[-\theta(t)/RC] \qquad (4.18)$$

where $\theta(t)$ is the time that describes the LCR transient in terms of the RC-only backbone circuit model. It is required that

$$\theta(t) \to t \qquad (L \to 0) \qquad (4.19)$$

By substituting $V_0(t)$ into Equation (4.2) it is transformed to

$$T_{LR}\left[\frac{\theta'(t)^2}{T_{RC}} - \theta''(t) \right] - \theta'(t) + 1 = 0 \qquad (4.20)$$

where the initial conditions are

$$\theta(+0) = 0 \qquad \theta'(+0) = 0 \tag{4.21}$$

and where the prime means the time derivative. At time $t = 0$ the capacitance discharge is suppressed by the inductance. The time derivative of the local time is zero. If we use $g(t) = \theta'(t)$, we have

$$\frac{dg}{(g - g_0)(g - g_1)} = \frac{dt}{T_{RC}} \tag{4.22}$$

where

$$g_0(-), \, g_1(+) = \frac{1 \pm \sqrt{1 - 4\tau}}{2\tau} \qquad \tau = \frac{T_{LR}}{T_{RC}} \tag{4.23}$$

and we have

$$\theta'(t) = g(t) = \frac{g_0 g_1 [1 - \exp(\gamma t)]}{g_0 - g_1 \exp(\gamma t)}$$

where $$\tag{4.24}$$

$$\gamma = \frac{g_1 - g_0}{T_{RC}}$$

Upon further integration we get a formula for the local time:

$$\theta(t) = -T_{RC} \log \left[\frac{g_1 \exp(\gamma t) - g_0}{g_1 - g_0} \right] + g_1 t \tag{4.25}$$

In the limit as $t \to \infty$ we have

$$\theta(t) \to g_0 t - (g_0/g_1) T_{RC} \tag{4.26}$$

where in the limit as $\tau \to 0$,

$$g_0 \to 1 + (Z/R)^2 \qquad (g_0/g_1) T_{RC} \to T_{LR} \tag{4.27}$$

with $Z = \sqrt{L/C}$. This equation was derived before. In the limit as $t \to 0$

$$\theta(t) \to t^2 / 2T_{LR} \tag{4.28}$$

If $L \to 0$ and $\tau \to 0$, the local time $\theta(t)$ approaches the global time t, and therefore the theory has physically meaningful limits.

4.3 Local Time in an Oscillatory *LCR* Circuit

If the inductance L of Figure 4.1(d) is increased from 0, the circuit goes successively through an overdamping, a critical damping, and an underdamping (oscillatory) response. The global time and local time concepts of the last section have clear meanings in the overdamping and in the critical damping regime. A natural question would be, how the underdamping transient is described using the local time concept. The node waveform $V_0(t)$ in each regime is given, by using the two time constants $T_{LR} = L/R$ and $T_{CR} = RC$, as follows:

- *Overdamping, $T_{LR} < T_{CR}/4$:*

$$V_0(t) = V_{DD} \frac{-\beta \exp(\alpha t) + \alpha \exp(\beta t)}{\alpha - \beta} \tag{4.29}$$

$$\alpha(+), \beta(-) = \frac{-1 \pm \sqrt{1 - 4(T_{LR}/T_{CR})}}{2T_{LR}}$$

- *Critical damping, $T_{LR} = T_{CR}/4$:*

$$V_0(t) = V_{DD} \exp(\alpha t)(1 - \alpha t) \qquad \alpha = -1/2T_{LR} \tag{4.30}$$

- *Underdamping, $T_{LR} > T_{CR}/4$:*

$$V_0(t) = V_{DD} \exp(\alpha_R t) \left[\cos(\alpha_I t) - (\alpha_R/\alpha_I) \sin(\alpha_I t)\right] \tag{4.31}$$

where

$$\alpha_R = -1/2T_{LR} \qquad \alpha_I = \sqrt{4(T_{LR}/T_{CR}) - 1}/2T_{LR} \tag{4.32}$$

The underdamping (oscillatory) response is shown in Figure 4.7. The response of the RC-only backbone circuit is given by

$$V(t) = V_{DD} \exp(-t/T_{CR}) \tag{4.33}$$

The local time t_0 is determined by solving

$$V(t_0) = V_0(t) \tag{4.34}$$

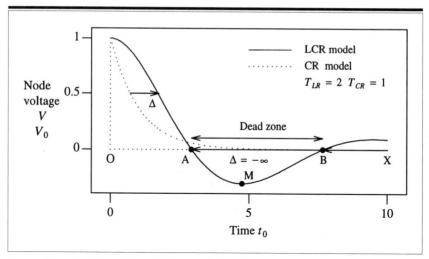

Figure 4.7 Definition of Time Shift of Oscillatory Response

Local times t_0 determined by this method are plotted versus global t in Figures 4.8(a) and (b) for several values of T_{LR}. The solid curve of Figure 4.8(a) is for an overdamping response, the dotted curve is for a critical-damping response, and the solid curve of Figure 4.8(b) is for the underdamping response.

In Figure 4.7 in the oscillatory regime, V_0 reaches zero at A. At this point the local time diverges to positive infinity. When the node voltage

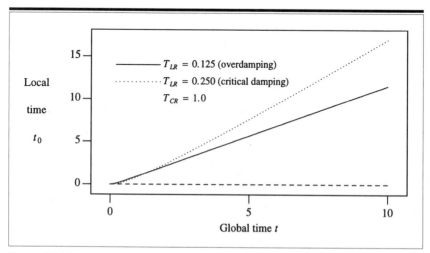

Figure 4.8(a) Local Time Versus Global Time for Overdamping and Critical-Damping Cases

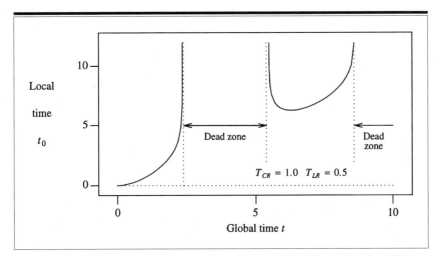

Figure 4.8(b) Local Time Versus Global Time for Underdamping Cases ($T_{LR} = 0.5$)

becomes negative, the local time cannot be determined from the formula discussed before. At point B the node voltage becomes positive again. Then the time shift can be defined by the formula, but the physical meaning of this local time is not clear. If L is higher than the critical value ($T_{LR} > T_{CR}/4$), the curve of local time versus global time consists of many fragmented components separated by dead zones, as shown by Figure 4.8(b). Only the first part of the oscillatory $V_0(t)$ waveform $V_0 > 0$ ($\overline{OA}$ of Figure 4.7) provides a well-defined local time.

Dead zones of the local time appear between zeros of Equation (4.31), which are given by

$$t(N) = \frac{[(1/2) + N]\pi - \chi}{\alpha_I}$$

where   (4.35)

$$\chi = \tan^{-1}(\alpha_R/\alpha_I)$$

and where N is an integer. The first few zeros for $T_{LR} = T_{CR} = 1$ are

$$t(1) = 2.4184 \qquad t(2) = 6.0460 \qquad t(3) = 9.6736 \qquad (4.36)$$

The dead zones are the ranges

$$t(N) < t < t(N + 1) \qquad\qquad (4.37)$$

where N is an *odd* integer. We attempt to extend the theory, in both mathematical and in physical terms, by redefinition of the local time. The new local time t_0 is defined by

$$t_0 = -T_{CR} \log [V_0(t)/V_{DD}] \qquad (4.38a)$$

as before, in the region $\overline{OA}$ of Figure 4.7. In the region $\overline{AB}$ of Figure 4.7, $V_0(t) < 0$. The local time t_0 could then be redefined by

$$t_0 = -T_{CR} \log [|V_0(t)|/V_{DD}] - j\pi T_{CR} \qquad (4.38b)$$

It acquires an imaginary part. As a complex function, the logarithm acquires an imaginary part $j\pi$ whenever $V_0(t)$ changes sign, on circling the origin of the complex V_0 -plane counterclockwise [41]. Similarly, on changing the sign of $V_0(t)$ at B, the local time t acquires an imaginary part $-2j\pi T_{CR}$. If the real part of the local time t_0 is plotted versus global time t, the dotted curve is inserted in the dead zone, as shown in Figure 4.8(c). The local time becomes a quasiperiodic function of the global time that diverges to $+\infty$ at the zeros of $V_0(t)$. The minimum of the quasiperiodic function increases with increasing t. The minima follow approximately

$$t_0 = (T_{CR}/2T_{LR})t - (T_{CR}/2) \log[1 + [1/(4(T_{LR}/T_{CR}) - 1)]] \qquad (4.39)$$

This relationship is plotted as the dashed line in Figure 4.8(c).

The complex local time was introduced as a mathematical fiction but it is a concept that has significant potential for theoretical development.

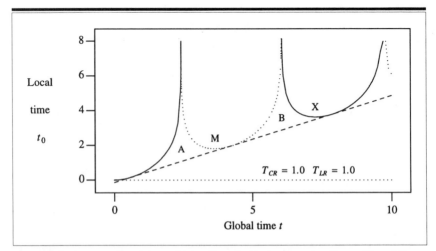

Figure 4.8(c) Local Time Versus Global Time for Underdamping Cases ($T_{LR} = 1.0$)

The complex local time is a consequence of the time-scale shrinkage caused by the inductance. As time scale shrinks beyond the limit of the critical damping, the curve of $V_0(t)$ intersects the time axis $[V_0(t) = 0]$ with nonzero angle. Then a solution of

$$V(t_0) = V_0(t) \tag{4.40}$$

must be sought that is an *analytic* extension of the log function. Every time $V_0(t)$ crosses the time axis, the argument of the log function goes around the branch point by π rad, and the local time adds an imaginary part $-j\pi T_{CR}$. The imaginary part is an *index* of the V_0 sign change, and as such an indicator of advancing time.

In Figure 4.8(c) the real part of the local time *decreases* in the regions A-M and B-X [Figures 4.7 and 4.8(c) have the same identification]. The physical meaning of decreasing local time (or reverse local-time flow) is that the circuit enters into a new operational regime, in which its initial condition is reset. To reset the initial condition is to *undo* the previous changes, and this is logically consistent with the backward flow of time. The energy required to establish a new initial condition has been stored outside the circuit, as the magnetic field energy. Inductance has the capability of removing the energy from the circuit and setting it outside. When the energy returns from the outside to the circuit, the approach to a steady state is reversed, and a new initial condition is set, away from the steady state. This interpretation is consistent with the fact that an *RC* circuit approaches the final steady-state regardless of the initial condition from which it starts off, because it has no outside energy source. Energy exchange between the circuit and the outside determines the direction of the time flow. This may well be a quite general property of physical systems, but it shows up especially clearly in electronic circuits, since they are simple enough to be analyzed in a closed form, but are still complex enough to reflect the properties of large and complex systems.

The energies stored in C and L are given, respectively, by

$$E_C = CV_0^2/2 = (CV_{DD}^2/2) \exp(2\alpha_R t) \, [\cos{(\alpha_i t)} - (\alpha_R/\alpha_i) \sin{(\alpha_i t)}]^2$$

$$E_L = LI^2/2 = \frac{CV_{DD}^2/2}{1 - (CR^2/4L)} \exp(2\alpha_R t) \, [\sin{(\alpha_i t)}]^2 \tag{4.41}$$

and V_0, E_C/E_0, and E_L/E_0 are plotted in Figure 4.9, where $E_0 = CV_{DD}^2/2$ is the energy at $t = 0$. At the time when V_0 vanishes at A, E_L is close to the maximum. In Figure 4.9 the time regions where the local time flows backward are indicated by the double arrows. In these regions the magnetic

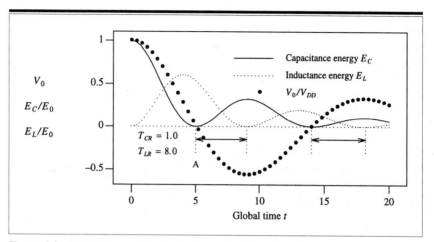

Figure 4.9 Capacitive and Inductive Energies of the Circuit

energy stored by the inductance decreases, and the circuit gets energy back from the outside, which resets the circuit's initial condition. If R is zero,

$$V_0(t) = V_{DD} \cos (t/\sqrt{LC})$$

and

$$E_L = (CV_{DD}^2/2)[\sin (t/\sqrt{LC})]^2$$

(4.42)

and E_L has maxima at the zeros of V_0.

4.4 Analysis of a Device Driving an Interconnect

Figure 4.10(a) is the equivalent circuit of a pulldown device connected to a capacitive load C_1 through an interconnect loop having inductance L_1 and resistance R_1. The device is so small that the inductive parasitic of the

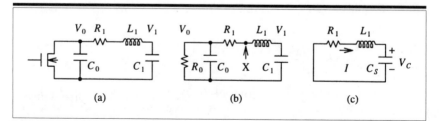

Figure 4.10 A Device Driving an Interconnect

loop including the device is negligible. In this section we model the conducting device by a resistor R_0, and the equivalent circuit is modified to Figure 4.10(b). C_0 and C_1 are charged to V_{DD} at $t = 0$. The capacitance discharge through R_0 is described by the circuit equations

$$C_0 \frac{dV_0}{dt} + \frac{V_0}{R_0} + C_1 \frac{dV_1}{dt} = 0$$

(4.43)

$$V_0 = V_1 + R_1 C_1 \frac{dV_1}{dt} + L_1 C_1 \frac{d^2 V_1}{dt^2}$$

Suppose that the node voltages depend on t as $\exp(St)$. Then S satisfies the secular equation

$$L_1 C_0 C_1 S^3 + [R_1 C_0 C_1 + (L_1 C_1 / R_0)]S^2$$

(4.44)

$$+ [C_0 + C_1 + (R_1 / R_0)C_1]S + (1/R_0) = 0$$

Let the (negative) solutions of the secular equation be α, β, and γ, in the order of increasing absolute value. If L_1 is small, the circuit is overdamped, and the solutions are all real and negative. The solution α gives the longest time constant $1/|\alpha|$, which is approximated well by Elmore's time constant

$$T_\alpha = 1/|\alpha| \approx T_E = R_0(C_0 + C_1) + R_1 C_1$$ (4.45)

The correlation between the accurately computed and the approximate time constant is shown in Figure 4.11(a), and it is good. Values of the inductance L_1 were randomly generated between 0 and 0.005.

The solution β gives the second longest time constant, which is that of voltage equalization of the two nodes. If the inductance is neglected, this time constant is approximated by

$$T_\beta = 1/|\beta| \approx R_1 C_0 C_1 / [C_0 + C_1 + C_1(R_1 / R_0)]$$ (4.46)

The correlation, as shown in Figure 4.11(b), is reasonable.

The shortest time constant is approximated as follows. For the fast transient, the device has no time to respond, and therefore R_0 can be set at ∞. Then the equivalent circuit is simplified to Figure 4.10(c). The capacitance C_S of Figure 4.10(c) is a series combination of C_0 and C_1, or $C_S = C_0 C_1 / (C_0 + C_1)$. For the fast transient the capacitance develops a small voltage.

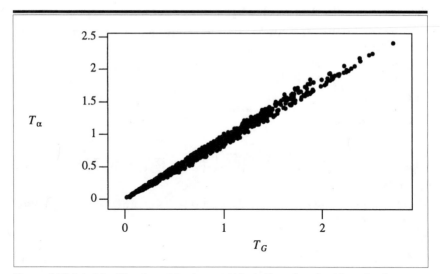

Figure 4.11(a)   Identification of the Longest Mode Time Constant for a Device Driving an Interconnect

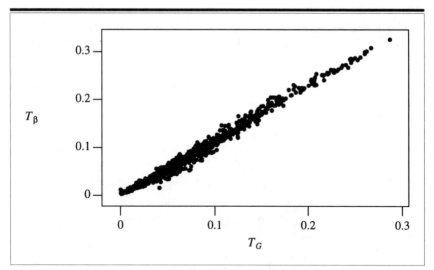

Figure 4.11(b)   Identification of the Second Longest Mode Time Constant for a Device Driving an Interconnect

For the first approximation we neglect the voltage. The loop current I of Figure 4.10(c) is then given by

$$I(t) = I(0) \exp[-(R_1/L_1)t] \tag{4.47}$$

The time constant is given by

$$T_\gamma = 1/|\gamma| \approx L_1/R_1 \qquad (4.48)$$

This is, however, an underestimate of T_γ, since C_S is in series. The equation of the simplified circuit of Figure 4.10(c) gives the secular equation

$$(L_1 C_S)S^2 + (R_1 C_S)S + 1 = 0 \qquad (4.49)$$

The solution of this equation having the largest absolute value is

$$S_- = -[(R_1 C_S) + \sqrt{(R_1 C_S)^2 - 4L_1 C_S}]/(2L_1 C_S) \approx \gamma. \qquad (4.50)$$

If L_1 is small, we have

$$1/|\gamma| \approx L_1/[R_1 - (L_1/R_1 C_S)] \qquad (4.51)$$

Figure 4.11(c) shows the correlation between the accurate and the guessed time constants. The good correlation indicates that the mechanism of the mode has been explained.

The trajectory of the complex frequency S when L is increased starting from 0 is shown in Figure 4.12. If L is small, the smallest complex frequency (in absolute value) is estimated from Elmore's time constant, and

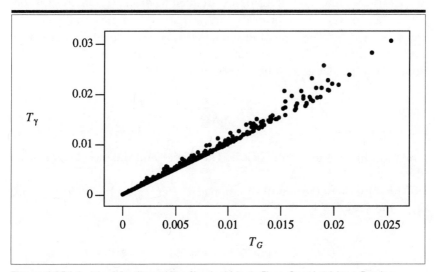

Figure 4.11(c) Identification of the Shortest Mode Time Constant for a Device Driving an Interconnect

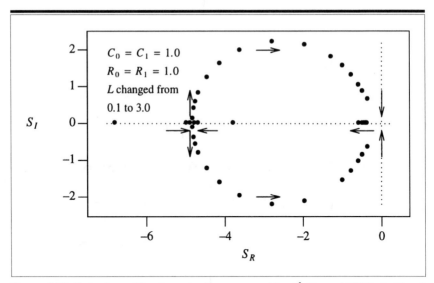

Figure 4.12 Trajectory of the Complex Frequency of the $1\frac{1}{2}$-Stage *LCR* Chain Circuit

the largest complex frequency from Equation (4.48). The third complex frequency is estimated from the voltage equalization time constant of the two-stage *RC* chain. As *L* increases, the smallest complex frequency does not change significantly. The second smallest complex frequency increases and the largest complex frequency decreases in magnitude. After they merge at the critical point, a pair of complex conjugate frequencies emerge. They trace out semicircles above and below the real axis, and approach the origin in the limit as $L_1 \rightarrow \infty$. In the limit as $L_1 \rightarrow \infty$, the roots of the secular equation are approximated by

$$\alpha \approx -1/\sqrt{R_0 C_0} \qquad \beta \approx j/\sqrt{L_1 C_1} \qquad \gamma \approx -j/\sqrt{L_1 C_1} \quad (4.52)$$

The meaning of this result is clear: The device and the interconnect have their own time constants.

The solution of the circuit equation is written, using the three solutions $\alpha, \beta,$ and γ, as

$$V_1(t) = A \exp(\alpha t) + B \exp(\beta t) + C \exp(\gamma t)$$
$$V_0(t) = D \exp(\alpha t) + E \exp(\beta t) + F \exp(\gamma t)$$
$\qquad\qquad\qquad (4.53)$

where

$$D = [1 + (R_1C_1)\alpha + (L_1C_1)\alpha^2]A$$
$$E = [1 + (R_1C_1)\beta + (L_1C_1)\beta^2]B \qquad (4.54)$$
$$F = [1 + (R_1C_1)\gamma + (L_1C_1)\gamma^2]C$$

and the coefficients A, B, and C are determined from the initial conditions

$$V_1(0) = V_{DD} \qquad V_1'(0) = 0 \qquad V_1''(0) = 0 \qquad (4.55)$$

as the roots of the simultaneous equations

$$A + B + C = V_{DD}$$
$$\alpha A + \beta B + \gamma C = 0 \qquad (4.56)$$
$$\alpha^2 A + \beta^2 B + \gamma^2 C = 0$$

They are

$$A = \frac{\beta\gamma(\gamma - \beta)}{\Delta} V_{DD}$$

$$B = \frac{\gamma\alpha(\alpha - \gamma)}{\Delta} V_{DD} \qquad (4.57)$$

$$C = \frac{\alpha\beta(\beta - \alpha)}{\Delta} V_{DD}$$

where

$$\Delta = \alpha\beta(\beta - \alpha) + \beta\gamma(\gamma - \beta) + \gamma\alpha(\alpha - \gamma) \qquad (4.58)$$
$$= (\alpha - \beta)(\beta - \gamma)(\gamma - \alpha)$$

We note that node X of Figure 4.10(b) does not have a physically meaningful voltage, since the node exists only in the circuit diagram and is assigned no capacitance (Section 1.11). This node is a pseudonode, which does not exist in the real hardware. The node 0 and 1 voltage waveforms

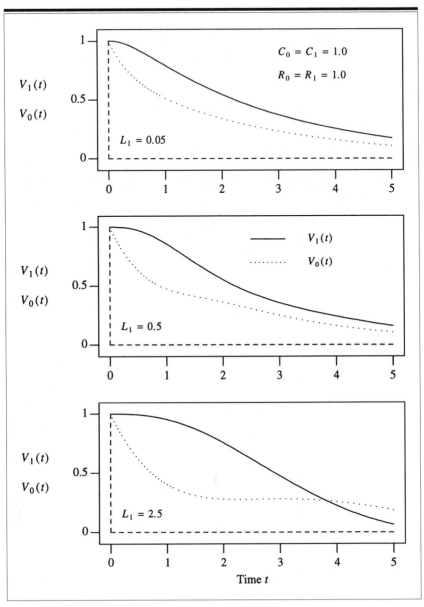

Figure 4.13 Node Voltage Waveforms of a Device Driving an Interconnect

for the overdamping (top), the slightly underdamping (middle), and the underdamping (bottom) cases (for the same set of values for R_0, R_1, C_0, and C_1 and three values of L_1) are shown in Figure 4.13.

For small t a power series expansion of the solution, of the form

$$\frac{V_1(t)(LCR)}{V_{DD}} = 1 - \frac{t^3}{6(R_0C_0)(L_1C_1)} + \frac{(R_0C_0)(R_1C_1) + (L_1C_1)}{24(R_0C_0)^2(L_1C_1)^2}t^4 + \cdots$$

$$\frac{V_0(t)(LCR)}{V_{DD}} = 1 - \frac{t}{(R_0C_0)} + \frac{t^2}{2(R_0C_0)^2} + \frac{(R_0C_0)(R_0C_1) - (L_1C_1)}{6(R_0C_0)^3(L_1C_1)}t^3 + \cdots$$

$$(4.59)$$

is derived. $V_1(t)(LCR)$ starts at the term proportional to t^3, since the initial condition requires $V'_1(0) = 0$ and $V''_1(0) = 0$. The power series is convenient for computing the node voltages if L_1 is *not* small. It diverges in the limit as $L_1 \to 0$, because the coefficient of the highest-order derivative of the differential equation vanishes.

The device driving the interconnect has two physically meaningful nodes. A local time can be assigned to each node, to observe the progress of the discharge relative to the *RC*-only backbone circuit. Let the node waveforms of the backbone RC circuit having the same parameter values R and C and the same initial conditions be $U_1(t)$ and $U_0(t)$. We use the results of Section 2.7:

$$U_1(t) = A' \exp(\alpha' t) + B' \exp(\beta' t) \qquad (4.60)$$

$$U_0(t) = C' \exp(\alpha' t) + D' \exp(\beta' t)$$

where $A', B', C', D', \alpha',$ and β' are given in Section 2.7; in particular,

$$\alpha' = (-\Gamma + \sqrt{\Delta})/2C_0C_1 \qquad \beta' = (-\Gamma - \sqrt{\Delta})/2C_0C_1 \quad (4.61)$$

where $\Gamma = (C_1/R_0) + (C_1/R_1) + (C_0/R_1)$ and $\Delta = \Gamma^2 - 4(C_0C_1/R_0R_1)$. Using the solution for the *LCR* circuit given in this section, we solve the node 1 local time t_1 for global time t:

$$U_1(t_1) = V_1(t)$$

or $\qquad\qquad\qquad\qquad\qquad\qquad\qquad\qquad\qquad\qquad\qquad (4.62)$

$$A \exp(\alpha t) + B \exp(\beta t) + C \exp(\gamma t)$$

$$= A' \exp(\alpha' t_1) + B' \exp(\beta' t_1)$$

The node 0 local time t_0 is found similarly, from the equation $U_0(t_0) = V_0(t)$.

Figure 4.14(a), (b), and (c) show the node local times t_1 and t_0 versus

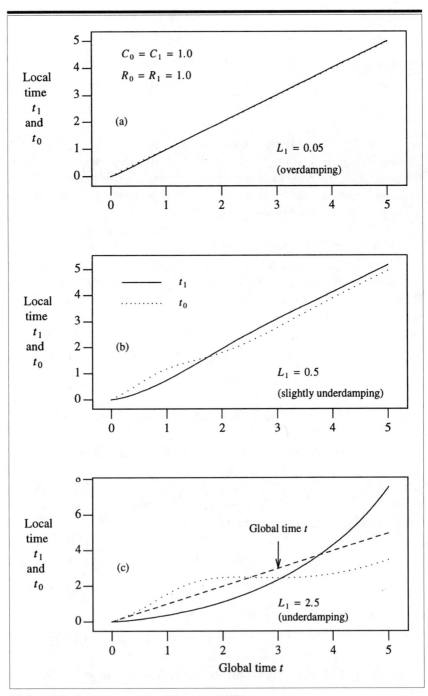

Figure 4.14 Node's Local Time Versus Global Time

the global time t, for the overdamping, slightly underdamping, and underdamping cases, respectively. In the overdamping regime the node local times t_1 and t_0 and the global time t are the same. In the underdamping regime the node 1 local time is delayed, and the node 0 local time is advanced, relative to the global time at the beginning. Node 1 is unable to discharge until current is established in L_1, and before the current is established, node 0 is effectively isolated from node 1. The node 0 discharge is fast, since R_0 discharges C_0 only. After current is established in L_1, the node 1 discharge rate does not decrease rapidly as V_1 decreases: The current established in L_1 continues to discharge node 1 at the high rate. The current from node 1 to node 0 reduces the node 0 discharge rate, and therefore the node 1 local time is advanced and the node 0 local time is delayed, relative to the global time [the global time is shown by the dashed line in Figure 4.14(c)]. In the underdamping regime the mechanism creates a significant difference between the local and the global times. In the underdamping regime the node voltage may be driven down below zero. Then it is impossible to define node local times.

The backbone RC circuit has two modes, whose complex frequencies are α' and β' (both real numbers). The node voltages are superpositions of the component modes. Modes of a circuit have their own identity, which are maintained all through the transient of the circuit. The mode local times are convenient for observing how the modes of the backbone RC circuit are modified by the inductance. As an alternative to the node local times, the *mode* local times t_α and t_β can be defined by solving the simultaneous equations

$$A' \exp(\alpha' t_\alpha) + B' \exp(\beta' t_\beta) = V_1(t)(LCR)$$

$$C' \exp(\alpha' t_\alpha) + D' \exp(\beta' t_\beta) = V_0(t)(LCR)$$

(4.63)

The mode local times versus the global time for four cases that cover the range from the overdamping to the strongly underdamping regime are shown in Figure 4.15(a)–(d). The local time of mode α', whose complex frequency is small, is almost the same as the global time. The local time of mode β', having a large complex frequency, has a maximum if L_1 is about 0.22, but if L_1 is larger than that, the local time diverges. The physical effect of the mode is to equalize the voltages of the two nodes. If significant inductance exists, the node 0 and 1 voltages approach equilibrium, but not monotonically: The inductance current may polarize nodes 1 and 0 in reverse polarity. The voltage equalization mode dies at the moment when the voltages equalize, and the local time diverges. After that, the

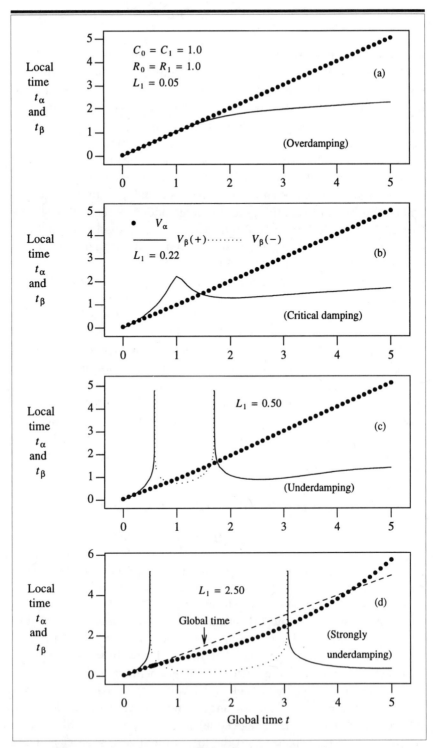

Figure 4.15 Mode Local Time Versus Global Time

magnetic energy of the inductance returns to the circuit, and the circuit resets the initial condition of the mode, now having the reverse polarity. Figure 4.15(b) and (c) show a transition from a peak to a divergence of the mode local time t_β. The behavior of t_β is found from the simultaneous equations as

$$t_\beta = \frac{1}{\beta'} \log \left[\frac{A' V_0(t)(LCR) - C' V_1(t)(LCR)}{A'D' - B'C'} \right] \tag{4.64}$$

By using

$\alpha'(+), \beta'(-) =$

$$\frac{-[R_0(C_0 + C_1) + R_1 C_1] \pm \sqrt{[R_0(C_0 + C_1) + R_1 C_1]^2 - 4(R_0 C_0)(R_1 C_1)}}{2(R_0 C_0)(R_1 C_1)}$$

$$A' = -\frac{\beta' V_{DD}}{\alpha' - \beta'}$$

$$B' = \frac{\alpha' V_{DD}}{\alpha' - \beta'} \tag{4.65}$$

$$C' = [1 + (R_1 C_1)\alpha']A'$$

$$D' = [1 + (R_1 C_1)\beta']B'$$

we have

$$t_\beta = \frac{1}{\beta'} \log \left[\frac{V_1(t)(LCR)}{V_{DD}} + \frac{V_1(t)(LCR) - V_0(t)(LCR)}{(R_1 C_1)\alpha' V_{DD}} \right]$$

$$= \frac{1}{\beta'} \log \left[1 + \frac{1}{(R_0 C_0)(R_1 C_1)\alpha'} \left[t - \frac{t^2}{2(R_0 C_0)} \right] - \frac{1}{6(R_0 C_0)(L_1 C_1)} \right.$$

$$\left. \times \left[1 + \frac{(R_0 C_0)^2 + (R_0 C_0)(R_0 C_1) - (L_1 C_1)}{(R_0 C_0)^2(R_1 C_1)\alpha'} \right] t^3 + o(t^4) \right] \tag{4.66}$$

If the argument of log is positive and if the terms up to t^2 are dominant, the maximum of t_β occurs at the maximum of $t - [t^2/2(R_0 C_0)]$, or $t = R_0 C_0$. It does not occur at small L_1, where the higher-order terms including L_1 in the denominator of the expansion coefficient become important. Figure 4.15(b) satisfies the condition for the existence of a peak in t_β. The location of the peak agrees with the simple estimate. If L_1 is increased, the argument of log vanishes, and the local time t_β diverges.

4.5 Generalization of the Theory to Many-Loop Circuits

The driver-and-interconnect combination has only one inductance loop. By studying a circuit that has two inductance loops, we gain an idea about some problems of introducing the local time concept into the study of complex circuits. We classified the RC equivalent circuits of digital logic circuits in Section 2.6, and concluded that the varieties of circuits having a small number of nodes are quite limited. In the classification we saw that the secular equation of an N-node circuit has N roots, and the circuit has N modes. To assign local times to nodes or to modes was a matter of free choice. In general, this is not so. This two-loop example shows clearly that the node local times are valid only if the inductances are small.

We consider the two-stage cascaded LCR chain circuit shown in Figure 4.16(a) that includes inductances L_1 and L_2. The node voltages V_1 and V_2 are used as the variables, and the circuit equations are

$$(L_1 C_1)\frac{d^2 V_1}{dt^2} + (L_1 C_2)\frac{d^2 V_2}{dt^2} + (R_1 C_1)\frac{dV_1}{dt} + (R_1 C_2)\frac{dV_2}{dt} + V_1 = 0$$

$$(L_2 C_2)\frac{d^2 V_2}{dt^2} + (R_2 C_2)\frac{dV_2}{dt} + V_2 = V_1 \tag{4.67}$$

If V_1 and V_2 depend on time as $\exp(St)$, then S satisfies a secular equation

$$A_4 S^4 + A_3 S^3 + A_2 S^2 + A_1 S + A_0 = 0 \tag{4.68}$$

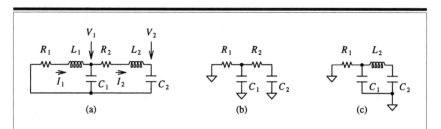

(a)   (b)   (c)

Figure 4.16 Two-Stage LCR Chain Circuit

where

$$A_4 = (L_1C_1)(L_2C_2)$$

$$A_3 = (L_1C_1)(R_2C_2) + (L_2C_2)(R_1C_1)$$

$$A_2 = (L_1C_1) + (L_1C_2) + (L_2C_2) + (R_1C_1)(R_2C_2) \qquad (4.69)$$

$$A_1 = (R_1C_1) + (R_1C_2) + (R_2C_2)$$

$$A_0 = 1$$

This equation can be solved by the Ferrari's method [25] as follows. We have

$$S^4 + a_3S^3 + a_2S^2 + a_1S + a_0 = 0 \qquad (4.70)$$

where $a_i = A_i/A_4$ for $i = 1, 2, 3$. On using a transform $S = x - (a_3/4)$, x satisfies

$$x^4 = -px^2 - qx - r \qquad (4.71)$$

where

$$p = a_2 - (3/8)a_3^2$$

$$q = (a_3^3/8) - (a_2a_3/2) + a_1 \qquad (4.72)$$

$$r = -(3a_3^4/256) + (a_2a_3^2/16) - (a_1a_3/4) + a_0$$

On adding $yx^2 + (y^2/4)$ to both sides, the right side becomes a perfect square if

$$y^3 - py^2 - 4ry + (4pr - q^2) = 0 \qquad (4.73)$$

By solving this cubic equation by Cardano's method [25] (Section 2.7) and substituting the one root into Equation (4.71), we have two second-order equations

$$x^2 - \sqrt{y - p}\,x + (y/2) + (q/2\sqrt{y - p}) = 0 \qquad (4.74)$$

$$x^2 + \sqrt{y - p}\,x + (y/2) - (q/2\sqrt{y - p}) = 0$$

By solving the two equations, we get the four roots of the fourth-order secular equation.

Let the four solutions be S_1–S_4. The solution of the circuit equation is given by

$$V_1 = A_1 \exp(S_1 t) + A_2 \exp(S_2 t) + A_3 \exp(S_3 t) + A_4 \exp(S_4 t) \qquad (4.75)$$

$$V_2 = B_1 \exp(S_1 t) + B_2 \exp(S_2 t) + B_3 \exp(S_3 t) + B_4 \exp(S_4 t)$$

where S_i, A_i, and B_i are in general complex numbers. Then the right-hand sides are computed as complex numbers, and the real parts give V_1 and V_2. If nodes 1 and 2 are charged to unit voltage at $t = 0$, then A_1–B_4 are determined from the initial conditions

$$V_1(+0) = V_2(+0) = 1 \qquad dV_1/dt|_{t=+0} = dV_2/dt|_{t=+0} = 0 \quad (4.76)$$

as

$$A_i = [1 + (R_2 C_2)S_i + (L_2 C_2)S_i^2]B_i \qquad i = 1, 2, 3, 4 \qquad (4.77)$$

$$B_i = \Delta_i/\Delta \quad i = 1, 2, 3 \qquad B_4 = 1 - B_1 - B_2 - B_3$$

where

$$\Delta = \begin{bmatrix} S_1 - S_4 & S_2 - S_4 & S_3 - S_4 \\ S_1^2 - S_4^2 & S_2^2 - S_4^2 & S_3^2 - S_4^2 \\ S_1^3 - S_4^3 & S_2^3 - S_4^3 & S_3^3 - S_4^3 \end{bmatrix} \qquad (4.78)$$

$$\Delta_1 = \begin{bmatrix} -S_4 & S_2 - S_4 & S_3 - S_4 \\ -S_4^2 & S_2^2 - S_4^2 & S_3^2 - S_4^2 \\ -S_4^3 & S_2^3 - S_4^3 & S_3^3 - S_4^3 \end{bmatrix}$$

Δ_2 and Δ_3 are obtained similarly, by substituting a column $(-S_4, -S_4^2, -S_4^3)$ for the second and the third column of Δ, respectively.

To gain insight into the mode, we choose $C_1 = C_2 = 1.0$, $R_1 = R_2 = 1.0$, and $L_1 = L_2 = L$, and the common inductance L is increased from a small value. The track of the complex frequency S is shown in Figure 4.17. If $L = 0$, the circuit of Figure 4.16(a) is reduced to that of Figure 4.16(b). The two time constants of the RC circuit are derived from the roots of the reduced secular equation

$$(R_1 C_1)(R_2 C_2)S^2 + [(R_2 C_2) + (R_1 C_1) + (R_1 C_2)]S + 1 = 0 \quad (4.79)$$

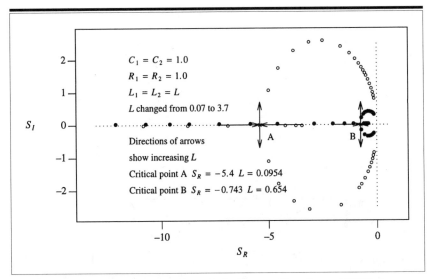

Figure 4.17 Complex Frequency Versus Inductance of Two-Stage *LCR* Chain

and for the pair of parameter values $S_{R+} = -0.382$ and $S_{R-} = -2.618$. Then $(S_{R+}, 0)$ and $(S_{R-}, 0)$ are the starting points of the tracks at $L = 0$. The other two starting points are both $(-\infty, 0)$, and they are the asymptotic solutions of the secular equation including inductance, in the limit as $L \to 0$. In general, the two large solutions of the original secular equation are of the order of $1/L$ in the limit as $L \to 0$. They satisfy the asymptotic secular equation $L_1 L_2 S^2 + (L_2 R_1 + L_1 R_2)S + R_1 R_2 = 0$, or $S_1 \to -R_1/L_1$ and $S_2 \to -R_2/L_2$. The complex frequencies trace out two loops. The number of loops in the trajectory equals the number of the roots that start from $(-\infty, 0)$ in the limit as $L \to 0$, and it equals the number of inductances. In an N-stage cascaded *LCR* chain N roots start from the infinity, N roots start from finite locations, and there are N loops in the trajectory. The N finite roots at $L = 0$ are the complex frequencies of the backbone RC circuit. As L increases, a pair of roots, one from a finite location and the other from $-\infty$, approach, and once they meet, they split into a pair of complex conjugate roots. Since the nonzero imaginary part means oscillating response of the mode, oscillation occurs as L increases and the first pair of complex roots emerge.

As L increases starting from zero, the pair of tracks of S starting from $(S_{R-}, 0)$ and $(-\infty, 0)$ meet at the first critical point A, at $L = L_{CA} = 0.0954$. For inductance less than L_{CA} all four complex frequencies are real and negative. The node voltage waveforms in this case are shown in Figure 4.18(a). The waveforms are closely similar to those of the backbone *RC*

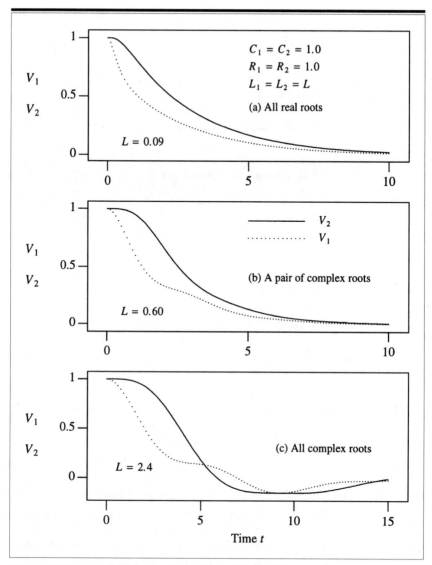

Figure 4.18 Node Waveforms of Two-Stage *LCR* Chain

circuit. If L increases still further, two complex frequencies having imaginary parts emerge. The node waveforms are shown in Figure 4.18(b). The node voltages decrease monotonically, but the rate of decrease oscillates. As L increases still further, the second critical point B of Figure 4.17 is reached. At $L = L_{CB} = 0.654$ the second pair of tracks originating from $(S_{R+}, 0)$ and $(-\infty, 0)$ meet. For $L > L_{CB}$ all the four complex frequencies

have nonzero imaginary parts, and the node waveforms become oscillatory, as shown in Figure 4.18(c).

There is a fundamental difference between the node and the mode local
times, which becomes clear from the study of a two-stage *LCR* chain circuit. The node local times t_1 and t_2 are defined by solving the equations

$$U_2(t_2) = A \exp(\alpha' t_2) + B \exp(\beta' t_2) = V_2(t) \tag{4.80}$$

$$U_1(t_1) = C \exp(\alpha' t_1) + D \exp(\beta' t_1) = V_1(t)$$

where $U_2(t)$ and $U_1(t)$ are the node voltage waveforms of the backbone
RC chain circuit (Section 2.7). If the nodes of the circuit are charged to V_{DD}, and if the discharge starts from the initial condition, we have

$$0 \leq U_1(t) \qquad U_2(t) \leq V_{DD} \tag{4.81}$$

$V_2(t)$ and $V_1(t)$ are the node voltage waveforms of the LCR circuit, and
may exceed this limit if the L, C, and R parameter values are arbitrarily chosen. Figures 4.19(a) and (b) show an example. Arrow A of
Figure 4.19(a) shows that $V_1(t) \rightarrow V_{DD}$ some time after the LCR circuit discharge starts. In this case C_1 is small. Node 1 discharges rapidly at the
beginning. Then the node voltage difference $V_2 - V_1 \ (> 0)$ builds the L_2
current up rapidly, since R_2 is small. Since C_2 is large, V_2 remains close to
V_{DD} during that time, and since C_1 is small, it is recharged quickly. The

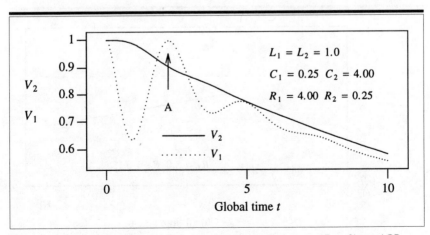

Figure 4.19(a) Node Voltage Waveforms of an Underdamped Two-Stage *LCR* Chain Circuit

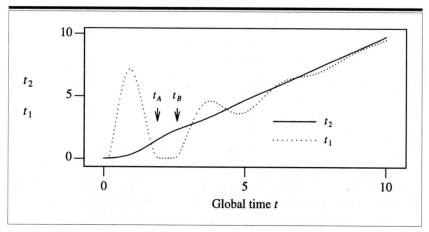

Figure 4.19(b)   Node Local Times t_2 and t_1 of an Underdamped Two-Stage *LCR* Chain Circuit

maximum node 1 voltage attained is higher than V_{DD} if the ratios C_2/C_1 and R_1/R_2 are high, as shown in Figure 4.20.

The two cases both have small C_1, large C_2, large R_1, and small R_2. In such cases two distinct transient processes take place simultaneously: (1) the resonant circuit consisting of L_1, C_1, and R_1 goes into a damped oscillation, and (2) the charges stored in C_1 and C_2 are discharged to ground by R_1. The simplified equivalent circuit is shown in Figure 4.16(c). As the discharge starts, the underdamped $L_2 C_1 C_2$ circuit goes into oscil-

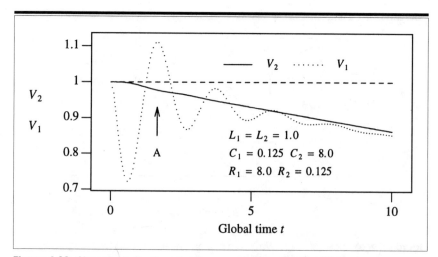

Figure 4.20   Strongly Underdamping Response of Two-Stage *LCR* Chain

lation. Since $C_2 \gg C_1$, the amplitude of oscillation of V_2 is about a factor C_1/C_2 smaller than the amplitude of V_1. The period of oscillation is given, approximately, by $T_P = 2\pi \sqrt{L_2 C_1}$ (C_2 does not affect the series-connected capacitance $C_2 \gg C_1$). The V_1 oscillation begins with a downward drive at $t = 0$, and therefore $t = 0$ is one of the maxima of V_1. Then V_1 has its second maximum at $t = T_P$. This is the maximum that can be higher than V_{DD}.

In the example of Figure 4.19(a) the node 1 local time t_1 bottoms at zero for the period $t_A < t < t_B$ of Figure 4.19(b). The current from node 2 totally undoes the initial rapid discharge of node 1, and the node 1 local time returns to zero. Since U_1 and U_2 are defined as functions of t for $t > 0$ only, if V_1 becomes higher than V_{DD} or less than 0, no meaningful node 1 local time can be defined. We observe that the node local time representation is limited to cases where the LCR circuit node waveforms are well-behaved, in the sense that the voltage does not leave the range from 0 to V_{DD}. If the range is exceeded, the definition of the node local time loses its meaning.

Node local times are parameters defined to compare two waveforms that are essentially alike. If the essential similarity is lost, they become meaningless, as is clear from the following analogy: The linear dimensions and the weight of a body are both measures of its volume, but the former depend on the shape, and the latter does not. In spite of this inconvenience, the node local time gives a clear idea about the node's discharge process, whether it is advanced or delayed relative to the discharge of the backbone RC circuit. We use the node local time for small inductance values only, where its use is advantageous. Many of the practical cases of digital circuits with small inductances satisfy this condition.

The mode local times t_α and t_β are determined by solving the simultaneous equations

$$A \exp(\alpha' t_\alpha) + B \exp(\beta' t_\beta) = V_2(t)$$

$$C \exp(\alpha' t_\alpha) + D \exp(\beta' t_\beta) = V_1(t)$$

(4.82)

The coefficients of the linear combination (A, B, C, and D) are determined to satisfy the initial conditions of the RC backbone circuit. The left-hand sides are linear combinations of the *modes* $\exp(\alpha' t_\alpha)$ and $\exp(\beta' t_\beta)$. The initial amplitudes of the modes are all 1 at $t_\alpha = t_\beta = 0$. The later transient developments are determined by the developments of the modes. Thus, in contrast with the definition of the node local time, the set of the simultaneous equations to determine the mode local times are valid all the times, even if the mode amplitudes should become negative or larger than 1. If

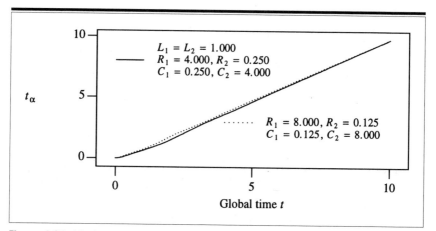

Figure 4.21 Mode α Local Times for Two Extreme Underdamping Conditions

an amplitude becomes larger than 1, the mode has only acquired more energy than at the beginning, and that is nothing unusual. For this reason the mode local times are more fundamental than the node local times.

The mode local times t_α of LCR chains having the same parameter values as those of Figure 4.19(a) and Figure 4.20 are shown by the solid and by the dotted curves of Figure 4.21, respectively. The mode α local time is almost always close to the global time, even in the extremely underdamped circuits. The time-averaged decay of the node voltages is determined by the mode that has the longest time constant. Figures 4.22(a) and (b) show the mode β local times of the examples of Figures 4.19(a) and 20, respectively. These times oscillate between the minima and infinity, and the minima increase slowly. The mode β local time deviates strongly from

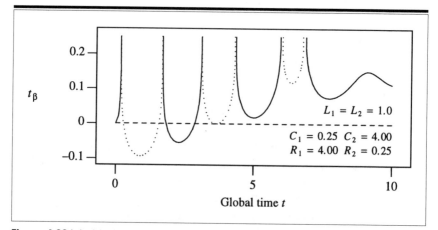

Figure 4.22(a) Mode Local Time t_β in the Underdamping Regime

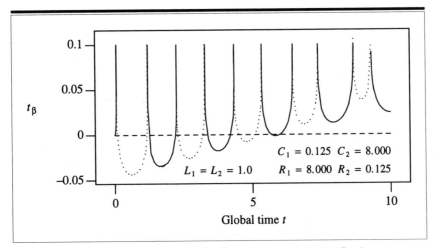

Figure 4.22(b) Mode Local Time t_β in the Extreme Underdamping Regime

the global time. The mode reduces the existing voltage difference between the nodes first, recharges the pair of the nodes to the opposite polarity, then reduces the reverse-polarity voltage difference, and so on. The negative mode local time means that the mode was excited more strongly by the energy returned from the inductance: The mode may acquire more energy than it had at $t = 0$. At the initial time mode α had energy. From this mode the energy is transferred to the inductance, and later to mode β.

We need to clarify an issue. Transfer of energy from one mode to the other takes place only in nonlinear circuits. The circuit we studied is linear. Why does the energy transfer take place? We interpreted the real LCR circuit transient in terms of the backbone RC circuit transient. The energy transfer among the modes of the RC circuit represents the mechanisms of discharge of the LCR circuit. Energy transfer among the LCR circuit modes never takes place.

Sometimes it is convenient to analyze the mode local times using the differential equations that are satisfied by them, as we did in Section 4.2. The equations of the circuit including inductance [Figure 4.16(a)] are

$$V_1 = V_2 + (R_2 C_2)\frac{dV_2}{dt} + (L_2 C_2)\frac{d^2 V_2}{dt^2}$$

$$0 = V_1 + (R_1 C_1)\frac{dV_1}{dt} + (R_1 C_2)\frac{dV_2}{dt} \qquad (4.83)$$

$$+ (L_1 C_1)\frac{d^2 V_1}{dt^2} + (L_1 C_2)\frac{d^2 V_2}{dt^2}$$

We substitute

$$V_1 = A_{11} \exp[\alpha'\theta_\alpha(t)] + A_{12} \exp[\beta'\theta_\beta(t)]$$

$$V_2 = A_{21} \exp[\alpha'\theta_\alpha(t)] + A_{22} \exp[\beta'\theta_\beta(t)]$$

(4.84)

and seek the differential equations satisfied by $\theta_\alpha(t)$ and $\theta_\beta(t)$. We have

$$H_{11}\theta_\alpha'' + H_{12}\theta_\beta'' = H_1 \qquad H_{21}\theta_\alpha'' + H_{22}\theta_\beta'' = H_2 \qquad (4.85)$$

where

$$
\begin{aligned}
H_1 &= \left(\frac{1 + T_2\alpha'}{\alpha'} + [(R_1C_1)(1 + T_2\alpha') + (R_1C_2)]\theta_\alpha' \right. \\
&\quad \left. + \alpha'[(L_1C_1)(1 + T_2\alpha') + (L_1C_2)](\theta_\alpha')^2 \right) \exp(\alpha'\theta_\alpha) \\
&\quad - \left(\frac{1 + T_2\beta'}{\beta'} + [(R_1C_1)(1 + T_2\beta') + (R_1C_2)]\theta_\beta' \right. \\
&\quad \left. + \beta'[(L_1C_1)(1 + T_2\beta') + (L_1C_2)](\theta_\beta^2) \right) \exp(\beta'\theta_\beta) \\
H_2 &= [-T_2 + T_2\theta_\alpha' + \alpha'(L_2C_2)(\theta_\alpha')^2] \exp(\alpha'\theta_\alpha) \\
&\quad - [-T_2 + T_2\theta_\beta' + \beta'(L_2C_2)(\theta_\beta')^2] \exp(\beta'\theta_\beta) \\
H_{11} &= -[(L_1C_1)(1 + T_2\alpha') + (L_1C_2)] \exp(\alpha'\theta_\alpha) \\
H_{12} &= [(L_1C_1)(1 + T_2\beta') + (L_1C_2)] \exp(\beta'\theta_\beta) \\
H_{21} &= -(L_2C_2) \exp(\alpha'\theta_\alpha) \\
H_{22} &= (L_2C_2) \exp(\beta'\theta_\beta)
\end{aligned}
$$

(4.86)

This set of equations is solved subject to the initial conditions

$$\theta_\alpha(0) = \theta_\beta(0) = 0 \qquad \theta_\alpha'(0) = \theta_\beta'(0) = 0 \qquad (4.87)$$

Figure 4.23 shows a result. This method requires no solution of the secular equation, and therefore it is often convenient.

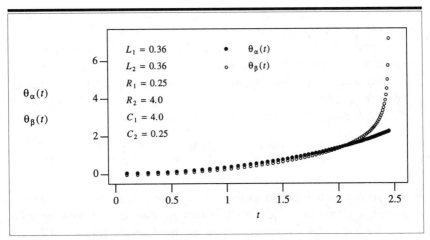

Figure 4.23 Mode Time Shifts Computed from the Differential Equations

4.6 Time Shift Between the Nodes

A node local time was introduced to observe the relative progress of the charge and discharge of an LCR circuit relative to the backbone RC circuit. The node local time and the excess charge and discharge currents to the node due to the inductances are closely related. The excess currents are, however, contributed by many modes, and their sum does not allow a simple physical interpretation. This is the reason why the node local time does not have universal validity. An alternative way to arrive at this conclusion is as follows. If inductances are added to the backbone RC circuit of Figure 4.24(b), the nodes that communicate with node 0 of the LCR circuit of Figure 4.24(a) via resistances R_i and inductances L_i should have effective times (not the local times defined before) that are all delayed relative to the time of node 0. This is because an inductance current requires

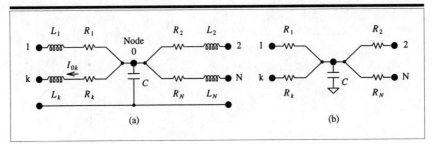

Figure 4.24 Circuits Including Inductance

time to build up. Let us rewrite the circuit equations to make the delay effects clearly observable.

A digital circuit node has a capacitance connected to it, whose other terminal is grounded. The nodes are connected together by resistances and inductances. Figure 4.24(a) and (b) show one such node 0, connected to N nodes labeled by $1, 2, \ldots, k, \ldots, N$. The node equation of the RC backbone circuit of Figure 4.24(b) is written as

$$C_0 \frac{dV_0(t)}{dt} = - \sum_{k=1}^{N} \frac{V_0(t) - V_k(t)}{R_k} \qquad (4.88)$$

where all the voltages refer to the *present* time t, and the dependence is explicitly written in the equation. If inductances are included as shown in Figure 4.24(a), however, the equations are modified as follows. It is convenient to maintain the general structure of Equation (4.88) to observe the physical meanings in the limit as $L_k \to 0$.

If inductance is included, the current I_{0k} of branch k (from node 0 to k) satisfies the equation

$$R_k I_{0k}(t) + L_k \frac{dI_{0k}(t)}{dt} = V_0(t) - V_k(t) \qquad (4.89)$$

If $I_{0k}(0) = 0$, the solution for $I_{0k}(t)$ is given by

$$I_{0k}(t) = \frac{1}{L_k} \int_0^t \exp[-(R_k/L_k)(t - \theta)] \cdot [V_0(\theta) - V_k(\theta)] \, d\theta \qquad (4.90)$$

$V_0(\theta)$ and $V_k(\theta)$ are expanded into power series in the vicinity of $\theta = t$ where the exponential term is significant. Then

$$I_{0k}(t) = \frac{1}{L_k} \sum_{n=0}^{\infty} \frac{1}{n!} \frac{d^n}{dt^n}[V_0(t) - V_k(t)]$$

$$\int_0^t \exp[-(R_k/L_k)(t - \theta)] \cdot (\theta - t)^n \, d\theta$$

$$= \frac{1}{L_k} \sum_{n=0}^{\infty} (-1)^n (L_k/R_k)^{n+1} \frac{d^n}{dt^n}[V_0(t) - V_k(t)] \qquad (4.91)$$

$$= \frac{V_0(t) - V_k(t)}{R_k} - \frac{L_k}{R_k^2}\left[\frac{dV_0(t)}{dt} - \frac{dV_k(t)}{dt} \right]$$

$$+ \frac{L_k^2}{R_k^3}\left[\frac{d^2V_0(t)}{dt^2} - \frac{d^2V_k(t)}{dt^2} \right] - \cdots$$

because the L_k's are all small, the factors R_k/L_k in the exponential are large, and therefore the lower limit of the integral can be extended to $-\infty$. Therefore the node equation is written as

$$C_0 \frac{dV_0(t)}{dt} = -\sum_{k=1}^{N} I_{0k}(t)$$

$$= -\sum_{k=1}^{N} \frac{V_0(t) - V_k(t)}{R_k}$$

$$+ \sum_{k=1}^{N} \frac{L_k}{R_k^2}\left[\frac{dV_0(t)}{dt} - \frac{dV_k(t)}{dt}\right]$$

$$- \sum_{k=1}^{N} \frac{L_k^2}{R_k^3}\left[\frac{d^2V_0(t)}{dt^2} - \frac{d^2V_k(t)}{dt^2}\right] + \cdots$$

(4.92)

or

$$\left(C_0 - \sum_{k=1}^{N} \frac{L_k}{R_k^2}\right)\frac{dV_0(t)}{dt} = -\sum_{k=1}^{N} \frac{V_0(t) - V_k[t - (L_k/R_k)]}{R_k}$$

$$+ \, o[(L_k/R_k)^2]$$

(4.93)

By adding inductances the node capacitance is reduced by the sum of (L_k/R_k^2), and the effective time of a node connected by L_k and R_k is delayed by L_k/R_k. Node k is delayed with respect to node 0, but node 0 is delayed with respect to node k as well. The complex relationships cannot be described simply by defining a local time for each node.

4.7 Modes of a Uniform Straight *LCR* Chain

Study of the modes of a uniform RC chain circuit of Section 2.8 provided an insight into the mode mechanisms. We repeat the same for a uniform LCR chain circuit shown in Figure 4.25. We assume $RC = 1$, and the inductance L is small. All the roots of the secular equation are real and

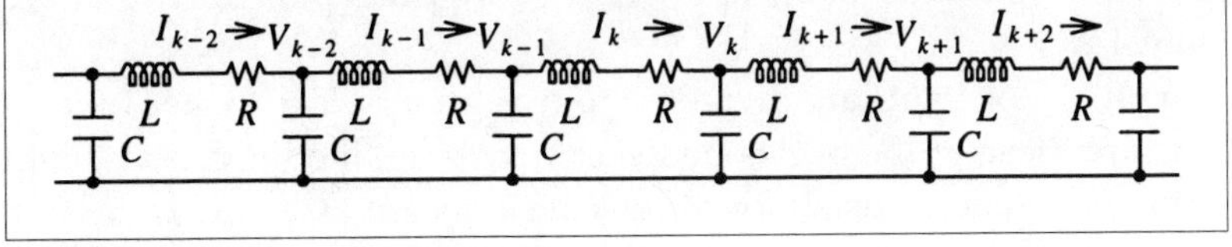

Figure 4.25 *LCR* Chain Circuit

negative. We use a parameter $T = L/R$ to characterize the effects of inductance. Only the voltages developed across the capacitances have valid physical meanings. The voltage profile of the chain is given by a sum of the modes of the circuit. If the number of the cascaded stages is N, there are $2N$ modes. The complex frequencies of the modes are determined from the following secular equations:

$$TS^2 + S + 1 = 0 \qquad (N = 1) \quad (4.94a)$$

$$T^2S^4 + 2TS^3 + (3T + 1)S^2 + 3S + 1 = 0 \qquad (N = 2) \quad (4.94b)$$

$$T^3S^6 + 3T^2S^5 + (5T^2 + 3T)S^4 + (10T + 1)S^3$$
$$+ (6T + 5)S^2 + 6S + 1 = 0 \qquad (N = 3) \quad (4.94c)$$

$$T^4S^8 + 4T^3S^7 + (7T^3 + 6T^2)S^6 + (21T^2 + 4T)S^5$$
$$+ (15T^2 + 21T + 1)S^4 + (30T + 7)S^3$$
$$+ (10T + 15)S^2 + 10S + 1 = 0 \quad (N = 4) \quad (4.94d)$$

$$T^5S^{10} + 5T^4S^9 + (9T^4 + 10T^3)S^8 + (36T^3 + 10T^2)S^7$$
$$+ (28T^3 + 54T^2 + 5T)S^6 + (84T^2 + 36T + 1)S^5$$
$$+ (35T^2 + 84T + 9)S^4 + (70T + 28)S^3$$
$$+ (70T + 35)S^2 + 15S + 1 = 0 \qquad (N = 5) \quad (4.94e)$$

$$T^6S^{12} + 6T^5S^{11} + (11T^5 + 15T^4)S^{10} + (55T^4 + 20T^3)S^9$$
$$+ (45T^4 + 110T^3 + 15T^2)S^8 + (180T^3 + 110T^2 + 6T)S^7$$
$$+ (84T^3 + 270T^2 + 55T + 1)S^6 + (252T^2 + 180T + 11)S^5$$
$$+ (70T^2 + 252T + 45)S^4 + (140T + 84)S^3$$
$$+ (21T + 70)S^2 + 21S + 1 = 0 \qquad (N = 6) \quad (4.94f)$$

Assuming $T = L/R = 0.05$, the equations are solved numerically, and the node time constants defined by $T_M = 1/|S|$ are plotted versus N in Figure 4.26.

The time constants are divided into two groups. The longer time constants are those of the backbone RC chain circuit, modified only slightly by the inductance. The shorter time constants are essentially due to the inductances. They vary only slightly with N, compared with the longer

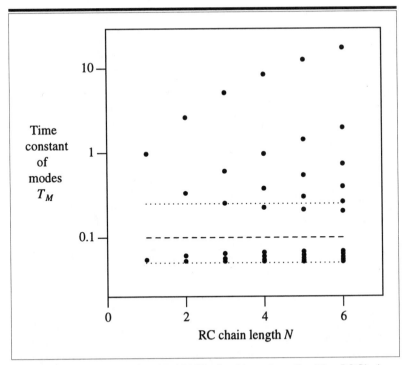

Figure 4.26 Mode Time-Constant Distribution Versus Length of the *RC* Chain

time constants. They are all clustered around $T = L/R$. They are the time constants by which the currents of the N sections of the series-connected R and L approach the values determined by the resistance alone. The variation with N is small because C presents an effective short circuit to the high-speed transients of all the LR loops, and that divides the long chain into many independent $CLRC$ loops.

The voltage profiles of the long-time-constant modes are the same as we observed in Section 2.8. The profiles of the short-time-constant modes are shown in Figure 4.27. The mode with the shortest time constant has about $\frac{1}{4}$ standing wave in the chain; with the second shortest, about $\frac{1}{2}$ standing wave; with the third shortest, about 1 standing wave; and so on. The relationship between the number of standing waves and the magnitude of the time constant is different from that of Section 2.8. This observation on the LCR chain circuit can be explained from the following theory.

The node voltage V_k and current I_k defined in Figure 4.25 satisfy a recurrence relationship

$$C \frac{dV_k}{dt} = I_k - I_{k+1} \qquad V_{k-1} - V_k = RI_k + L \frac{dI_k}{dt} \qquad (4.95)$$

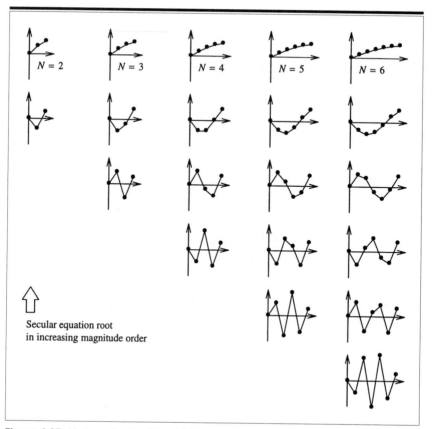

Figure 4.27 Voltage Profile of the *LCR* Modes

If all the variables depend on time as $\exp(St)$, we have

$$CSV_k = I_k - I_{k+1} \qquad V_{k-1} - V_k = (R + LS)I_k \qquad (4.96)$$

or

$$(R + LS)CSV_k = V_{k-1} + V_{k+1} - 2V_k \qquad (4.97)$$

Suppose that the modes are indexed by a sine function and a parameter α such that V_k has the dependence

$$V_k = V_0 \sin(\alpha k) \qquad (4.98)$$

In a long uniform chain the node voltage profile depends on the location k only through a periodic function. Since $V_{k-1} + V_{k+1} = 2V_0 \sin(\alpha k) \cos(\alpha)$

we have a quadratic equation satisfied by S:

$$LCS^2 + RCS - 2[\cos(\alpha) - 1] = 0 \qquad (4.99)$$

which is solved as

$$S = \frac{R}{2L}\left[-1 \pm \sqrt{1 - 8\frac{(L/C)}{R^2}[1 - \cos(\alpha)]}\right] \qquad (4.100)$$

In the limit of small L the negative sign gives a solution

$$S_- = -\frac{R}{L}\left[1 - 2\frac{(L/C)}{R^2}[1 - \cos(\alpha)]\right] \qquad (4.101)$$

and the positive sign

$$S_+ = -\frac{2}{CR}[1 - \cos(\alpha)] = -\frac{\alpha^2}{CR} \qquad (\alpha \to 0) \qquad (4.102)$$

S_+ gives the longer time constants of the circuits, and S_- gives the shorter time constants. The parameter α specifies how frequently the potential profile oscillates in the chain, that is, the number of the standing waves in the chain. If $\frac{1}{4}$ of a standing wave exists in the N-link chain, $\alpha = \pi/2N$. Then the longest time constant is estimated as

$$T_{\max} = 1/S_{+\min} = CR(2N/\pi)^2 \approx CRN(N + 1)/2 \qquad (4.103)$$

and this is Elmore's time constant discussed before. The dependence of the time constant on α is different for S_+ and S_-. For S_+, larger α gives a shorter time constant. A mode having a smaller number of standing waves has a longer time constant. For S_-, larger α gives a longer time constant. This explains why Figure 2.18 and Figure 4.27 show different time constant dependence.

4.8 Elmore's Formula for the *LCR* Chain Circuit

The delay time and the rise-fall time of a straight RC chain circuit are given by Elmore's formula. We seek a similar formula for a straight LCR chain circuit. Derivation of the following results may be found in reference [7] or the original paper [27]. In the equivalent circuit of Figure 4.28 the

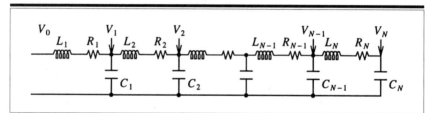

Figure 4.28 Derivation of Elmore's Formula for an *LCR* Chain Circuit

Laplace-transformed response function $G_N(S)$ that relates the Laplace-transformed input and output node voltages $V_0(S)$ and $V_N(S)$ is written as

$$G_N(S) = \frac{V_N(S)}{V_0(S)} = \frac{1 + a_1 S + a_2 S^2 + \cdots}{1 + b_1 S + b_2 S^2 + \cdots} \qquad (4.104)$$

The delay time $T_D(\text{Elmore})$ and the rise-fall time $T_{R/F}(\text{Elmore})$ are given by

$$T_D(\text{Elmore}) = b_1 - a_1 \qquad (4.105)$$
$$T_{R/F}(\text{Elmore}) = [b_1^2 - a_1^2 + 2(a_2 - b_2)]^{1/2}$$

The problem is to calculate the first few terms of the power series expansion of the response function. The Laplace transforms of the node voltages satisfy the recurrence formulas

$$V_{N-1}(S) = [1 + (R_N C_N)S + (L_N C_N)S^2]V_N(S)$$

$$V_{N-2}(s) = [1 + (R_{N-1} C_{N-1})S + (L_{N-1} C_{N-1})S^2]V_{N-1}(S) \qquad (4.106)$$
$$+ [(R_{N-1} C_N)S + (L_{N-1} C_N)S^2]V_N(S)$$

$$\vdots$$

The algebraic operations to eliminate $V_{N-1}(S), \ldots, V_1(S)$ are straightforward but very tedious. Here we show the results only. We find

$$a_1 = a_2 = \cdots = 0$$
$$T_D(\text{Elmore}) = b_1 = R_1(C_1 + C_2 + \cdots + C_N) \qquad (4.107)$$
$$+ R_2(C_2 + C_3 + \cdots + C_N) + \cdots + R_N C_N$$

This is identical to the conventional Elmore's formula. Addition of inductance to the RC backbone circuit does not affect the Elmore delay time. Since inductance delays the response at the beginning of the transient but speeds up the response in the later part, the overall effect of inductance is to maintain the same delay time. This conclusion is subject to the assumptions underlying Elmore's definition of the delay time of a large-amplitude signal. Since in a digital circuit model the delay at the beginning matters more, T_D(Elmore) in an inductive circuit becomes an underestimate.

The coefficient b_2 consists of two terms:

$$b_2 = b_2(LC) + b_2(RC) \tag{4.108}$$

where

$$b_2(LC) = L_1(C_1 + C_2 + \cdots + C_N) \tag{4.109}$$
$$+ L_2(C_2 + C_3 + \cdots C_N) + \cdots + L_N C_N$$

and

$$b_2(RC) = \sum_{i,j,k,l} (R_i C_j)(R_k C_l) \tag{4.110}$$

For small N,

$$b_2(RC) = (R_1 C_1)(R_2 C_2) \qquad (N = 2)$$
$$b_2(RC) = (R_1 C_1)(R_2 C_2) + (R_1 C_1)(R_2 C_3) \qquad (N = 3) \tag{4.111}$$
$$+ (R_1 C_1)(R_3 C_3) + (R_1 C_2)(R_3 C_3) + (R_2 C_2)(R_3 C_3)$$

If all the possible product terms of the form $(R_i C_j)(R_k C_l)$ are represented as in Figure 4.29, the required terms of the sum are shown by the filled circles. The diagram can be easily extended to larger N [1].

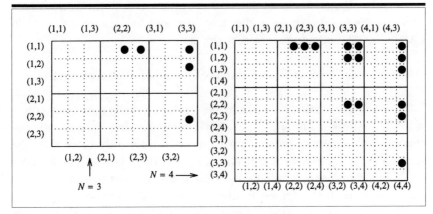

Figure 4.29 Terms of the Form $(R_i C_j)(R_k C_l)$ That Must Be Summed

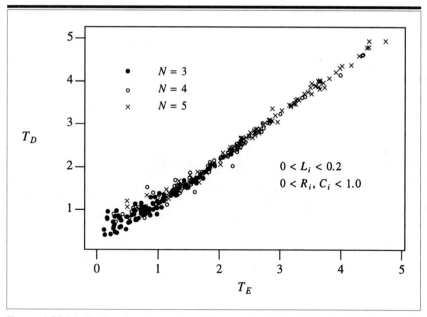

Figure 4.30(a)   Testing the Accuracy of Elmore's Delay Formula for the *LCR* Chain Circuit

The accuracy of Elmore's formula can be checked by numerical simulation. As was done in Section 2.9, the LCR parameter values are generated randomly, and the delay is computed by the two methods. The delay determined by the numerical simulation is plotted versus the delay computed by Elmore's formula, as shown in Figure 4.30(a). In this trial the resistance and the capacitance parameter values were generated uniformly in the range between 0 and 1, and the inductance parameters in the range between 0 and 0.2. In this comparison the difference in the logic threshold voltage was taken into consideration as follows. Since V_{DD}/e was used as the threshold in the derivation of Elmore's formula, but in CMOS $V_{DD}/2$ is commonly used, we have

$$T_E = (\log 2) \times T_D(\text{Elmore}) = 0.691 T_D(\text{Elmore}) \quad (4.112)$$

The overall correlation is quite good. If N is small, there are points above the straight line $T_D = T_E$. The inductance-capacitance time constant $\sqrt{L_i C_i}$ of some stages of the chain is significantly more than the resistance-capacitance time constant $R_i C_i$. For these points the effects of the inductances are certainly underestimated by Elmore's formula.

The formula for the rise-fall time can be tested similarly. We need to compute the accurate rise-fall times. The node waveform $V_N(t)$ is deter-

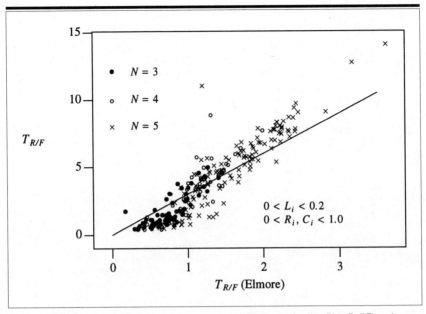

Figure 4.30(b)   Testing the Accuracy of Elmore's Formula for the Rise-Fall Time in an *LCR* Chain Circuit

mined by numerical analysis. At the point where $V_N(T_D) = V_{DD}/2$, a tangent is drawn, and its slope is determined using the least-squares method. We obtain the rise-fall time by dividing V_{DD} by the slope. Although this definition is not the same as that used to derive Elmore's formula, the difference is not significant. The parameter values were generated randomly in the ranges indicated in Figure 4.30(b). The following approximate formula holds between the simulated rise-fall time and Elmore's estimate:

$$T_{R/F} \approx 3.0 \cdot T_{R/F}(\text{Elmore}) \tag{4.113}$$

Microstates, Submicrostates, and Local Times

5.1 Introduction

In the preceding chapter we have explored the need for the notions of local time in the understanding of integrated circuits where inductance must be taken into account. In this chapter we develop a theory that relates the three basic concepts: microstates, submicrostates, and local times. As a product of this integration, we will have a *microstate sequence* including indication of the evolution of submicrostates supported by the microstate. Indication of activities of the submicrostates is reversal in the direction of mode local time flow. The enhanced microstate sequence describes exchange of hardware and energy between the digital circuit and its environment, and as such, it is a complete description of the operation of a digital circuit at high speed. The concept of microstate and microstate sequence are assumed known. The basic microstate theory is presented in my previous book [1].

5.2 Forward and Backward Flow of Local Time

In Chapter 4 we saw that a real and positive mode local time can be defined, but may diverge to positive infinity. Following the divergence, the local time decreases from positive infinity and reaches a minimum. Beyond the minimum it increases again. We had to explain what is meant by *decreasing* (or backward flow of) local time. What we understood is summarized as follows. In a circuit including inductance, the current established in the inductance continues to discharge the capacitance at the high rate in the later phase of the transient, thereby completing the discharge earlier than $t = \infty$; we say the mode that discharges the capacitance *dies* before $t = \infty$. As the current is established in the inductance, the electrical energy is converted to magnetic energy, thereby transferring the energy from the circuit to the outside. An *RC*-only circuit is, by contrast, entirely self-contained: Outside energy storage does not exist in an RC circuit. The energy of the LCR circuit stored in the outside space returns during the later phase of the transient. It recharges the capacitances to a new initial condition, from which the circuit approaches the steady state again. Thus the backward local time flow is interpreted as the reestablishment of a new initial condition of the RC circuit by using the energy that came back from the outside.

Inductors as components that remove energy from the circuit may sound strange, but the following observation helps to clarify the point. The time-dependent electric field in a capacitor need not be considered as storing energy outside the circuit, because the rate of change of the electric field is proportional to the displacement current density in the capacitor. This displacement current is as much a current as the conduction current of the rest of the circuit. Indeed, in a resistive conductor both currents coexist. At the capacitor location the current spreads out over the cross section, but it still flows within the circuit. In an inductor the magnetic energy is stored outside the *current path*. If it has a magnetic core the magnetic energy is stored there, that is still *outside* the connected conductor of the circuit. In this sense the RC-only circuit contains all its energy within the connected part of the circuit, whereas in the LCR circuit some energy moves back and forth between the connected part and the outside.

An LCR circuit can be in one of two states: a state whose local time flows (1) forward or (2) backward. The circuit operation may be understood as the sequence of the events that describe the changes in the direction of the local time flow. Let us call that sequence of the events the *submicrostate* sequence. The events that characterize the boundaries of the local times of different submicrostates include the following:

1. The local time diverges to positive infinity (a mode's death, the end of the forward time flow, and the beginning of the backward time flow).

2. The local time flow changes direction (opposite of forward), indicating the reestablishment of the initial condition.

These two are the most conspicuous submicrostate events.

In a circuit that has more than one mode, the local times of some modes flow forward, and of others backward, at the same global time. A diagram indicating the change of the direction of the local time flow provides the

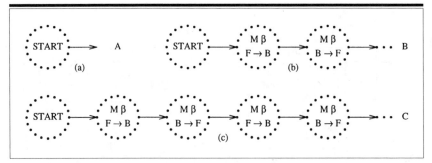

Figure 5.1 Mode Submicrostate Sequence

most fundamental information for understanding the circuit operation. Conforming to the notation used to describe the microstate sequence [1], we introduce the notations of Figure 5.1. Each symbol bears indications of the mode and the change of the direction of the mode local time flow. At the beginning of the discharge process local times flow forward. In the RC circuit model the direction of the time flow never changes (sequence A). In the example of Figure 5.1(b) the mode β local time changes direction, followed by a second change that brings the local time flow back to forward. As the circuit approaches the final steady state, the local-time flow is forward. In Figure 5.1(c) the mode β time changes its direction of flow periodically, showing that a quasistable and generally quasiperiodic oscillation takes place.

5.3 Microstates and Submicrostates—I

The operation of a digital CMOS circuit is described by the method I introduced in my previous book [1], the microstate sequence. A microstate sequence indicates the circuit's connectivity change caused by the switching devices. The basic concepts of microstate and microstate sequence need not be repeated here. If necessary, the reader is referred to my first book. If the microstate sequence is supplemented by the sequence of the submicrostates of LCR circuits, the resulting composite of microstate and submicrostate sequences represents the circuit operation in detail: In the microstate-submicrostate sequence the energy transfer between the circuit and the outside is shown, in addition to the connectivity changes produced by the switching devices. A microstate event connects or disconnects a part of the circuit. The disconnected part becomes the outside of the circuit. A microstate event is an indication of a hardware exchange to or from the outside, and a submicrostate event is indication of an energy exchange. Therefore it is natural to integrate them together. Furthermore,

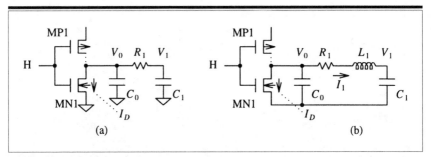

Figure 5.2 Inductance in Pulldown Switching Circuit

microstate event and submicrostate event interact: The energy stored outside the circuit is able to change the circuit connectivity. Submicrostate changes, however, have meaning only within a single microstate, or a given circuit connectivity.

In the circuit of Figure 5.2(a) the input voltage has made a rapid low-to-high transition. MP1 turns off, MN1 turns on, and the NFET begins to discharge C_0 and C_1. If MN1 is modeled by a collapsible current generator, the pulldown process is described by the circuit equations

$$C_1 \frac{dV_1}{dt} = \frac{V_0 - V_1}{R_1} \qquad C_0 \frac{dV_0}{dt} + I_D - \frac{V_1 - V_0}{R_1} = 0 \qquad (5.1)$$

where I_D is the MN1 current in the saturation region. If time dependence of the form $\exp(St)$ is assumed, S satisfies the secular equation

$$S[C_0 C_1 R_1 S + (C_0 + C_1)] = 0 \qquad (5.2)$$

and $S = 0$ or $S = -1/t_{EQ}$, where $t_{EQ} = C_0 C_1 R_1 / (C_0 + C_1)$ is the voltage equalization time constant for nodes 0 and 1. For $S = 0$ it might be thought that there is no solution that contains t, but this is not the case. The equation satisfied by V_1,

$$C_0 C_1 R_1 \frac{d^2 V_1}{dt^2} + (C_0 + C_1) \frac{dV_1}{dt} + I_D = 0 \qquad (5.3)$$

has a solution of the form

$$V_1(t) = \alpha_1 t + \alpha_0$$

where (5.4)

$$\alpha_1 = -I_D / (C_0 + C_1)$$

where α_0 is a constant. This type of solution plays the role of the mode whose S is zero. It is derived from an infinite series expansion

$$\exp(St) = 1 + St + o[(St)^2] \tag{5.5}$$

by considering the limit as $S \to 0$ for finite t. If the initial conditions are $V_0(+0) = V_1(+0) = V_{DD}$, we have

$$V_1(t)(RC) = (V_{DD} + X_1) - X_0 t - X_1 \exp(-t/t_{EQ}) \tag{5.6a}$$

$$V_0(t)(RC) = (V_{DD} - X_2) - X_0 t + X_2 \exp(-t/t_{EQ}) \tag{5.6b}$$

where X_0, X_1, and X_2 are determined from the initial condition $V_0(+0) = V_{DD}$ and $V_1(+0) = V_{DD}$ as

$$X_0 = \frac{I_D}{C_0 + C_1} \qquad X_1 = \frac{R_1 C_0 C_1 I_D}{(C_0 + C_1)^2} \qquad X_2 = \frac{R_1 C_1^2 I_D}{(C_0 + C_1)^2} \tag{5.7}$$

At $t = t_{GND}$, V_0 becomes zero. The time t_{GND} is given by

$$t_{GND} = t_{EQ}\theta[-C_0/C_1, 1 - [(C_0 + C_1)^2 V_{DD}/R_1 C_1^2 I_D]] \tag{5.8}$$

where $\theta(\kappa, \chi)$ is an implicit function defined by solving $\exp(-\theta) + \kappa\theta = \chi$ for θ [1]. For $t > t_{GND}$ we have

$$V_0(t) = 0 \qquad V_1(t) = V_1(t_0) \exp[-(t - t_{GND})/R_1 C_1] \tag{5.9}$$

and MN1 is in the triode region.

Similarly the pulldown process of the LCR circuit is described by

$$C_1 \frac{dV_1}{dt} = I_1$$

$$L_1 \frac{dI_1}{dt} = V_0 - V_1 - R_1 I_1 \tag{5.10}$$

$$C_0 \frac{dV_0}{dt} = -(I_D + I_1)$$

The set of equations can be solved either in closed form or numerically, subject to the initial condition $V_0(+0) = V_{DD}$, $V_1(+0) = V_{DD}$, and $I_1'(+0) =$

0. If t is small, the node voltages depend on t as

$$V_0 \approx V_{DD} - (I_D/C_0)t \qquad V_1 = V_{DD} - [I_D/(6L_1C_0C_1)]t^3 \quad (5.11)$$

as is shown by a direct substitution. This relationship is used to start the numerical integration. In the closed-form analysis the circuit equation can be integrated once and we obtain

$$L_1C_0C_1 \frac{d^2V_1}{dt^2} + R_1C_0C_1 \frac{dV_1}{dt} + (C_0 + C_1)V_1$$

$$+ I_Dt - (C_0 + C_1)V_{DD} = 0 \quad (5.12)$$

The solution of this equation is given by

$$\beta(+),\, \gamma(-) = -\frac{R_1}{2L_1} \pm \frac{R_1}{2L_1} \sqrt{1 - [4L_1(C_0 + C_1)/(C_0C_1R_1^2)]}$$

$$B = -\frac{1}{\gamma - \beta} \frac{I_D}{C_0 + C_1} \left(1 + \frac{R_1C_0C_1\gamma}{C_0 + C_1}\right)$$

$$C = \frac{1}{\gamma - \beta} \frac{I_D}{C_0 + C_1} \left(1 + \frac{R_1C_0C_1\beta}{C_0 + C_1}\right)$$

$$(5.13)$$

$$V_1(LCR) = V_{DD} + X_1 - X_0t + B \exp(\beta t) + C \exp(\gamma t)$$

$$V_0(LCR) = V_{DD} - X_2 - X_0t - (C_1/C_0)B \exp(\beta t)$$

$$- (C_1/C_0)C \exp(\gamma t)$$

where β and γ can be complex numbers if $R_1^2 < 4L_1(C_0 + C_1)/C_0C_1$. Then V_0 and V_1 are computed as complex numbers, and the real parts are taken. At $t = t_{GND}$, MN1 goes into the triode region. This t_{GND} is numerically different from that of the RC circuit, but there is no confusion in using the same symbol. After that time we have

$$V_1(LCR) = B \exp(\beta t) + C \exp(\gamma t) \quad (5.14)$$

where

$$B = \frac{\gamma V_1(t_{GND}) - V_1'(t_{GND})}{(\gamma - \beta) \exp(\beta t_{GND})} \qquad C = \frac{V_1'(t_{GND}) - \beta V_1(t_{GND})}{(\gamma - \beta) \exp(\gamma t_{GND})} \quad (5.15)$$

where

$$\beta(+),\ \gamma(-) = -\frac{R_1}{2L_1} \pm \frac{R_1}{2L_1}\sqrt{1 - 4(L_1/C_1)(1/R_1^2)} \quad (5.16)$$

and where $V_1(t_{GND})$ and $V_1'(t_{GND})$ are the values at the end of the first microstate.

Figure 5.3 shows the node voltage waveforms of the *RC* and the *LCR* circuits. The response is moderately underdamping, and therefore the node voltages oscillate. At the beginning, when the L_1 current has not yet been built up, the node 0 voltage of the LCR circuit decreases faster than that of the RC circuit. Once the L_1 current is established, the node 0 voltage is partly sustained by the inductance current, and the rate of the voltage decrease becomes smaller. V_0 reaches zero at $t_{GND} = 3.5$, and thereafter V_0 is held at zero by MN1 in the triode region. A microstate change MN1: $S \to T$ takes place at t_{GND}. After that, the L_1 current overdischarges C_1, and V_1 becomes negative. Since $V_0 = 0$, the $L_1C_1R_1$ closed loop continues a damping oscillation.

In Equations (5.6a) and (5.6b), $\exp(-t/t_{EQ})$ is a mode of the RC circuit model. We call this mode β. The other mode has zero complex frequency. We use $V_{DD} - [I_D/(C_0 + C_1)]t$ as the expression for mode α. Let the modes of the RC circuit be $\xi_A = V_{DD} - [I_D/(C_0 + C_1)]t_\alpha$ and $\xi_B = \exp(-t_\beta/t_{EQ})$, where t_α and t_β are the mode local times. We have, from Equations (5.6a)

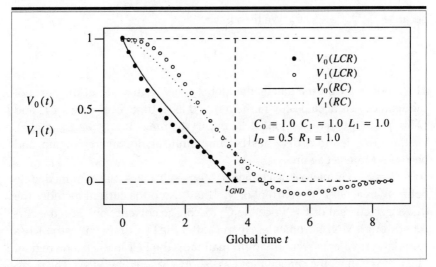

Figure 5.3 Voltage Waveforms of the Driver-Interconnect Combination

and (5.6b), $V_1(RC) = \xi_A - X_1\xi_B + X_1$ and $V_0(RC) = \xi_A + X_2\xi_B - X_2$. From the definition of the mode local time, we can solve the following simultaneous equations for ξ_A and ξ_B:

$$\xi_A - X_1\xi_B = V_1(t)(LCR) - X_1 \tag{5.17}$$
$$\xi_A + X_2\xi_B = V_0(t)(LCR) + X_2$$

Using the solutions of the simultaneous equations, ξ_A and ξ_B, we have the mode local times as

$$t_\alpha = [(C_0 + C_1)/I_D](V_{DD} - \xi_A) \qquad t_\beta = -t_{EQ}\,\log(\xi_B) \tag{5.18}$$

If $\xi_B < 0$ we define the mode β local time by $t_\beta = -t_{EQ}\,\log(-\xi_B)$, so that the local time acquires a nonzero imaginary part, as we discussed in Section 4.3. Among the simultaneous equations ξ_B is eliminated to obtain

$$(X_1 + X_2)\xi_A = X_2 V_1(t)(LCR) + X_1 V_0(t)(LCR) \tag{5.19}$$

This becomes

$$(X_1 + X_2)[V_{DD} - [I_D/(C_0 + C_1)]t_\alpha]$$
$$= (X_1 + X_2)[V_{DD} - [I_D/(C_0 + C_1)]t] \tag{5.20}$$

because $X_2/X_1 = C_1/C_0$. We then have

$$t_\alpha = t \tag{5.21}$$

The mode α local time equals the global time. In this circuit the effect of inductance is reflected only in the mode β local time. In Figure 5.4 t_α and t_β are plotted versus the global time t. In the range $0 \leq t \leq t_{GND} (= 3.5)$, we have $t_\alpha = t$, and t_β diverges, becomes finite again, diverges again, and becomes finite yet again.

At $t = t_{GND}$ a microstate change MN1:S $\rightarrow$ T takes place. Immediately before V_0 reaches 0, the current of MN1 is I_D, and the current includes the discharge current of C_0. Since the C_0 discharge current vanishes, the current through MN1 becomes less than I_D after the FET drain voltage reaches 0, and the smaller current (than I_D) maintains the FET in the triode region. After the change, the capacitance C_0 is short-circuited by MN1. The number of nodes decreases by 1 (the pseudonode at the junction of R_1 and L_1

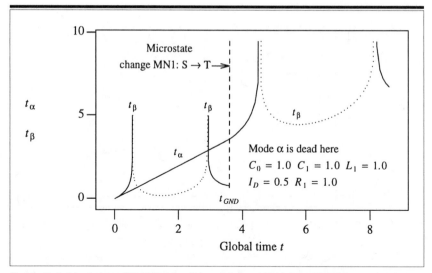

Figure 5.4 Mode Local Times of the Driver-Interconnect Combination

does not count). The number of modes decreases by 1 as well. The remaining mode is called mode β. The new mode local time t_β is determined from

$$V_1(t_{GND})(LCR)\ \exp[-(t_\beta - t_{GND})/(R_1 C_1)] = V_1(t)(LCR) \quad (5.22)$$

The local time t_β is plotted versus the global time t in Figure 5.4 ($t > t_{GND}$). At $t = t_{GND}$, we have $t_\beta = t_{GND}$ because the mode local time has meaning only within a microstate. A boundary between microstates is a boundary between local-time definitions, where the origin of the local time is reset at the global time. Figure 5.5 shows the mode local times t_α and t_β versus the global time t of the circuit in the overdamping regime. The mode β local time does not diverge: Generally, no significant anomaly of local-time-like divergence takes place in the overdamping regime.

It is often convenient to use a numerical analysis of the set of circuit equations. If the equations are to be solved numerically, the following problem must be taken into consideration. If V_0 decreases and falls below zero, the collapsible current generator changes from I_D to $-I_D$. This is because the collapsible current generator always forces current to the source (to ground). Because of this current discontinuity, V_0 oscillates around 0 at a period twice the integration step. This is not clean, but is a convenient indicator that the FET is in the triode region. If the numerical data are used to draw the curves, they are certainly not neat. Since the small oscillation around $V_0 = 0$ of the node waveforms indicates clearly

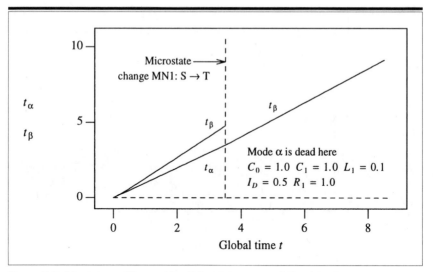

Figure 5.5 Mode Local Times of the Driver-Interconnect Combination

that the NFET is in the triode region, there are spurious features in the following drawings.

Figure 5.6 shows the node voltage waveforms for a significantly under-damped case. The arrow S indicates the NFET turn-on at the beginning, which takes place within a negligibly short time. The NFET pulls node 0 down, and the node voltage arrives at the ground potential at the time indicated by arrow A. The NFET goes into the triode region. The microstate sequence of this transition is described by the two microstate

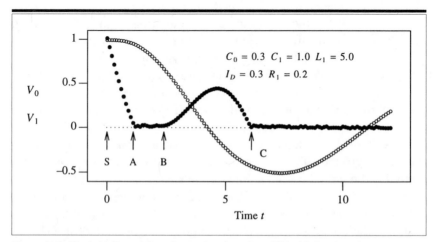

Figure 5.6 Node Voltage Waveform of an Inverter with Inductive Load

changes, MN1:N → S and MN1:S → T, as is shown in Figure 5.8 below. The small oscillation around the time axis indicates that MN1 ends up in the triode region.

The current in the inductance is proportional to the integral of the voltage applied across the terminals. If the voltage does not change sign, the current keeps increasing in the same direction. At the beginning of the transient the current is small, MN1 pulls node 0 down quickly, and it goes into the triode region. The inductance current keeps increasing. Some time later the inductance current becomes higher than the maximum current for MN1 to stay in the triode region (I_D). Then MN1 moves out from the triode region. The correspondence between the node waveforms and the microstate-submicrostate sequence is shown by the same indicators S, A, B, ... in Figures 5.6 and 5.8. At B the current from the inductance becomes so high that MN1 is driven out from the triode region. Then the increasing V_0 and decreasing V_1 cross, and after that the inductance current decreases.

The mode local times between the microstate events B and C are determined from the general solution of the circuit equations (5.10) as

$$V_1(t)(LCR) = A - X_0 t + B \exp(\beta t) + C \exp(\gamma t)$$

$$V_0(t)(LCR) = A - (X_1 + X_2) - X_0 t - (C_1/C_0) \qquad (5.23)$$

$$[B \exp(\beta t) + C \exp(\gamma t)]$$

where

$$\beta(+), \gamma(-) = -\frac{R_1}{2L_1} \pm \frac{R_1}{2L_1} \sqrt{1 - [4L_1(C_0 + C_1)/C_0 C_1 R_1^2]}$$

$$I_D(t_{DG}) = -C_1[dV_1(t)/dt]_{t=t_{DG}} \qquad (5.24)$$

$$= -C_1[\beta B \exp(\beta t_{DG}) + \gamma C \exp(\gamma t_{DG}) - X_0]$$

where $I_D(t)$ is the FET current assuming $V_0(t) = 0$. This relation is valid only at $t = t_{DG}$. A, B, and C are the constants determined from the initial conditions at $t = t_{DG}$, when MN1 moves out the triode region. The latter are, at $t = t_{DG}$,

$$V_1(t)(LCR) = V_1(t_{DG})(LCR) \text{ (last microstate)}$$

$$V_0(t) = 0 \qquad (5.25)$$

$$I_D(t_{DG}) = I_D$$

After some algebra we obtain

$$V_1(t)(LCR) = A - X_0(t - t_{DG}) + B \exp[\beta(t - t_{DG})]$$
$$+ C \exp[\gamma(t - t_{DG})]$$

$$V_0(t)(LCR) = A - (X_1 + X_2) - X_0(t - t_{DG}) - (C_1/C_0)$$
$$[B \exp[\beta(t - t_{DG})] + C \exp[\gamma(t - t_{DG})]]$$

(5.26)

where the redefined constants A, B, and C are given by

$$A = \frac{C_1}{C_0 + C_1} V_1(t_{DG}) + X_1$$

$$B = \frac{C_1\gamma[C_0(C_0 + C_1)V_1(t_{DG}) - R_1C_0C_1I_D] + C_0(C_0 + C_1)I_D}{C_1(C_0 + C_1)^2(\gamma - \beta)}$$

$$C = -\frac{C_1\beta[C_0(C_0 + C_1)V_1(t_{DG}) - R_1C_0C_1I_D] + C_0(C_0 + C_1)I_D}{C_1(C_0 + C_1)^2(\gamma - \beta)}$$

(5.27)

The RC circuit solution satisfying the initial conditions at $t = t_{DG}$

$$V_0(t)(RC) = 0$$
$$V_1(t)(RC) = V_1(t_{DG})(LCR) \text{ (last microstate)}$$

(5.28)

is derived as

$$V_1(t)(RC) = \frac{C_1}{C_0 + C_1} V_1(t_{DG}) + X_1 - X_0(t - t_{DG})$$
$$+ \left[\frac{C_0}{C_0 + C_1} V_1(t_{DG}) - X_1\right] \exp[-(t - t_{DG})/t_{EQ}]$$

$$V_0(t)(RC) = \frac{C_1}{C_0 + C_1} V_1(t_{DG}) - X_2 - X_0(t - t_{DG})$$
$$- \left[\frac{C_1}{C_0 + C_1} V_1(t_{DG}) - X_2\right] \exp[-(t - t_{DG})/t_{EQ}]$$

(5.29)

The simultaneous equations that determine the local times t_α and t_β are

$$-X_0(t_\alpha - t_{DG}) + \left[\frac{C_0}{C_0 + C_1} V_1(t_{DG}) - X_1 \right] \exp[-(t_\beta - t_{DG})/t_{EQ}] = P$$

$$-X_0(t_\alpha - t_{DG}) - \left[\frac{C_1}{C_0 + C_1} V_1(t_{DG}) - X_2 \right] \exp[-(t_\beta - t_{DG})/t_{EQ}] = Q$$

$$(5.30)$$

where

$$P = V_1(t)(LCR) - \frac{C_1}{C_0 + C_1} V_1(t_{DG}) - X_1$$

$$(5.31)$$

$$Q = V_0(t)(LCR) - \frac{C_1}{C_0 + C_1} V_1(t_{DG}) + X_2$$

By solving the simultaneous equations we determine t_α and t_β as functions of the global time t. We have

$$t_\alpha = t \tag{5.32}$$

as before. The local times versus the global time are plotted in Figure 5.7.

After MN1 goes into the triode region at $t = t_{END}$ for the second time, at the microstate event C of Figures 5.6 and 5.8, the mode β local time is determined by the same procedure as in the second microstate, or the

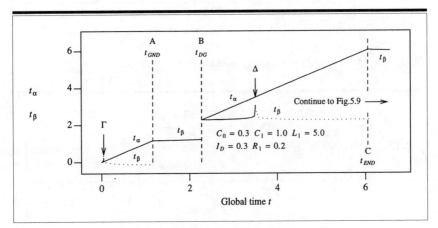

Figure 5.7 Mode Local Times t_α and t_β Versus Global Time

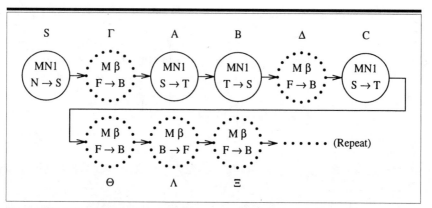

Figure 5.8 Microstate Sequence of Extremely Underdamped Driver-Interconnect Combination

interval between the microstate events A and B. Figure 5.9 shows the mode β local time versus the global time after the microstate event C. The FET current plotted in Figure 5.10 never exceeds I_D (the value at 0.3). There are no more microstate changes. There are submicrostate changes, in that the local-time flow direction of the second mode changes periodically even

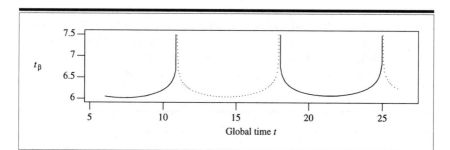

Figure 5.9 Mode Local Time t_β Versus Global Time

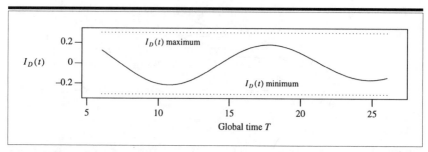

Figure 5.10 FET Current Versus Time

after C. The entire microstate-submicrostate sequence of this circuit is shown in Figure 5.8.

A closed-form solution for the local time t_β in the last phase of the transient is derived as

$$
\begin{aligned}
t_\beta = t_{END} &+ (R_1^2 C_1/2L_1)(t - t_{END}) \\
&- (R_1 C_1) \log\left| \sqrt{1 + [A/V_1(t_{END})]^2} \right. \\
&\left. \cos\left[\sqrt{(1/L_1 C_1) - (R_1/2L_1)^2}\,(t - t_{END}) + \phi \right] \right|
\end{aligned}
\tag{5.33}
$$

where A is determined from

$$
A = [(C_1 R_1/2L_1)V_1(t_{END}) - I_D]/[C_1 \sqrt{(1/L_1 C_1) - (R_1/2L_1)^2}] \tag{5.34}
$$

and $\phi = -\tan^{-1}[A/V_1(t_{END})]$. From this expression we learn that the mode local time increases slowly if $R_1/\sqrt{L_1/C_1}$ is small, as in the present example. Its divergence is logarithmic, and it has very narrow peaks, as is observed in Figure 5.7.

5.4 Microstates and Submicrostates—II

In the source-follower pullup circuit driving a load through an interconnect, as shown in Figure 5.11(a), the NFET MN1 is modeled by a collapsible current generator, and the current is given by

$$
I_D(t) = G_m[V_{DD} - V_{TH} - V_0(t)] \tag{5.35}
$$

where the back-bias effect is neglected; V_{TH} is the constant FET threshold voltage, and G_m is the transconductance. The FET gate is pulled up to the high level V_{DD} at $t = 0$. The circuit equations are

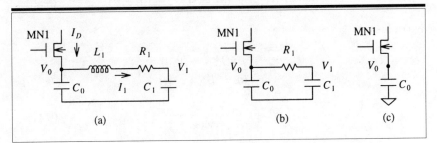

Figure 5.11 Source-Follower Pullup Transient

$$C_0 \frac{dV_0}{dt} = I_D - I_1$$

$$I_1 = C_1 \frac{dV_1}{dt} \tag{5.36}$$

$$V_0 - V_1 = R_1 I_1 + L_1 \frac{dI_1}{dt}$$

The equation satisfied by V_1 is derived by eliminating V_0 and I_1:

$$L_1 C_0 C_1 \frac{d^3 V_1}{dt^3} + (R_1 C_0 C_1 + G_m L_1 C_1) \frac{d^2 V_1}{dt^2}$$

$$+ (C_0 + C_1 + G_m R_1 C_1) \frac{dV_1}{dt} + G_m V_1 = G_m (V_{DD} - V_{TH}) \tag{5.37}$$

The special solution of this equation is $V_1 = V_{DD} - V_{TH}$. The homogeneous part of the equation gives the secular equation

$$L_1 C_0 C_1 S^3 + (R_1 C_0 C_1 + G_m L_1 C_1) S^2$$

$$+ (C_0 + C_1 + G_m R_1 C_1) S + G_m = 0 \tag{5.38}$$

By using the three solutions α, β, and γ of this equation, V_1 and V_0 are given by

$$V_1(t)(LCR) = (V_{DD} - V_{TH}) + A \exp(\alpha t) + B \exp(\beta t) + C \exp(\gamma t)$$

$$V_0(t)(LCR) = (V_{DD} - V_{TH}) + [1 + (R_1 C_1)\alpha + (L_1 C_1)\alpha^2]A \exp(\alpha t)$$

$$+ [1 + (R_1 C_1)\beta + (L_1 C_1)\beta^2]B \exp(\beta t)$$

$$+ [1 + (R_1 C_1)\gamma + (L_1 C_1)\gamma^2]C \exp(\gamma t) \tag{5.39}$$

From the initial conditions

$$V_1(0) = 0 \qquad V_1'(0) = 0 \qquad V_0(0) = 0 \tag{5.40}$$

we obtain

$$A = \frac{\beta\gamma(V_{DD} - V_{TH})}{(\alpha - \beta)(\gamma - \alpha)}$$

$$B = \frac{\gamma\alpha(V_{DD} - V_{TH})}{(\beta - \gamma)(\alpha - \beta)} \tag{5.41}$$

$$C = \frac{\alpha\beta(V_{DD} - V_{TH})}{(\gamma - \alpha)(\beta - \gamma)}$$

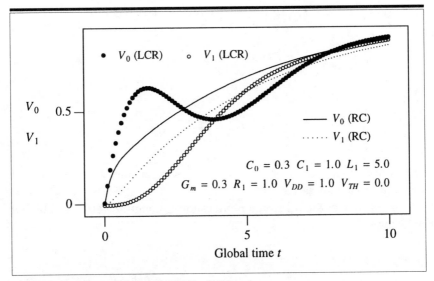

Figure 5.12 Source-Follower Pullup Node Waveforms

An example of the underdamped node waveforms is shown in Figure 5.12 by the filled (V_0) and the open (V_1) circles.

From the set of the initial conditions and from the expressions for $V_1(t)$ and $V_0(t)$ we conclude $V_1''(0) = 0$. Then $V_1(t)$ as a power series in small t starts from the term proportional to t^3. We have

$$\frac{V_1(t)}{V_{DD} - V_{TH}} = \frac{1}{6}\left(\frac{G_m}{C_1}\right)\frac{1}{(L_1 C_0)}t^3$$

$$- \frac{1}{24}\frac{1}{(L_1 C_0)^2}\left[G_m^2\left(\frac{L_1}{C_1}\right) + G_m R_1\left(\frac{C_0}{C_1}\right)\right]t^4 + \cdots$$

$$\frac{V_0(t)}{V_{DD} - V_{TH}} = \left(\frac{G_m}{C_0}\right)t - \frac{1}{2}\left(\frac{G_m}{C_0}\right)^2 t^2$$

$$+ \frac{1}{6}\left(\frac{G_m}{C_0}\right)\frac{1}{(L_1 C_0)}\left[G_m^2\left(\frac{L_1}{C_0}\right) - 1\right]t^3$$

$$+ \frac{1}{24}\frac{1}{(L_1 C_0)^2}\left[-G_m^4\left(\frac{L_1}{C_0}\right)^2 + 2G_m^2\left(\frac{L_1}{C_0}\right) + G_m R_1\right]t^4$$

$$+ \cdots$$

$$(5.42)$$

where the first two terms of $V_0(t)$ do not contain inductance. L_1 may be considered as an open circuit at $t = 0$. Then the equivalent circuit at $t = 0$

is as shown in Figure 5.11(c), and the circuit equation is

$$C_0 \frac{dV_0}{dt} = G_m(V_{DD} - V_{TH} - V_0)$$

or   (5.43)

$$V_0 = \frac{G_m(V_{DD} - V_{TH})}{C_0} t + \cdots$$

Thus the first two terms of the power series expansion of the solution of the LCR model are derived.

The solution of the RC circuit model shown in Figure 5.11(b) is obtained by using

$$\alpha'(+), \beta'(-) =$$
$$\frac{-(C_0 + C_1 + G_m R_1 C_1) \pm \sqrt{(C_0 + C_1 + G_m R_1 C_1)^2 - 4G_m R_1 C_0 C_1}}{2C_0 C_1 R_1}$$

(5.44)

as

$$V_1(RC) = (V_{DD} - V_{TH}) + A' \exp(\alpha' t) + B' \exp(\beta' t)$$

$$V_0(RC) = (V_{DD} - V_{TH}) + [1 + (R_1 C_1)\alpha'] A' \exp(\alpha' t) \quad (5.45)$$

$$+ [1 + (R_1 C_1)\beta'] B' \exp(\beta' t)$$

From the initial condition

$$V_1(t)(RC) = V_0(t)(RC) = 0 \quad\quad\quad (5.46)$$

we have

$$A' = \frac{\beta'(V_{DD} - V_{TH})}{\alpha' - \beta'} \qquad B' = -\frac{\alpha'(V_{DD} - V_{TH})}{\alpha' - \beta'} \quad (5.47)$$

The node voltage waveforms for the same parameter values in an RC-only circuit are shown in Figure 5.12 by the solid (V_0) and the dotted (V_1) curve. We solve simultaneous equations to determine the mode local times t_α and t_β as functions of the global time t:

$$A' \exp(\alpha' t_a) + B' \exp(\beta' t_\beta) = V_1(t)(LCR) - (V_{DD} - V_{TH})$$

$$[1 + (R_1 C_1)\alpha']A' \exp(\alpha' t_a) + [1 + (R_1 C_1)\beta']B' \exp(\beta' t_\beta)$$
$$= V_0(t)(LCR) - (V_{DD} - V_{TH})$$
$$(5.48)$$

These equations are solved for

$$\exp(\alpha' t_\alpha) = 1 - \frac{V_1(t)(LCR)}{V_{DD} - V_{TH}} + \frac{V_0(t)(LCR) - V_1(t)(LCR)}{(R_1 C_1)\beta'(V_{DD} - V_{TH})} \qquad (5.49)$$

$$\exp(\beta' t_\beta) = 1 - \frac{V_1(t)(LCR)}{V_{DD} - V_{TH}} + \frac{V_0(t)(LCR) - V_1(t)(LCR)}{(R_1 C_1)\alpha'(V_{DD} - V_{TH})}$$

Figure 5.13 shows t_α versus global time. In the limit as $t \to \infty$, we observe $t_\alpha/t = \text{const} > 1$. The smallest-magnitude complex frequency α of the LCR circuit is approximated by

$$\alpha(L_1) = -(G_m/C_X) - (G_m^2 R_1 C_0 C_1/C_X^3) - (G_m^3 L_1 C_1/C_X^3) \quad (5.50)$$

where $C_X = C_0 + C_1(1 + G_m R_1)$. In the limit as $t \to \infty$ we have

$$t_\alpha = [\alpha(L_1)/\alpha(0)]t + \text{const} = (1 + \Delta\alpha)t + \text{const} \quad (5.51)$$

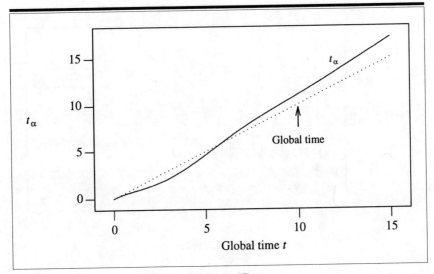

Figure 5.13 Mode α Local Time Versus Global Time

where

$$\Delta\alpha = G_m^2 \frac{L_1}{C_1} \frac{1}{[1 + G_m R_1 + (C_0/C_1)]^2} \qquad (5.52)$$

Since $1/G_m$ is the output impedance of the source follower, this formula is structurally similar to the asymptotic form derived in Section 4.2. For small t, $V_1(t)(LCR)$ and $V_0(t)(LCR)$ are given by the power series, and in particular $V_1(t)(LCR) \approx o(t^3)$. We then have, up to the first power of t, by setting $V_1(t)(LCR) = 1$,

$$\exp(\beta' t_\beta) = 1 + \frac{V_0(t)(LCR)}{(R_1 C_1)\alpha'(V_{DD} - V_{TH})} \qquad (5.53)$$

$$= 1 + \frac{1}{(R_1 C_1)\alpha'} \left(\frac{G_m}{C_0}\right) t + o(t^2)$$

or

$$t_\beta = \frac{1}{\beta'} \log\left[1 + \frac{1}{(R_1 C_1)\alpha'} \left(\frac{G_m}{C_0}\right) t\right] \qquad (5.54)$$

where $\alpha' < 0$. As t increases, the argument of log becomes zero, and t_β diverges. The divergence of t_β at Γ in Figure 5.14 can be understood from

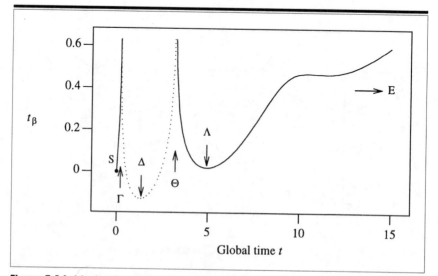

Figure 5.14 Mode β Local Time Versus Global Time

this formula. Similarly

$$t_\alpha = \frac{1}{\alpha'} \log \left[1 + \frac{1}{(R_1 C_1)\beta'} \left(\frac{G_m}{C_0} \right) t \right] \tag{5.55}$$

We note that t_α does not diverge, as is observed in Figure 5.13. This is because

$$\left| \frac{\beta'}{\alpha'} \right| = \frac{(C_0 + C_1 + G_m R_1 C_1)^2}{G_m^2 (C_0/G_m)(R_1 C_1)} = \frac{(1.6)^2}{0.09} \approx 28 \tag{5.56}$$

(The numerator is Elmore's formula if G_m is replaced by $1/R_0$.) Then before the argument of the logarithm in t_α vanishes by virtue of the negatively increasing term proportional to t, the neglected higher-order terms become important.

Figure 5.14 shows t_β versus global time. The source-follower pullup transient has no microstate change, as shown in the microstate-submicrostate sequence of Figure 5.15. The final microstate M is the boundary between the nonconducting and the saturation region of MN1. The circuit goes through four submicrostate changes. In contrast with the pulldown transition of the last section, there are only a finite number of submicrostate changes. The pullup device presents successively higher internal impedance as the circuit approaches the final M state. The circuit moves out of the underdamping regime before reaching the final steady state, and there is no ringing oscillation in the later phase of the transient. In the first submicrostate, t_β flows forward, and likewise in the last submicrostate, which is the final approach to the steady state. As we observed from many examples in Sections 4.4, 4.5, 5.3, and 5.4, addition of inductance does not significantly affect the local time of the mode having the longest time constant, which governs the overall charge-discharge process in the circuit.

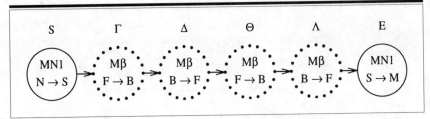

Figure 5.15 Microstate Sequence of Underdamped Source Follower

Many interesting features appear in the local times of the modes, whose role is to equalize the voltages of the nodes. Thus the LCR circuit theory becomes a neat extension of the RC circuit theory, and the basic theoretical connection is clearly observable. This is a nice feature of the microstate-submicrostate theory.

5.5 Microstates and Submicrostates—III

All the switching devices of a single logic gate are laid out close together. The inductance of a loop connecting the devices is negligible. This is true most of the time, but there is an exception. A transmission gate may be placed a long distance away from the driving gate, and the inductance of the interconnect can be significant. The equivalent circuit for the combination is shown in Figure 5.16. L_1 is the inductance of a loop closed by the lumped capacitances C_0 (driver end) and C_1 (transmission-gate end).

Let the voltages developed across C_0, C_1, and C_2 be V_0, V_1, and V_2, respectively, and the currents in the interconnect loops of L_1, of MN0, and of MN1 be I_1, I_0, and I_2, respectively. The device has its own shunt capacitance, and it closes a loop (both MN0 and MN1). The current polarities are defined in Figure 5.16. We study a nontrivial case. MN0 is small and is turned on before, C_2 is large and is charged to V_{DD}, and MN1 is a large transmission gate that is turned on at $t = 0$. The circuit equations are

$$C_0 \frac{dV_0}{dt} = I_1 - I_0 \qquad C_1 \frac{dV_1}{dt} = I_2 - I_1 \qquad C_2 \frac{dV_2}{dt} = -I_2$$

$$\text{(5.57)}$$

$$V_1 - V_0 = R_1 I_1 + L_1 \frac{dI_1}{dt}$$

and

$$I_0 = I_D \quad (V_0 > 0) = -I_D \quad (V_0 < 0) \qquad \text{(5.58)}$$

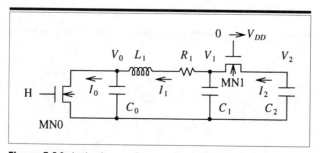

Figure 5.16 Inductance Connecting a Transmission Gate to a Driver

where we assume that V_0 is not driven significantly in the negative polarity. $I_0 = -I_D$ is an approximation subject to restrictions. We have

$$I_2 = G_m(V_{DD} - V_1) \ (V_2 > V_1) = -G_m(V_{DD} - V_2) \quad (V_2 < V_1) \quad (5.59)$$

where we note that I_0 at $V_0 = 0$ and I_2 at $V_1 = V_2$ are not defined. This is convenient in the numerical analysis for producing the sign of the triode state, as discussed in Section 5.3. Figure 5.17 shows an example of numerical analysis. At S $(t = t_0 = 0)$, MN1 is turned on. At A $(t = t_1)$, MN1 goes into the triode region because it is a large NFET. MN1 stays in the triode region after that. MN0 stays in the triode region until B $(t = t_2)$. The small oscillation close to the horizontal axis and the small offset of the V_2 and V_1 curves indicate that MN0 and MN1 are in the triode region, respectively. At B $(t = t_2)$, MN0 moves out from the triode region because the current established in L_1 exceeds I_D, the maximum current of MN0. MN0 stays in the saturation region until C $(t = t_3)$, when the inductance current has decayed and MN0 goes back to the triode region. The numerical analysis can be carried out using a very simple computer program, and it is convenient for getting an overall idea of the transient. The oscillation associated with the triode region, however, makes it difficult to determine the local times. The closed-form circuit analysis to derive local times is elementary but is tedious. Only the results are listed below.

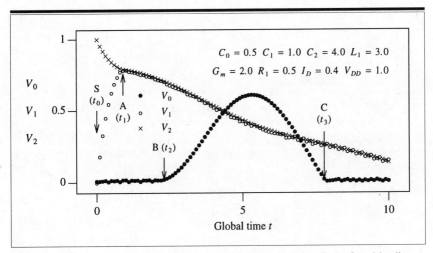

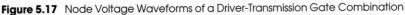

Figure 5.17 Node Voltage Waveforms of a Driver-Transmission Gate Combination

- *Microstate 1 (S → A):*

$$V_0(t)(LCR) = 0$$

$$V_1(t)(LCR) = [R_1 G_m / (1 + R_1 G_m)] V_{DD} + A \exp(\alpha t) + B \exp(\beta t)$$

$$V_2(t)(LCR) = V_{DD} - [G_m V_{DD} / C_2 (1 + R_1 G_m)] t$$

$$+ (G_m A / C_2 \alpha)[\exp(\alpha t) - 1] + (G_m B / C_2 \beta)[\exp(\beta t) - 1]$$

$$(5.60)$$

where the initial conditions to determine A and B are

$$V_1(+0) = 0 \quad \text{and} \quad V_1'(+0) = G_m V_{DD} / C_1 \qquad (5.61a)$$

whence

$$A = - \frac{[(1 + R_1 G_m) + (R_1 C_1)\beta] G_m V_{DD}}{C_1(\beta - \alpha)(1 + R_1 G_m)}$$

$$(5.61b)$$

$$B = \frac{[(1 + R_1 G_m) + (R_1 C_1)\alpha] G_m V_{DD}}{C_1(\beta - \alpha)(1 + R_1 G_m)}$$

and where

$$\alpha(+), \beta(-) =$$

$$\frac{-(L_1 G_m + R_1 C_1) \pm \sqrt{(L_1 G_m + R_1 C_1)^2 - 4 L_1 C_1 (1 + R_1 G_m)}}{2 L_1 C_1}$$

$$(5.62)$$

- *Microstate 2 (A → B):*

$$V_0(t)(LCR) = 0$$

$$V_1(t)(LCR) = V_2(t)(LCR) = A \exp[\alpha(t - t_1)] + B \exp[\beta(t - t_1)]$$

$$(5.63)$$

where

$$A = \frac{(C_1 + C_2)\beta V_1(t_1) + I_1(t_1)}{(C_1 + C_2)(\beta - \alpha)}$$

$$(5.64)$$

$$B = - \frac{(C_1 + C_2)\alpha V_1(t_1) + I_1(t_1)}{(C_1 + C_2)(\beta - \alpha)}$$

and where

$$\alpha(+), \beta(-) = \frac{-R_1(C_1 + C_2) \pm \sqrt{[R_1(C_1 + C_2)]^2 - 4L_1(C_1 + C_2)}}{2L_1(C_1 + C_2)}$$

(5.65)

and $V_1(t_1)$ and $I_1(t_1)$ are the values at the end of microstate 1.

- *Microstate 3 (B → C):*

$$V_1(t)(LCR) = V_2(t)(LCR) = A - \frac{I_D}{C_0 + C_1 + C_2}(t - t_2)$$

$$+ B \exp[\beta(t - t_2)] + C \exp[\gamma(t - t_2)]$$

$$V_0(t)(LCR) = A - \frac{(C_1 + C_2)R_1 I_D}{C_0 + C_1 + C_2}$$

(5.66)

$$- \frac{I_D}{C_0 + C_1 + C_2}(t - t_2)$$

$$- \frac{C_1 + C_2}{C_0}[B \exp[\beta(t - t_2)] + C \exp[\gamma(t - t_2)]]$$

where

$$A = V_1(t_2) - \frac{C_0^2 R_1 I_D}{(C_0 + C_1 + C_2)^2} + \frac{C_0}{C_0 + C_1 + C_2}[R_1 I_D - V_1(t_2)]$$

$$B = - \frac{[C_0/(C_0 + C_1 + C_2)]L_1 \gamma I_D + [R_1 I_D - V_1(t_2)]}{L_1(C_1 + C_2)\beta(\gamma - \beta)}$$

(5.67)

$$C = \frac{[C_0/(C_0 + C_1 + C_2)]L_1 \beta I_D + [R_1 I_D - V_1(t_2)]}{L_1(C_1 + C_2)\gamma(\gamma - \beta)}$$

in which

$$\beta(+), \gamma(-) =$$
$$\frac{-C_0(C_1 + C_2)R_1 \pm \sqrt{[C_0(C_1 + C_2)R_1]^2 - 4C_0(C_1 + C_2)L_1(C_0 + C_1 + C_2)}}{2C_0(C_1 + C_2)L_1}$$

(5.68)

and $V_1(t_2)$ is the value at the end of microstate 2.

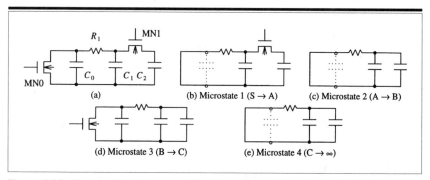

Figure 5.18 Connectivity Change of the Driver-Transmission Gate Combination

- *Microstate 4 (C→):* This is the same as microstate 2. The backbone *RC* circuit of Figure 5.18(a) changes connectivity as shown in Figure 5.18(b), (c), (d), and (e).

- *Microstate 1 (S → A):* See Figure 5.18(b). The set of equations that determine the mode local times t_α and t_β are

$$V_1(t)(LCR) = \frac{G_m R_1 V_{DD}}{1 + G_m R_1} \left[1 - \exp\left(-\frac{1 + G_m R_1}{C_1 R_1} t_\beta \right) \right]$$

$$V_2(t)(LCR) = V_{DD} - \frac{G_m V_{DD}}{C_2(1 + G_m R_1)} t_\alpha$$

$$- \frac{C_1 (G_m R_1)^2 V_{DD}}{C_2(1 + G_m R_1)^2} \left[1 - \exp\left(-\frac{1 + G_m R_1}{C_1 R_1} t_\beta \right) \right]$$

$$(5.69)$$

and we obtain

$$t_\alpha = \frac{C_2}{G_m} (1 + G_m R_1) \left[1 - \frac{V_2(t)(LCR)}{V_{DD}} - \frac{C_1}{C_2} \frac{G_m R_1}{1 + G_m R_1} \frac{V_1(t)(LCR)}{V_{DD}} \right]$$

$$t_\beta = -\frac{C_1 R_1}{1 + G_m R_1} \log \left| 1 - \frac{V_1(t)(LCR)(1 + G_m R_1)}{G_m R_1 V_{DD}} \right| \qquad (5.70)$$

- *Microstate 2 (A → B):* See Figure 5.18(c). We have

$$t_\alpha = t_1 - R_1(C_1 + C_2) \log[V_1(t)(LCR)/V_1(t_1)] \qquad (5.71)$$

- *Microstate 3 (B → C):* See Figure 5.18(d). We have

$$t_\alpha = t_2 + (1/I_D)[(C_1 + C_2)[V_1(t_2) - V_1(t)(LCR)] - C_0 V_0(t)(LCR)]$$

$$t_\beta = t_2 - \frac{C_0(C_1 + C_2)R_1}{C_0 + C_1 + C_2} \log \left| \frac{V_1(t)(LCR) - V_0(t)(LCR) - V_X}{V_1(t_2) - V_X} \right|$$

$$(5.72)$$

where $V_X = R_1(C_1 + C_2)I_D/(C_0 + C_1 + C_2)$.

- *Microstate 4 (C → ∞):* See Figure 5.18(e). We have

$$t_\alpha = t_3 - R_1(C_1 + C_2) \log[V_1(t)(LCR)/V_1(t_3)] \qquad (5.73)$$

The mode that equalizes V_1 and V_2 in the first microstate has the local time t_β. As V_1 increases and V_2 decreases, the mode dies, and the local time diverges at Γ of Figures 5.19 and 5.20. At this point, still $V_2 > V_1$, as we see from Figure 5.17, and the source-drain identification of MN1 remains the same. The mode is recharged. The mode β local time flows backward during the period between Γ and A.

In the second microstate the inductance current is still increasing, but it is low enough to maintain MN0 in the triode region. Since R_1 is small but L_1 is large, the discharge is controlled by the current of L_1. The local time

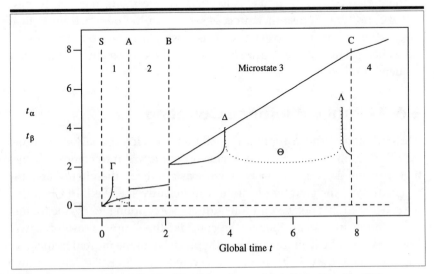

Figure 5.19 Mode Local Times of the Driver-Transmission Gate Combination

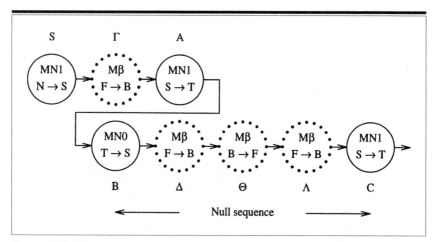

Figure 5.20 Microstate-Submicrostate Sequence of the Driver-Transmission Gate Combination

referred to the backbone RC circuit model time constant, $R_1(C_1 + C_2)$, does not move forward rapidly during the period between A and B.

As MN0 changes to the saturation region at C, V_0 begins to increase. The time constant of the mode β, given by $R_1 C_0 (C_1 + C_2)/(C_0 + C_1 + C_2)$, is that of the voltage equalization between nodes 0 and 1. The mode β that forces current from node 1 to node 0 dies at Δ (close to the time when $V_1 = V_0$ is reached) and is then recharged. The backward local time flow in the period between Δ and Θ is to effect the reestablishment of the new initial condition. The similar process repeats in the period from Θ to Λ to C. After the microstate event C, I_0 is less than I_D and no more microstate changes take place. Figure 5.20 shows the microstate-submicrostate sequence of this circuit.

5.6 Mutual Inductance Coupling

If mutual inductance coupling is included, the circuit analysis becomes significantly more complicated. Inclusion of mutual inductance assumes that a magnetic coupling mechanism exists outside the circuit, and the mechanism of the coupling is quite insufficiently specified. In Chapter 1 we concluded that delayed mutual inductance coupling marks the limit of extension of the circuit theory. We consider the simplest case of a two-stage cascaded LCR circuit with negligible delay in the mutual inductance coupling. The equivalent circuit is shown in Figure 5.21. The loops shown by the dotted rectangles have mutual inductance M.

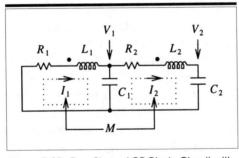

Figure 5.21 Two-Stage *LCR* Chain Circuit with
Mutual Inductance Coupling

A prerequisite of writing the circuit equations is to clarify a confusing
issue: The polarity of the voltage induced by the magnetic induction and
the sign of the mutual inductance. In Figure 5.22(a) the two current loops
1 and 2 are broken open at A, B and C, D, respectively. They have pairs of
terminals A, B and C, D, and A and C are identified by the dots. *A* and *C*
are the *positive* terminals of the loop inductances. The voltage V_1 developed across AB (A positive), the voltage V_2 developed across CD (C positive), the current I_1 flowing in loop 1 from A to B, and the current I_2
flowing in loop 2 from C to D are related by

$$V_1 = L_1 I_1' + M I_2' \qquad V_2 = M I_1' + L_2 I_2' \qquad (5.74)$$

where the primes indicate time derivatives. The equivalent circuit of the
coupled inductances—a transformer—is shown in the inset. The terminal

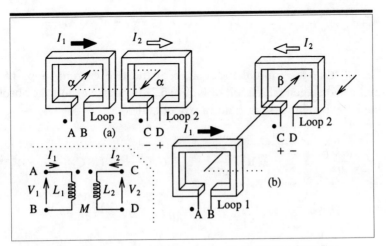

Figure 5.22 Voltage Induced by the Mutual Inductance Coupling

identification and the location of the dot are important in the following discussions, to establish correlation between the hardware of Figure 5.22(a) or (b) and the equivalent circuit.

Current I_1 is forced into loop 1 from the positive terminal A. The current creates a magnetic field that is shown by arrow α of Figure 5.22(a) and that links loop 2. The current and magnetic field directions follow Ampere's law: The current in the direction of advance of a right-hand screw creates a magnetic field in the direction of rotation of the screw. If I_1 increases, the magnetic field (arrow α) increases, and the magnetic flux that links loop 2 from the back to the front increases. The induced electric field in loop 2 is in such a direction as to create a loop 2 current that tends to maintain the magnetic flux linked to loop 2 unchanged. The current in loop 2 is shown by the open arrow. The current generates voltage across the gap C–D with polarity such that D is positive relative to C. The voltage is negative according to the convention, and accordingly $M < 0$. If the loops are arranged as shown in Figure 5.22(b), the increasing magnetic flux in loop 2 in the direction of arrow β creates current shown by the open arrow. The current generates voltage at the gap CD such that C is positive relative to D. The induced voltage is positive and $M > 0$, according to the convention.

The set of circuit equations of the two-stage LCR circuit of Figure 5.21 are

$$C_1 \frac{dV_1}{dt} = I_1 - I_2 \qquad\qquad R_1 I_1 + L_1 \frac{dI_1}{dt} + V_1 = M \frac{dI_2}{dt}$$

$$\tag{5.75}$$

$$C_2 \frac{dV_2}{dt} = I_2 \qquad\qquad -V_1 + R_2 I_2 + L_2 \frac{dI_2}{dt} + V_2 = M \frac{dI_1}{dt}$$

We recall that the mutual inductance of the two current loops of the straight two-stage LCR circuit of Figure 5.21 is negative, and the absolute value is M. By eliminating I_1 and I_2 we obtain

$$(L_1 C_1) \frac{d^2 V_1}{dt^2} + (L_1 - M)C_2 \frac{d^2 V_2}{dt^2} + (R_1 C_1) \frac{dV_1}{dt} + (R_1 C_2) \frac{dV_2}{dt} + V_1 = 0$$

$$-MC_1 \frac{d^2 V_1}{dt^2} + (L_2 - M)C_2 \frac{d^2 V_2}{dt^2} + (R_2 C_2) \frac{dV_2}{dt} - V_1 + V_2 = 0$$

$$\tag{5.76}$$

Assuming that the time dependence of V_1 and V_2 is of the form $\exp(St)$, complex frequency S satisfies the secular equation

$$(L_1L_2 - M^2)C_1C_2S^4 + [(L_2C_2)(R_1C_1) + (L_1C_1)(R_2C_2)]S^3$$
$$+ [(R_1C_1)(R_2C_2) + (L_1C_1) + (L_1C_2) + (L_2C_2) - 2(MC_2)]S^2 \quad (5.77)$$
$$+ [(R_1C_1) + (R_2C_2) + (R_1C_2)]S + 1 = 0$$

In this equation all the coefficients are positive, since $L_1L_2 \geq M^2$ from electromagnetic theory (Section 1.13). Since $L_1 + L_2 \geq 2\sqrt{L_1L_2} \geq 2|M|$, the coefficient of S^2 is positive as well.

To gain insight into the effects of the mutual inductance coupling, we choose a set of parameter values $C_1 = C_2 = 1.0$, $R_1 = R_2 = 1.0$, $L_1 = L_2 = 1.0$ and study the locus of the complex frequency S as M is changed. M is in the range between $-\sqrt{L_1L_2}$ and $\sqrt{L_1L_2}$. As M decreases from the maximum value $M = 1$, the complex frequency varies as shown in Figure 5.23. The first critical point A is reached when $M = 0.900$ and $S_R = -4.98$. The second critical point B is reached when $M = -0.38$ and $S_R = -0.795$.

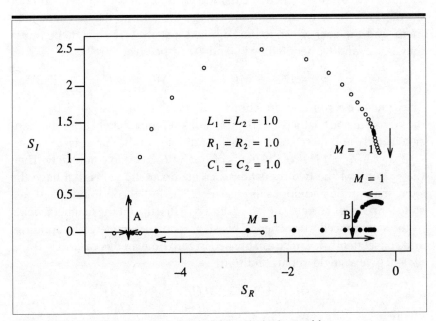

Figure 5.23 Complex Frequency Versus Mutual Inductance M

The set of the equations are solved as

$$V_2(t) = \sum_{n=1}^{4} A_n \exp(S_n t) \qquad V_1(t) = \sum_{n=1}^{4} B_n \exp(S_n t) \quad (5.78)$$

where S_1–S_4 are the four solutions of the secular equation. By substituting $V_1(t)$ and $V_2(t)$ into Equation (5.76), A_n and B_n are related by

$$B_n = q_n A_n$$

where $\hspace{9cm}$ (5.79)

$$q_n = \frac{(L_2 - M)C_2 S_n^2 + (R_2 C_2)S_n + 1}{(MC_1)S_n^2 + 1} A_n$$

The coefficients A_n are determined from the initial conditions

$$V_1(0) = V_2(0) = V_{DD} \, (= 1) \quad \text{and} \quad I_1(0) = I_2(0) = 0 \quad (5.80)$$

The four initial conditions set the four coefficients A_1–A_4. The initial conditions for the currents require explanation. The currents of L_1 and L_2 create a magnetic field, which stores energy. If $\Delta = L_1 L_2 - M^2 > 0$, a fraction of the magnetic flux from L_1 does not link L_2, and vice versa. Thus there is a region where the magnetic fields due to the L_1 and L_2 currents do not overlap. The currents must supply energy to build up the magnetic field there, and the energy cannot be supplied instantly. The magnetic fluxes that link L_1 and L_2, denoted by Φ_1 and Φ_2, respectively, are

$$\Phi_1 = L_1 I_1 + M I_2 \qquad \Phi_2 = M I_1 + L_2 I_2 \qquad (5.81)$$

They must be continuous functions of time. If $I_1(-0) = I_2(-0) = 0$ (before the transient starts), then $I_1(+0) = I_2(+0) = 0$ immediately after the transient sets in, and these are the initial conditions for the currents.

If $\Delta = L_1 L_2 - M^2 = 0$, the fields of L_1 and L_2 overlap completely. Then the currents of the two inductances create magnetic fields that have the same magnitude and opposite polarities. The initial condition specifies a state that carries current $[I_1(+0) \neq 0 \text{ and } I_2(+0) \neq 0]$, but the magnetic fields in the state cancel everywhere. This state requires no energy to build up, and consequently it can be established in zero time. In this case $\Phi_1 = 0$ and $\Phi_2 = 0$ give a single initial condition

$$I_1(+0)/I_2(+0) = -(M/L_1) = -(L_2/M)$$

where $\hspace{9cm}$ (5.82)

$$L_1 L_2 - M^2 = 0$$

Then the circuit starts from an initial condition that carries currents. In this special case of close coupling, the secular equation is reduced to a cubic equation

$$[(L_2C_2)(R_1C_1) + (L_1C_1)(R_2C_2)]S^3$$

$$+ [(R_1C_1)(R_2C_2) + (L_1C_1) + (L_1C_2) + (L_2C_2) - 2(MC_2)]S^2 \quad (5.83)$$

$$+ [(R_1C_1) + (R_2C_2) + (R_1C_2)]S + 1 = 0$$

and by using the three solutions S_1, S_2, and S_3, we obtain $V_2(t)$ and $V_1(t)$ as

$$V_2(t) = \sum_{n=1}^{3} A_n \exp(S_n t) \qquad V_1(t) = \sum_{n=1}^{3} B_n \exp(S_n t) \quad (5.84)$$

where A_n and B_n are related by Equation (5.79). The three coefficients A_1–A_3 are determined from the three initial conditions

$$V_1(+0) = V_2(+0) = 0, \quad (5.85)$$

$$I_1(+0)/I_2(+0) = -M/L_1 = -L_2/M.$$

The set of simultaneous equations that determine A_n for $\Delta > 0$ are

$$\sum_{n=1}^{4} A_n = 1 \qquad \sum_{n=1}^{4} S_n A_n = 0 \qquad \sum_{n=1}^{4} q_n A_n = 1 \quad (5.86)$$

and

$$\sum_{n=1}^{4} (C_1 q_n + C_2) S_n A_n = 0 \quad (5.87)$$

where the second and the fourth conditions are $I_2(+0) = 0$ and $I_1(+0) = 0$, respectively. Since S_n are generally complex numbers, the set of equations must be solved in the complex number domain. V_2 and V_1 of Equation (5.78) are taken as the real parts of the resulting complex numbers. For a typical set of the parameter values L_1, L_2, R_1, R_2, C_1, and C_2 and values of M in the range $-\sqrt{L_1 L_2} < M < \sqrt{L_1 L_2}$, the results are shown in Figure 5.24(a)–(e).

The circuit equations can be solved as power series for small t. We set

$$V_2(t) = 1 + a_2 t^2 + a_3 t^3 + \cdots \quad (5.88)$$

$$V_1(t) = 1 + b_2 t^2 + b_3 t^3 + \cdots$$

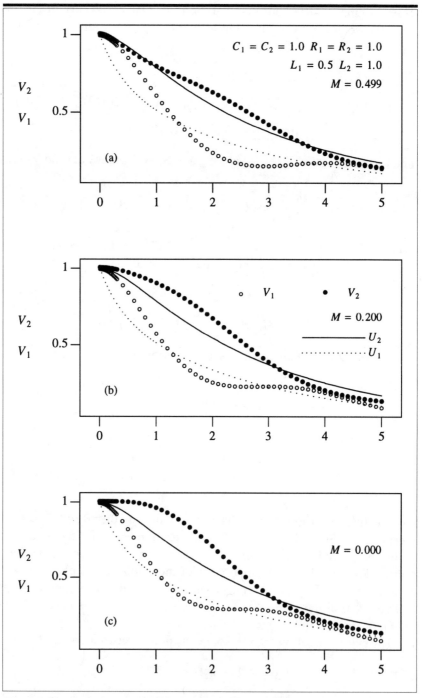

Figure 5.24 Node Voltage Waveforms of Two-Stage *LCR* Chain Coupled by Mutual Inductance

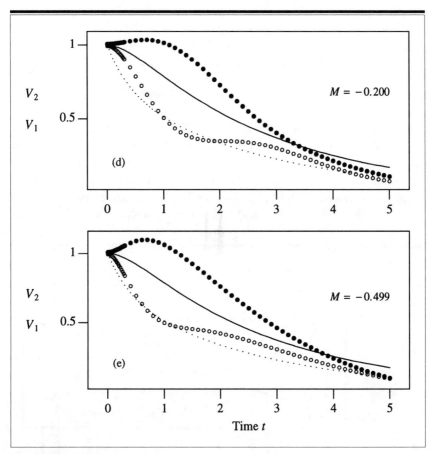

Figure 5.24 Continued

and substitute them into the circuit equations. We assume $\Delta = L_1 L_2 - M^2 > 0$. We have

$$a_2 = -\frac{M}{2C_2\Delta} \qquad a_3 = \frac{M(R_1 L_2 + R_2 L_1)}{6C_2\Delta^2}$$

$$b_2 = -\frac{L_2 - M}{2C_1\Delta} \qquad b_3 = \frac{R_1 L_2(L_2 - M) - R_2 M(L_1 - M)}{6C_1\Delta^2}$$

(5.89)

$V_2(t)$ has a maximum of $M < 0$, as is observed in Figure 5.24(d) and (e). The location of the maximum, t_{max}, is found from $V_2'(t_{max}) = 0$ as

$$t_{max} = -\frac{2a_2}{3a_3} = \frac{2\Delta}{R_1 L_2 + R_2 L_1} \approx 0.6 \qquad (5.90)$$

for the parameter values of the example.

The solid and the dotted curves of Figure 5.24(a)–(e) are the responses of the backbone RC circuit, given by

$$U_2(t) = A \exp(\alpha't) + B \exp(\beta't)$$

$$U_1(t) = C \exp(\alpha't) + D \exp(\beta't)$$

(5.91)

where A, B, C, D, α', and β' are listed in Section 2.7. The initial response of the LCR circuit is delayed with respect to the backbone RC circuit, but as the loop current is built up, the LCR circuit response catches up, and eventually overtakes the RC circuit response. This is reflected in the *node local times* shown in Figure 5.25(a) and (b). As the inductances exchange energy with the circuit, the node local times oscillate. Node local times

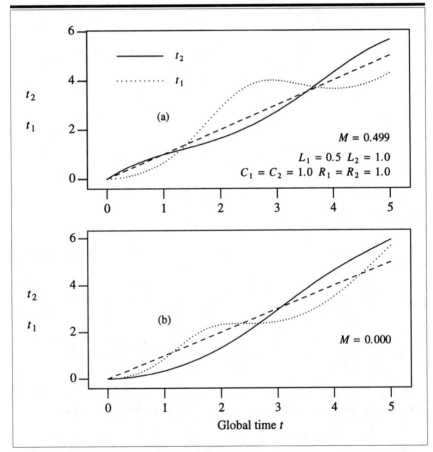

Figure 5.25 Node Local Times Versus Global Time in a Two-Stage *LCR* Chain with Mutual Inductance

cannot be defined for $M < 0$. Then, immediately after the beginning of the transient, V_2 becomes higher than V_{DD}, since the mutual inductance coupling drives node 2 up before the discharge begins. The validity of the node local time in this problem is limited to $M > 0$. In this limited range the node local times give the relative delay or advance of the response.

The mode local times are determined by solving the simultaneous equations

$$A \exp(\alpha' t_\alpha) + B \exp(\beta' t_\beta) = V_2(t)(LCR)$$
$$C \exp(\alpha' t_\alpha) + D \exp(\beta' t_\beta) = V_1(t)(LCR)$$
(5.92)

As we saw before, the mode α local time is close to the global time, but the mode β local time diverges periodically to infinity, indicating energy exchange with the outside of the circuit. [See Figures 5.26(a) and (b).]

We note that mutual inductance coupling should include signal propagation delay, but that cannot be modeled by the value of M alone (Section 1.12). The two loops that interact may have some distance between them, so that the signal is delayed by the distance divided by the wave velocity. As we observed in Section 1.13, the transformer model including delay must be supplemented by a shunt resistance to allow for the radiation loss. Since the loss cannot be determined from information on the circuit alone, this is the natural limit of extension of the circuit theory.

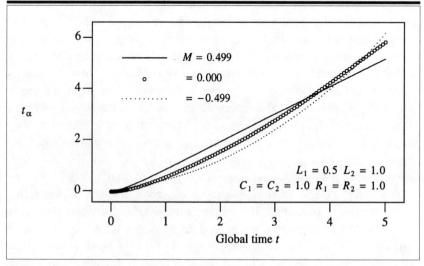

Figure 5.26(a) Mode α Local Time Versus Global Time *t* of a Two-Stage *LCR* Chain with Mutual Inductance

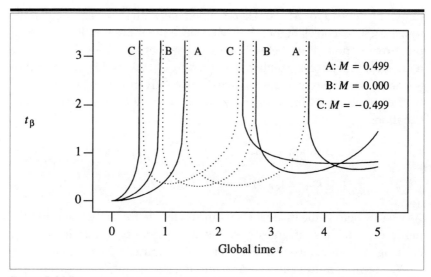

Figure 5.26(b) Mode β Local Time Versus Global Time in Two-Stage *LCR* Chain with Mutual Inductance

The direction of flow of time, or *time's arrow*, has been of interest to philosophers, but in recent years physicists have considered the subject as well [42, 43]. Before concluding this discussion of discrete circuit theory including inductance, I wish to mention the following viewpoint, to persuade the reader that backward flow of time is not at all an odd feature of my circuit theory. When an inductance sends energy out, the local time flows forward, and when the energy comes back, the local time *may* flow backward. For the local time to change direction, the outside magnetic energy must not be dissipated. The inductance acts as a gate between the circuit and the outside space (or the core), where the magnetic energy is stored. This energy has no other route of exchange. Thus the energy exchange is carried out in a quite restricted way. The restrictions are to satisfy requirements that the reverse time flow be a *natural* process of the circuit, and not due to an independent intervention of a subject outside the circuit. An outside intervention such as taking energy away from one capacitor and giving it to another might be viewed as if the local time flowed backward, but that is not a natural phenomenon. If there are many independent self-inductances, each of them goes through the restricted mode of energy exchange. If mutual inductance is included with zero delay, the inductances as a whole exchange energy with the circuit in a restricted way, and the magnetic energy is still in a finite region of the outside space. If delay is included in the mutual inductance interaction, however, the "core" for each inductance expands to the entire space. The resulting unlimited core volume and unlimited overlap of the cores introduce something qualitatively new. The large outside space may be *inhabited* by a *subject* that has the capability to interfere with the circuit inside, or it may encounter a dissipative medium equivalent to a resistor. This conclusion is inevitable, since there is one and only one space in which everything claims existence.

From this observation we may conclude as follows. In a simple system like a circuit including inductance, it is no surprise at all that the local time seen by the circuit often flows backward. In a larger system, the energy transferred out from the circuit may not return completely, or may not return in an orderly way. If some energy does not return, we see that the local time predominantly flows forward. If energy returns in disorder as seen from the circuit, we infer the intervention of an outside subject. Forward and backward flow of time is not a strange concept at all, in a simple object like an electronic circuit. This viewpoint is consistent with the fact that time practically never flows backward in any large system.

5.7 Circuits Having no Backbone *RC* Circuit

We have studied digital circuits that have a *backbone* circuit including R and C only. Inductances were added to the backbone circuit, and the effects were studied. There are, however, circuits that do not have meaningful backbone circuits or modes. We study two examples that create some difficulties in the theory.

The LC tuned oscillator is the first example. Figure 5.27(a) shows a typical LC oscillator circuit. The circuit generates an oscillatory waveform by the frequency-dependent feedback through the tuned transformer, whose primary inductance is L_1, secondary inductance is L_2, and mutual inductance is M. A relationship $L_1 L_2 \geq M^2$ holds. The resistance R_2 represents the *load* to the oscillator. The voltages and currents V_1, V_2, V_D, I_1, and I_2 are defined in Figure 5.27(a).

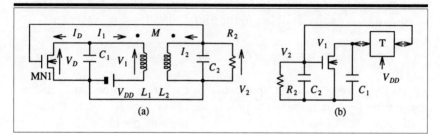

Figure 5.27 *LC* tuned oscillator

The circuit equations are

$$V_D = V_{DD} + V_1$$

$$I_D + I_1 + C_1 \frac{dV_1}{dt} = 0 \qquad I_2 + \frac{V_2}{R_2} + C_2 \frac{dV_2}{dt} = 0 \qquad (5.93)$$

and the transformer gives the relationships

$$V_1 = L_1 \frac{dI_1}{dt} + M \frac{dI_2}{dt} \qquad V_2 = M \frac{dI_1}{dt} + L_2 \frac{dI_2}{dt} \qquad (5.94)$$

The time derivatives of I_1 and I_2 are solved for as

$$\frac{dI_1}{dt} = \frac{L_2 V_1 - M V_2}{\Delta} \qquad \frac{dI_2}{dt} = \frac{L_1 V_2 - M V_1}{\Delta} \qquad (5.95)$$

where $\Delta = L_1 L_2 - M^2$ is assumed not zero. The NFET current I_D is modeled using the collapsible-current-generator model as

$$I_D = B(V_2 - V_{TH})$$

if

$$V_D = V_1 + V_{DD} > 0$$

and

$$V_2 > V_{TH} \qquad (5.96)$$

$$I_D = -B(V_2 - V_1 - V_{DD} - V_{TH})$$

if

$$V_D = V_1 + V_{DD} < 0$$

and

$$V_2 - V_1 - V_{DD} > V_{TH}$$

where V_{TH} is the NFET threshold voltage. If the gate voltage is less than the threshold voltage, the NFET is not conducting in any polarity. The set of equations are solved subject to the initial conditions

$$V_1(0) = -V_{DD}, \qquad V_2(0) = 0, \qquad I_1(0) = I_2(0) = 0 \qquad (5.97)$$

This is the initial condition established in the circuit immediately following the stepwise increase in the power supply voltage from 0 to V_{DD}.

To start oscillation MN1 must be an amplifying device in the feedback path at the beginning. V_{TH} of MN1 must be negative (in the example,

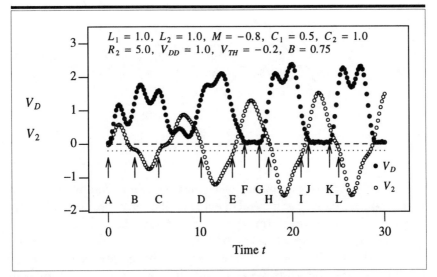

Figure 5.28 Waveforms of the *LC* Transformer-Coupled Oscillator

$V_{TH} = -0.2$). A simple computer program is used to solve the set of circuit equations. A result is shown in Figure 5.28. The parameter values are given in the figure.

The sign of M is negative to shift the phase by 180 deg from the transformer primary to the secondary. Immediately after the power supply turn-on, V_D increases, since C_1 is charged by the current from L_1. Now V_2 increases starting from zero, since $dI_1/dt < 0$ and $M < 0$. The induced secondary voltage charges C_2. At the beginning V_D and V_2 are in phase. V_D and V_2 must be out of phase in the steady-state oscillation, however, since MN1 works as an inverting amplifier in the feedback path. This transition takes place because the current established in L_1 decreases or increases, according as $V_{DD} - V_D$ is positive or negative. At the time when $V_D = V_{DD}$ is reached, the L_1 current is still of a polarity to charge C_1, but its magnitude begins to decrease. After that time $dI_1/dt > 0$. Since $M < 0$, a negative voltage is induced in the transformer secondary. V_2 becomes negative, and that turns MN1 off. Then the L_1 current is entirely used to charge C_1. The current established in L_1 keeps V_D up. Since $dI_1/dt \approx V_1/L \approx (V_D - V_{DD})/L_1$, as V_1 increases, V_2 decreases. V_D and V_2 are now out of phase. The microstate sequence of the oscillation buildup is shown in Figure 5.29. The transition from in phase to out of phase takes place near the microstate events B and C.

When a steady-state oscillation is established, the periodic microstate sequence consists of the two *null* sequences, MN1:S $\rightarrow$ T, T $\rightarrow$ S and MN1:

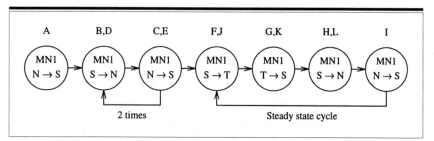

Figure 5.29 Microstate Sequence of *LC* Transformer-Coupled Oscillator

$S \rightarrow N$, $N \rightarrow S$. This is necessary, since the oscillator circuit must go back to the same initial state at the end of every oscillation cycle. The node voltage and the current waveforms are shown in Figure 5.30.

An *LC* oscillator circuit has no backbone RC circuit. The circuit never works without inductance. Still, the energy exchange between the circuit and the outside (the field) works in the same way. If the transformer is disconnected, the circuit of Figure 5.27(a) becomes the RC-only circuit of Figure 5.27(b), where the box *T* effects the energy exchange among C_1, C_2, and the power supply. In this RC circuit the approach to the steady state of the circuit is

$$V_1 \rightarrow 0 \qquad V_2 \rightarrow 0 \tag{5.98}$$

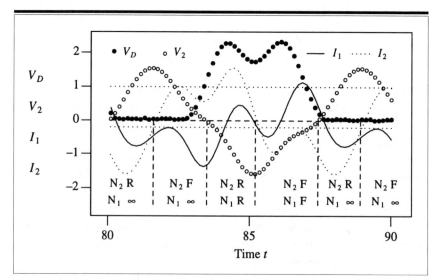

Figure 5.30 Voltage and Current Waveforms of the Transformer-Coupled Oscillator

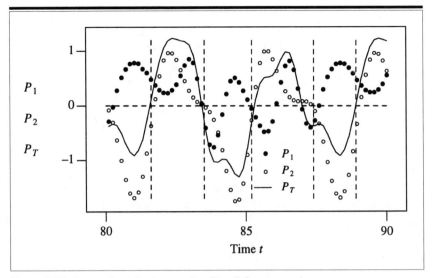

Figure 5.31 Energy Transfer Among the Circuit Components

If $V_1 = 0$, the node 1 local time has reached infinity. If V_1 and V_2 are heading to zero from either polarity (if their magnitudes decrease), the node local time flows forward. If not, the node local time flows backward. Figure 5.30 shows the directions of the local-time flow determined from this criterion. Figure 5.31 shows the powers P_1 and P_2 that are pumped into the transformer from the primary and from the secondary, respectively. They are defined by

$$P_1 = I_1 V_1 \qquad P_2 = I_2 V_2 \qquad (5.99)$$

and their sum $P_T = P_1 + P_2$ is the total power that is pumped into the transformer. In Figures 5.30 and 5.31 the broken vertical lines are drawn at the same locations. If we compare those figures, we see that the node local times flow forward when the power is pumped into the transformer. When the transformer sends the stored energy back to the circuit, the local time flows backward. The transformer may be considered as the outside, or the field, with which the backbone RC circuit of Figure 5.27(b) exchanges energy. That circuit is only a model of the oscillation mechanism, but our fundamental contention that the direction of flow of time is determined by the direction of energy flow solidly stands.

The second example is shown in Figure 5.32. A current generator establishes I_A in the inductor L. Then a switch turns from A to B and delivers

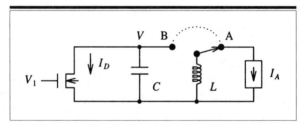

Figure 5.32 A Circuit That Has No *RC* Backbone Circuit

the current to a parallel combination of an NFET and a capacitor C. The NFET is a collapsible current generator that sinks current I_D to ground. The switch is turned to B at $t = 0$. The circuit equation is

$$I_D + C\frac{dV}{dt} + \frac{1}{L}\int_*^t V\,dt = 0 \qquad \text{or} \qquad \frac{d^2V}{dt^2} = -\frac{V}{LC} \qquad (5.100)$$

where the lower limit of the integral is unspecified. This equation is solved as

$$V(t) = V_{max}\sin(t/\sqrt{LC}) \qquad (5.101)$$

Since $C(dV/dt)_{t=0} = I_A - I_D$, we have $V_{max} = \sqrt{L/C}(I_A - I_D)$. We assume $I_A > I_D$. The voltage $V(t)$ is plotted in Figure 5.33(a). We consider the period $0 \le t \le 2t_0$, where $t_0 = \pi\sqrt{LC}$ is the half period.

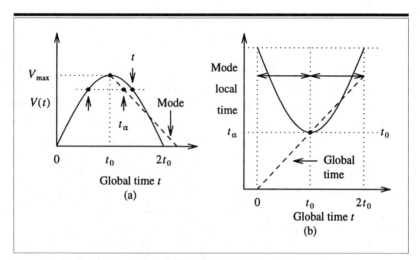

Figure 5.33 Mode Local Time of the Circuit of Figure 5.32

The mode of this circuit is

$$V_\alpha(t) = V_0 - (I_D/C)t \qquad (5.102)$$

as we saw in Section 5.3, where V_0 is the initial voltage of C. The problem is, $V_0 = 0$ at $t = 0$. There is no proper RC circuit transient to which to refer the LCR circuit transient so as to determine the local time. The source of this difficulty is that at $t = 0$ the RC circuit holds no energy. All the energy comes from the current of the inductor. The inductor delivers energy to the circuit and sets the initial condition later, at $t = t_0$. The interval $0 \le t \le t_0$ is the phase in which the inductor sets the initial condition of the circuit. At $t = t_0$ the RC circuit is charged, and the mode is defined as

$$V_\alpha(t) = V_{max} - (I_D/C)(t - t_0) \qquad (5.103)$$

This mode is shown by a dashed line in Figure 5.33(a). The mode local time t_α is now defined by solving $V(t) = V_\alpha(t_\alpha)$ as

$$t_\alpha = t_0 + \sqrt{LC}\,[(I_A - I_D)/I_D][1 - \sin(t/\sqrt{LC})] \qquad (5.104)$$

Figure 5.33(b) shows t_α versus t. Since the mode V_α is an object that has its own identity, it can be defined at t_0 instead of at $t = 0$. The local time is valid over the entire period $0 \le t \le 2t_0$. During the period $0 \le t \le t_0$ the local time t_α flows backward, and it has its minimum at the point where the initial condition is set. This is consistent with the fact that the circuit has not been set for further development of transients before the inductor is connected.

5.8 Inductance and Black Boxes

In circuit theory an independent functional block of circuits that has a small number of input and output terminals is singled out and is called a *black box*. The excitation-response relationship between the inputs and the outputs is specified, but the circuit implementation within the black box is conventionally not specified. If a black box is included in a circuit, the black box can be considered as the *outside* of the circuit, in that exchange of the signal and the energy takes place between the circuit and the black box. Thus a black box has some similarity to an inductance. In an inductance the electrical energy is converted to magnetic field energy, and is stored in the space outside. The magnetic energy cannot be accessed by any conventional circuit components except the inductance itself. Similarly, when energy goes into a black box, it becomes unavailable to the rest of the circuit.

This viewpoint is relevant to our concept of local time and global time. The forward and the backward local-time flow create an oscillating output, and this is the basic mechanism of a feedback oscillator. There are other kinds of oscillators, however. A cascaded inverting amplifier loop having an odd number of stages oscillates in a mode called a ring oscillation. In this circuit everything shown on the equivalent circuit diagram is device, resistance, or capacitance, and therefore energy cannot be placed outside of the circuit. The analogy between a black box and an inductance resolves this apparent contradiction.

Let us consider an example. The circuit shown within the dotted rectangle of Figure 5.34(a) is a black-box circuit that adds a *negative* capacitance to the capacitance C, thereby speeding up the response of the node voltage V. Because of this interesting characteristic, the properties of this circuit have been studied in detail [6]. If the amplifier in the black box has gain μ (noninverting), and if the response time is negligibly short, the effective capacitance the black box presents is given by

$$C_{\text{EFF}} = C_F(1 - \mu) \tag{5.105}$$

If $\mu > 1$, C_{EFF} becomes negative. The negative capacitance partially compensates the positive node capacitance C, and the node response speeds up. In a more detailed circuit model, the noninverting amplifier has signal delay time τ. The circuit equation of Figure 5.34(a) is

$$V_0 = V + R\left[C\frac{dV}{dt} - C_F\frac{d}{dt}[\mu V(t - \tau) - V] \right] \tag{5.106}$$

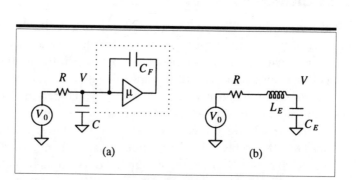

(a)   (b)

Figure 5.34   A Black Box and the Outside of a Circuit

If τ is small, a Taylor expansion of $V(t - \tau)$ up to the term proportional to τ can be used. We have

$$V_0 = C_F \mu \tau \frac{d^2V}{dt^2} + R[C + C_F(1 - \mu)] \frac{dV}{dt} + V \quad (5.107)$$

We assume that $V_0(t)$ is a step function: $V_0(t) = 0$ if $t < 0$, and $= 1$ if $t \geq 0$. The closed-form solution is

$$V(t) = 1 + A \exp(\alpha t) + B \exp(\beta t) \quad (5.108)$$

where

$$A = \beta/(\alpha - \beta) \qquad B = -\alpha/(\alpha - \beta) \quad (5.109)$$

and where the complex frequencies α and β are the solutions of the secular equation

$$C_F \mu \tau S^2 + R[C + C_F(1 - \mu)]S + 1 = 0 \quad (5.110)$$

If α and β are complex numbers, so are A and B. Then $V(t)$ is the real part of the solution. The circuit responses are shown in Figure 5.35. If gain μ of the amplifier is changed while maintaining all the other parameter values constant, the response changes from overdamping (squares) to critical

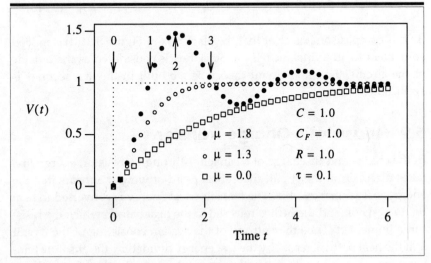

Figure 5.35 Response of a Negative-Capacitance Circuit

damping (open circles) and to underdamping (filled circles). The over-damping response at $\mu = 0$ is described by a simple time-constant process

$$V(t) = 1 - \exp[-1/R(C + C_F)] \tag{5.111}$$

This is the RC circuit model that determines the global time of the circuit. If the negative-capacitance circuit is activated, the response becomes underdamping, and the local time diverges at arrow 1. The equilibrium state of this circuit is $V(t) = 1$. A transient that approaches the equilibrium state is described by the forward flow of time, and a transient that departs from it by the backward flow of time. Between arrow 1 and arrow 2 the local time decreases, and it attains the minimum at arrow 2. From arrow 2 to arrow 3 the local time increases again. The same process repeats. Between arrow 0 and arrow 1 energy is sent from the circuit to the black box. Between arrow 1 and arrow 2 the energy returns, and a new initial condition, charging up C to voltages higher than 1, is set.

This negative capacitance circuit is structurally identical to the LCR circuit shown in Figure 5.34(b). The circuit equation is

$$V_0 = V + RC_E \frac{dV}{dt} + L_E C_E \frac{d^2V}{dt^2} \tag{5.112}$$

and the following correspondences exist:

$$C_E \leftrightarrow C + C_F(1 - \mu) \qquad L_E \leftrightarrow C_F \mu \tau / [C + C_F(1 - \mu)] \tag{5.113}$$

By this comparison it is clear that the black box of Figure 5.34(a) is indeed equivalent to an inductance. If the black box is considered as the outside of the circuit, the same circuit theory as we built in the last sections is applicable.

5.9 Quasistatic Charge Transfer

In the charge and discharge of a capacitor through a resistor, energy loss takes place. From the circuit-theoretical point of view, it appears that the energy loss is unavoidable. The lost electrical energy is converted to heat by the resistor, and the heat is released to the heat bath in which the resistor is immersed. Thus the entropy of the system consisting of the circuit and the heat bath increases by the lost energy divided by the absolute temperature of the heat bath. From the thermodynamical point of view, however, the entropy increase can be prevented by carrying out the process of

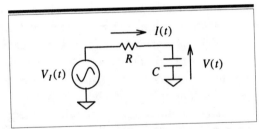

Figure 5.36 Quasistatic Charge Transfer Between Power Supply and Capacitor

charge transfer *quasistatically* [44]. In this section we observe how that can be achieved, by using a simple example shown in Figure 5.36.

In Figure 5.36 the AC power supply generates a voltage $V_I(t)$ given by

$$V_i(t) = V_I \sin (\omega t) \qquad (5.114)$$

The circuit equations satisfied by the node voltage $V(t)$ and current $I(t)$ are

$$\frac{dV(t)}{dt} + \frac{1}{RC} V(t) = \frac{V_I}{RC} \sin (\omega t) \quad \text{and} \quad I(t) = C \frac{dV(t)}{dt} \qquad (5.115)$$

The solution of this equation is given by

$$V(t) = V_I \frac{(\omega RC) \exp(-t/RC) + \sin (\omega t) - (\omega RC) \cos (\omega t)}{1 + (\omega RC)^2}$$

$$(5.116)$$

In the limit as $t \to \infty$ the exponential term decays and

$$V(t) \to V_I \frac{\sin (\omega t) - (\omega RC) \cos (\omega t)}{1 + (\omega RC)^2} \qquad (5.117)$$

The rate of energy loss in the resistor is given by

$$RI(t)^2 = \frac{RC^2 V_I^2 \omega^2 [\cos (\omega t) + (\omega RC) \sin (\omega t)]^2}{[1 + (\omega RC)^2]^2} \qquad (5.118)$$

The total energy loss per oscillation period $2\pi/\omega$ is given by

$$E = \int_{t_0}^{t_0 + (2\pi/\omega)} RI(t)^2 \, dt = \pi C V_I^2 \frac{\omega RC}{1 + (\omega RC)^2} \qquad (5.119)$$

Let us consider the physical meaning of these closed-form results. The resistor R is the model of a switching device, C is the load capacitance, and $V_i(t)$ is the power supply that generates a clocklike power supply voltage. According as $V_i(t) > V(t)$ or $V_i(t) < V(t)$, $I(t)$ flows to the capacitor (charge process) or to the power supply (discharge process). Every time the time-dependent power supply completes a cycle, we may say that one *logic operation* is carried out, and a certain amount of energy is lost by the circuit. The energy required by this *logic circuit* for each logic operation is given by Equation (5.119). From this closed-form result we note the following:

1. $E \to 0$ if $R \to \infty$.

2. $E \to 0$ if $C \to 0$.

3. $E \to 0$ if $R \to 0$.

4. $E \to 0$ if $\omega \to \infty$.

5. $E \to 0$ if $\omega \to 0$.

In cases 1 and 2 there is no connection from the power supply to the load (the power supply is standing alone). In case 3 the power supply is loaded purely by capacitance, so there is no loss. In case 4 the voltage $V(t)$ does not swing at all. That leaves case 5, where the logic circuit operates, and still the energy consumed by the circuit per unit operation is zero. This is the only nontrivial case.

In this case the logic circuit switches, but infinitesimally slowly. This leads to an interesting conclusion: that energy is required to *expedite* the logic operation, but not to execute it. If we note that $T = 2\pi/\omega$ is the time required to execute a logic operation, Equation (5.119) gives in the limit as $\omega \to 0$

$$T \cdot E = 2\pi^2 C V_i^2 \cdot (RC) \qquad (5.120)$$

This relationship may be regarded as the uncertainty principle between energy and time. In this limit

$$V(t) \to V_i \sin(\omega t) = V_i(t) \qquad (\omega \to 0) \qquad (5.121)$$

and this means that the voltage developed across the resistor R tends to zero, and the current $I(t)$ tends to zero as well. Charging and discharging of the capacitor C is carried out at an infinitesimally small rate, and this is the requirement for a quasistatic switching process in thermodynamics.

In the circuit of Figure 5.36, the logic operation is first carried out by drawing energy from the AC power supply, and then the inverse operation returns the expended energy to the power supply. In the Figure R is shown as a single resistor, but it can be a series-parallel combination of any number of resistors, each of which represent an electron triode that works as a switch. When the logic operation is finished, the output node voltage is sampled to obtain the result of the logic operation, and then $V_i(t)$ is reduced quasistatically to return the spent energy to the power supply.

C H A P T E R   S I X

Complexity of Interconnects

6.1 Introduction

A component of an integrated circuit has at most only four terminals. An integrated circuit is complex because many simple components are connected together by interconnects. The interconnects are featureless at low frequencies, but they present quite complex responses at medium and especially at high frequencies. At high frequencies the discrete circuit model of interconnects used in Chapters 4 and 5 is often inadequate. In high-frequency integrated circuit simulation a long interconnect is divided into many segments, and each segment is approximated by a lumped equivalent circuit. Although this method is convenient, it may not provide adequate understanding of the working mechanisms of the interconnects at high frequencies. A theory of distributed-parameter interconnects that is based on the basic equations and rigorous analytical methods is necessary, but only very limited work has been carried out. The mathematical difficulties are considerable, but modern computing power can be used to compute closed-form expressions that have well-understood physical meanings. Using a computer this way may appear conservative, but that is quite a superficial impression. Once a closed-form expression is derived, a lot of physical interpretation can be derived from it. In this chapter we develop a theory of isolated and interacting interconnects using the distributed circuit model, and using closed-form analyses. Certain basic features of the theories of the preceding sections are preserved; for instance, the circuit response is expressed as the sum of modes. In that respect, this chapter is a followup of Chapters 4 and 5.

6.2 Complexity of Interconnects

A transmission line that connects a microwave transmitter to a receiver is a uniform, isolated, and often quite low-loss line, whose properties can be understood by a simple analysis based on electromagnetic theory [45–47]. The signal source is characterized by an internal impedance and an ideal voltage source, and the destination by an equivalent impedance. The destination impedance is often very close to the matched termination resistance, to maximize the signal power. If the line is branched on the way, a matching circuit consisting of resistors is used to prevent reflection. The signal is sinusoidal with frequency close to the central frequency of the band. The analysis is in the frequency domain. The interconnects on an integrated circuit chip satisfy none of these special conditions of simplification.

In MOS IC interconnects, a shunt resistance reduces the signal amplitude. Therefore, in any *large-amplitude* logic circuit, the load on the transmission lines must be capacitive. If a capacitive load is connected to the end of a transmission line, the signal is reflected, and the entire energy of the wave is sent back to the driver. In a lossless transmission line the energy of the reflected signal comes back to the driver unattenuated, and then it may be reflected again. If the driver internal impedance does not absorb the reflected wave, the transient lasts a long time, and that complicates the transmission mechanism.

Branching of interconnects occurs frequently. Where branching occurs depends on the circuit layout requirements, and it is impossible to standardize the interconnect tree structure. At the branch point no matching circuit is inserted, so a fraction of the wave energy is reflected. Reflected waves from many branch points, from the line ends, and from the drivers interfere to create a complex, messy response. The signals on a digital signal interconnect must be analyzed in the time domain, so that the analysis results give easily measurable voltage and currents, rather than hard-to-measure frequency response and phases.

If there are more than one transmission line in close proximity, they interact with each other. This is almost inevitable in densely packed integrated circuits. Two lines interact by capacitive coupling through the nonconducting space above the lines, by electromagnetic coupling through the space between the line and the substrate, and by sharing the ground return current path through the substrate. If the lines are modeled as *RC* transmission lines and if the substrate is perfectly conductive, the third mechanism is absent.

Physically meaningful voltage is developed only across a capacitance.

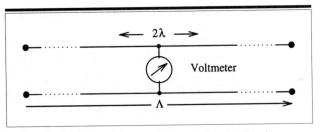

Figure 6.1 Effective Capacitance seen by a Voltmeter

We discussed this in Chapters 1 and 4, but it is reiterated here, since the individual capacitances of a distributed parameter circuit are infinitesimally small, and it might for some purposes be taken as zero. In a distributed parameter circuit, the capacitance of the node to which a voltmeter is connected depends on its response time (Figure 6.1). If the voltmeter requires time Δt to provide a reading, the effective capacitance of a distributed RC line is the parallel capacitance within distance λ, which is determined from

$$(R\lambda)(C\lambda) = \Delta t \quad \text{or} \quad \lambda = \sqrt{\Delta t/RC}$$

where R and C are the series resistance and parallel capacitance of the line per unit line length, respectively. The capacitance value is then $C_{EFF} = 2\,C\lambda = 2\sqrt{C\,\Delta t/R}$. If the total length of the line, Λ, is less than λ, the distributed circuit works effectively as a single node with capacitance $C\Lambda$. A similar relationship exists in an LC line as well. In this case the voltmeter affects the section having length λ determined from

$$(L\lambda)(C\lambda) = (\Delta t)^2 \quad \text{or} \quad \lambda = \Delta t/\sqrt{LC}$$

where L and C are the series inductance and parallel capacitance of the line per unit length, respectively. The requirement of voltage measurability leads to a conclusion that a signal must be carried by differential voltage developed across a pair of wires. This is true in an RC transmission line as well.

In this chapter we develop RC, LCR, and LC transmission-line theory at high frequencies. The theory of very complex RC transmission-line configurations has not yet been developed. Traditionally this problem has been left to numerical simulation. It may be thought that an idealized LC transmission-line theory is enough. In real IC interconnects, however, the series resistance creates the major part of the transmission delay.

6.3 Initial Conditions of *RC* Transmission Lines

The first subject of this chapter is signal transmission through RC lines. Signal transmission is analyzed assuming the simplest possible excitation, that a uniformly charged line is discharged by suddenly grounding the left end. Immediately following the grounding, a voltage profile is established in the line, and that becomes the initial condition. Determination of the initial voltage profile requires physical insight, since it is often quite confusing.

Suppose that the left end of line 0 of the double RC transmission line of Figure 6.2(a) was grounded a long time ago, line 1 is uniformly charged to V_{DD}, and the left end is grounded at time $t = 0$ by suddenly closing the switch. What is the voltage profile established in the line immediately after $t = 0$? Note that the voltages are referred to the ground at the left end of the line. The parameter values R_0, R_1, and C_0 are per unit length of the uniform distributed RC line. Within a short time Δt, only the capacitance within the portion of the line having length λ from the left end given by

$$(R_0 + R_1)\lambda \cdot C_0\lambda = \Delta t \quad \text{or} \quad \lambda = \sqrt{\Delta t/[C_0(R_0 + R_1)]} \quad (6.1)$$

is discharged by grounding the left end of line 1, thereby short-circuiting the line capacitances. We note, however, that only the voltage *difference* of the line, $V_1(x,t) - V_0(x,t)$, was reduced by the discharge. For $V_1(x,t)$ and $V_0(x,t)$ themselves, we need a different consideration. Since they may be the variables used to write the circuit equation, their initial conditions must be clearly understood. In the equivalent circuit of Figure 6.2(a) drawn by the solid lines, the part of the RC line to the right of the first capacitance is still charged to V_{DD} after the short time Δt. The charged

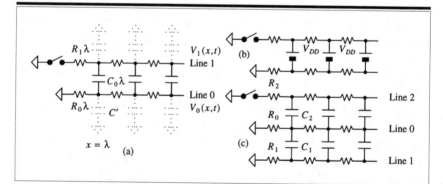

Figure 6.2 Mechanisms of Establishing the Initial Condition

capacitors may be regarded as batteries, as shown in Figure 6.2(b). Replacement of the capacitance by an equivalent battery lessens the confusion of this problem. Since the left ends of both lines are grounded, $R_0\lambda$ and $R_1\lambda$ must sustain voltages such that

$$V_1(\lambda,\Delta t) - V_0(\lambda,\Delta t) = V_{DD}$$

and   (6.2)

$$V_1(\lambda,\Delta t)/(R_1\lambda) = -V_0(\lambda,\Delta t)/(R_0\lambda)$$

where the second equation is the current conservation law: Net current should not flow from the ground at the left end to the line on the right. Therefore

$$V_1(\lambda,\Delta t) = \frac{R_1}{R_0 + R_1}V_{DD}$$

and   (6.3)

$$V_0(\lambda,\Delta t) = -\frac{R_0}{R_0 + R_1}V_{DD}$$

and since λ and Δt can be reduced to zero, the initial conditions hold for any x, as

$$V_1(x,+0) = R_1V_{DD}/(R_0 + R_1)$$ (6.4)

$$V_0(x,+0) = -R_0V_{DD}/(R_0 + R_1) \qquad (x > 0)$$

Before the switch turns on, $V_0(x,-0) = 0$ and $V_1(x,-0) = V_{DD}$. It might seem strange that V_1 and V_0 for $x > 0$ change instantly after the switch closure at the left end. This is an inevitable consequence of the equivalent circuit of Figure 6.2(a), since the lines stretching to the right have no capacitance to the ground. The change of V_1 and V_0 immediately after turn-on of the switch should not be interpreted as a signal arrival. The only meaningful signal is carried by the voltage difference $V_1 - V_0$. Since the internal nodes of the lines have no capacitance to the ground, the node voltage relative to the ground is unmeasurable. If the extra capacitances shown by the dotted lines are added to the equivalent circuit, V_1 and V_0 take time to change after the switch closure. If we identify which signal carries information and which does not, we are able to use a simplified equivalent circuit such as that of Figure 6.2(a) without the insignificant capacitances shown by the dotted lines.

Let us consider a more complex case. In Figure 6.2(c) line 2 is charged to V_{DD}, and lines 0 and 1 were grounded some time ago at the left end. This is the simplest case of a pair of interacting lines. If the left end of line 2 is grounded at $t = 0$, C_2 may be replaced by a battery whose voltage is V_{DD}, and C_1 by a battery having zero voltage. The resistances R_0 and R_1 are effectively in parallel and have the effective value $R_0R_1/(R_0 + R_1)$ immediately after $t = 0$. Then we have

$$\frac{V_2(x0,0)}{V_1(x,\ x0)} = -\frac{R_2}{R_0R_1/(R_0 + R_1)} \tag{6.5}$$

or

$$V_2(x0,0) = \frac{R_2 V_{DD}}{R_2 + [R_0R_1/(R_0 + R_1)]}$$

and $\tag{6.6}$

$$V_1(x,\ x0) = -\frac{[R_0R_1/(R_0 + R_1)]V_{DD}}{R_2 + [R_0R_1/(R_0 + R_1)]}$$

This result will be used in the next section.

6.4 Common-Ground *RC* Transmission Lines

If the semiconductor substrate has high conductivity, a uniform and isolated interconnect fabricated on it is represented by the equivalent circuit of Figure 6.3 at low speeds where inductance effects are negligible. In this chapter we assume that all the equivalent circuits are *distributed* equivalent circuits, representing uniform and continuous interconnects. The mathematical methods are treated, for example, in reference [48].

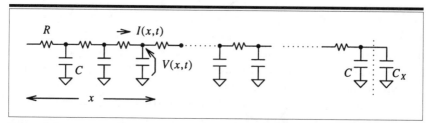

Figure 6.3 Common-Ground, Uniform *RC* Transmission Line

The voltage and the current profiles of the RC line, $V(x,t)$ and $I(x,t)$, respectively, defined in Figure 6.3 satisfy the circuit equations

$$C \frac{\partial V(x,t)}{\partial t} = - \frac{\partial I(x,t)}{\partial x} \quad \text{and} \quad \frac{\partial V(x,t)}{\partial x} = - RI(x,t) \quad (6.7)$$

where R and C are the series resistance and the parallel capacitance per unit length of the line. By eliminating $I(x,t)$ we obtain

$$CR \frac{\partial V(x,t)}{\partial t} = \frac{\partial^2 V(x,t)}{\partial x^2} \quad \text{and} \quad I(x,t) = - \frac{1}{R} \frac{\partial V(x,t)}{\partial x} \quad (6.8)$$

To find a special solution of this equation, we assume an exponential time dependence. We have

$$\frac{\partial^2 V(x)}{\partial x^2} = - CR|S|V(x)$$

where $\qquad\qquad\qquad\qquad\qquad\qquad\qquad\qquad\qquad\qquad\qquad (6.9)$

$$V(x,t) = V(x) \exp(-|S|t)$$

where $|S|$ is the reciprocal of the time constant of a mode. By solving this equation, we obtain

$$V(x) = A \sin (Kx) + B \cos (Kx)$$

where

$$K = \sqrt{CR|S|} \qquad\qquad\qquad\qquad (6.10)$$

or

$$S = -(K^2/CR)$$

The general solution is given by a superposition of all the special solutions as

$$V(x,t) = \int_0^\infty [A(K) \sin (Kx)$$
$$+ B(K) \cos (Kx)] \exp[-(K^2/CR)t] \, dK \quad (6.11)$$

The coefficients $A(K)$ and $B(K)$ are determined from the initial and the boundary conditions. The integral over K may become a sum over a discrete index in some cases, but in general it is convenient to write it in an integral form.

Suppose that the RC line having length Λ is uniformly charged to V_{DD}, the right-side end is open-circuited, and the left-side end is grounded at $t = 0$. The current at the right-side end is zero, or $I(\Lambda,t) = 0$. The solution is given by a sum of terms, each of which satisfies the boundary conditions, of the form

$$\exp(-K_n^2 t/CR)\,\sin\,(K_n x)$$

where (6.12)

$$K_n = (2n + 1)(\pi/2\Lambda) \qquad n = 0, 1, \ldots$$

From the initial condition,

$$V(x,0) = V_{DD} \quad (0 < x \le \Lambda)$$

where (6.13)

$$V(x,0) = \sum_{n=0}^{\infty} A_n \sin\,(K_n x)$$

By expanding $V(x,0)$ into a Fourier sine series [49] and by comparing the expansion coefficients we have

$$V(x,t) = \frac{4V_{DD}}{\pi} \sum_{n=0}^{\infty} \exp\left[-\frac{1}{CR}\left[\frac{(2n+1)\pi}{2\Lambda}\right]^2 t\right]$$
$$\frac{1}{2n+1} \sin\left[\frac{(2n+1)\pi}{2\Lambda}x\right]$$

(6.14)

$V(x,t)$ is plotted for several values of t in Figure 6.4, assuming $RC = 1.0$. The discharge process consists of two phases. During the initial phase the wavefront propagates in the RC line, until it reaches the right end. Until the wavefront reaches the end, the voltage at the right end remains unchanged, at V_{DD}. The wavefront during this period progresses proportionally to $\sqrt{t}$. As the wavefront reaches the right end, the voltage there, $V(\Lambda,t)$, begins to decrease [1].

If the right end of the RC line is loaded by a lumped capacitance C_X as shown in Figure 6.3, we still have

$$V(x,t) = \int_0^{\infty} A(K)\,\exp[-(K^2/CR)t]\sin\,(Kx)\,dK \qquad (6.15)$$

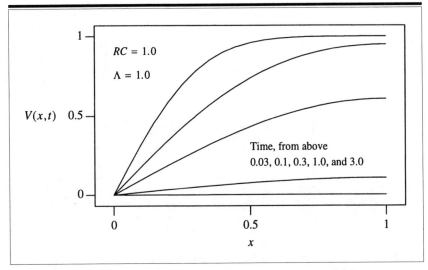

Figure 6.4 Time Dependence of the Potential Profile in a Uniform *RC* Line

and from this solution $I(x,t)$ is derived as

$$I(x,t) = -(1/R)[\partial V(x,t)/\partial x] \tag{6.16}$$

$$= -(1/R) \int_0^\infty KA(K) \exp[-(K^2/CR)t] \cos (Kx) \, dK$$

The boundary condition at the end of the line is

$$I(\Lambda,t) = C_X[\partial V(\Lambda,t)/\partial t] \tag{6.17}$$

By substitution we have

$$\int_0^\infty KA(K) \exp[-(K^2/CR)t] \, [\cos (K\Lambda)$$

$$- (C_X/C)K \sin (K\Lambda)] \, dK = 0 \tag{6.18}$$

Then K takes discrete values that are the solutions of a transcendental equation

$$\tan (\xi) = \gamma/\xi \quad \text{where} \quad \xi = K\Lambda \text{ and } \gamma = C\Lambda/C_X \tag{6.19}$$

If the nth root of the equation is ξ_n, then $K_n = \xi_n/\Lambda$. The root ξ_n is given by the intersection of the curves of the tangent function and a hyperbola,

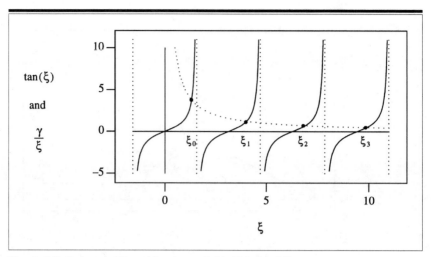

Figure 6.5 Determination of the Acceptable Values of K

as shown in Figure 6.5. The numerical values of the first several roots are plotted versus γ in Figure 6.6.

Using the values of the K_n's, the voltage profile in the RC line is written as

$$V(x,t) = \sum_{n=0}^{\infty} A_n \exp[-(K_n^2/CR)t] \sin(K_n x) \qquad (6.20)$$

The expansion coefficients A_n are determined from the initial condition

$$V(x,0) = V_{DD} \quad (0 < x \leq \Lambda)$$

where $\qquad\qquad\qquad\qquad\qquad\qquad\qquad\qquad\qquad\qquad\qquad$ (6.21)

$$V(x,0) = \sum_{n=0}^{\infty} A_n \sin(K_n x)$$

Determination of A_n is not a simple matter, however, since the sinusoids in the sum are not in a simple, harmonic relationship. Equation (6.21) is *not* a conventional Fourier series, and therefore the expansion coefficients A_n cannot be determined by using the orthogonality of the sine functions. In the following sections several problems of this type are discussed. We need to devise a new method. The following numerical methods are useful.

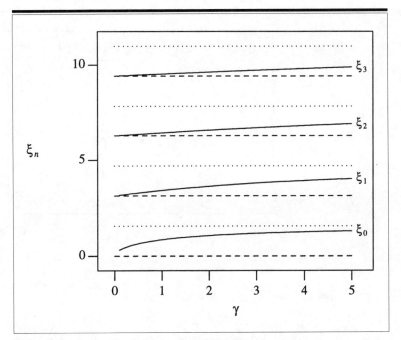

Figure 6.6 Roots of the Transcendental Equation Versus γ

From the general form of the Fourier expansion coefficient in Equation (6.14), we guess the functional dependence of A_n on n as

$$A_n = a/(1 + bn) \tag{6.22}$$

and compute the sum of Equation (6.21). The parameters a and b are adjusted so that the absolute error determined by

$$\text{error} = \sum_{n=1}^{20} |V_{DD} - V(0.05Ln)| \tag{6.23}$$

is minimized (we divided the line into 20 sections, but a different number may be chosen). In the following example the capacitance loading at the right end of the line, C_X, was the same as the capacitance of the entire line: $C_X = C\Lambda$. For $\gamma = 1$ the parameter values $a = 1.52$ and $b = 2.00$ were found to be the optimum.

The initial potential profile in this case is shown by the filled circles in Figure 6.7. The approximation can be improved by the following consideration. The *dip* of the first approximation of the initial potential profile at

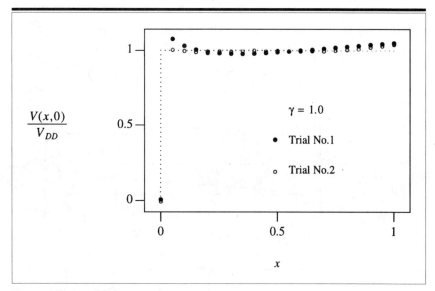

Figure 6.7 Coefficient Determination of Pseudo Fourier Series

about the middle of the line is caused by the overestimate of the Fourier coefficients in the middle spatial-frequency range. By suppressing the amplitude of the 3rd to 25th harmonics the dip can be reduced significantly. By reducing A_2 to A_{25} by a factor of 0.91 and reoptimizing the parameters to $a = 1.525$ and $b = 0.94$, the approximation can be improved, as shown by the open circles of Figure 6.7. Using the expansion coefficients, the evolution of the potential profile is determined as shown in Figure 6.8. During the first phase the line discharges from the left end. During the second phase the potential profile within the chain is a uniform ramp. This is because the line is effectively acting as a resistor that has small internal capacitive parasitics compared with the capacitive load C_X at the line's end. The voltage waveforms at the right-side end of the chain of the loaded and the unloaded chains are plotted in Figure 6.9. If the digital signal transmission delay is determined at $V(\Lambda,t)/V(\Lambda,0) = 0.5$, the loaded RC chain delay is about 3 times the unloaded RC line delay, if $C\Lambda = C_X$. This is because only half of the RC line capacitance $(C\Lambda)$ is effectively loading the end of the line.

The method of expansion coefficient determination is fast and simple, but is dependent on intuition. A systematic method is as follows. After setting up the first approximation of the Fourier coefficients as we did before, the initial voltage profile is computed. We observe that $V(\Lambda,0) < V(0,0)$ in Figure 6.7. This is corrected by increasing the Fourier coefficient

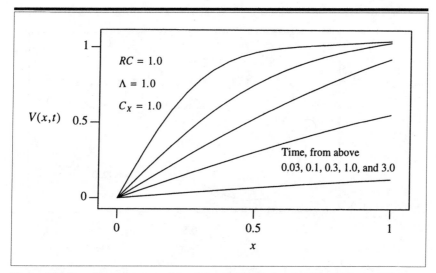

Figure 6.8 Time Dependence of the Potential Profile in a Loaded Uniform *RC* Line

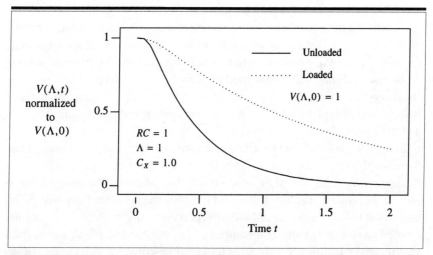

Figure 6.9 Discharge of the Loaded and the Unloaded *RC* Line at the Ends

at the fundamental spatial frequency A_0, since the fundamental mode contains only one quarter of a sinusoid, and it may be regarded as a linear ramp. Then we have the profile shown by the solid curve of Figure 6.10. Since $V(\Lambda/2,0) < V(0,0)$ [$\approx V(\Lambda,0)$], the Fourier coefficient A_1 is increased to match the voltage profile at the center of the line. This mode contains half of a sinusoid, and if A_1 is adjusted, the *droop* of the solid curve at the center can be removed. This procedure is repeated. While carrying out the

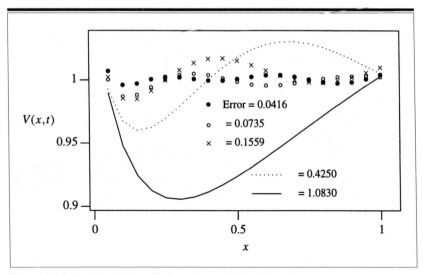

Figure 6.10 Systematic Error Reduction on Fourier Expansion

successive approximations, the parameters a and b are adjusted to minimize the error. After adjusting the first five Fourier coefficients, we obtain the initial potential profile shown by the filled circles of Figure 6.10, whose accuracy is about 0.2%. This method can be computerized. As we saw from the two examples, determination of the pseudo Fourier expansion coefficients of Equation 6.21 requires a computer, but mathematically it is a simple procedure.

Since the node voltage profile was expressed as the sum of *modes* of the form $\exp[-(K_n^2/CR)t]\sin(K_n x)$, the mode coefficients A_n versus the parameter $S_n = K_n^2/CR$ of Equation (6.20) may be called the spectral intensity of the voltage profile. Figure 6.11 shows the spectral intensity of the unloaded (filled circles) and loaded (open circles, $\gamma = C\Lambda/C_X = 1.0$) RC lines. The dominant spectral component of the loaded RC line is shifted to the lower-frequency side (the longer-time-constant side). The difference is that the loaded-line discharge is slower than the unloaded-line discharge because of the lower frequency shift of the spectral components. Spectral decomposition of the voltage profile indicates how the discharge is slowed down.

An interconnect may have branches. We give an example of the analysis of a branched RC line, whose equivalent circuit is shown in Figure 6.12. A uniform RC line having length Λ_A has a branch at the end, to the two RC lines having lengths Λ_B and Λ_C. We assume that the ends of the branched

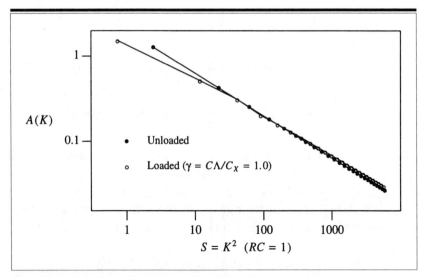

Figure 6.11 Spectra of the Modes of Loaded and Unloaded *RC* Lines

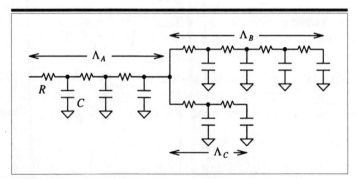

Figure 6.12 Branched *RC* Line

lines B and C are not loaded by any extra capacitance. Evolution of the potential profile of the branched RC line can be expressed as the sum of the modes, as before. The modes have a form

$$\exp[-(K^2/CR)t] \times [A(K) \sin(Kx) + B(K) \cos(Kx)] \quad (6.24)$$

The RC line was uniformly charged to V_{DD} before discharge, and the left end is grounded at $t = 0$. The voltage profiles in the three segments A, B, and C satisfying their respective boundary conditions at the ends are

$$V_A(x_A,t) = \int_0^\infty A(K) \exp[-(K^2/CR)t] \sin (Kx_A) \, dK$$

$$V_B(x_B,t) = \int_0^\infty B(K) \exp[-(K^2/CR)t] \cos[K(\Lambda_B - x_B)] \, dK \quad (6.25)$$

$$V_C(x_C,t) = \int_0^\infty C(K) \exp[-(K^2/CR)t] \cos[K(\Lambda_C - x_C)] \, dK$$

where x_A, x_B, and x_C are the distances measured from the left end of the line A, and from the branch point of the lines B and C, along those respective lines. These formulas reflect the fact that $V_A(0,t) = 0$ and that V_B and V_C have voltage maxima at the right ends. Since the voltages of the three lines must match at the branch point,

$$B(K) = [\sin (K\Lambda_A)/\cos (K\Lambda_B)]A(K) \qquad (6.26)$$

$$C(K) = [\sin (K\Lambda_A)/\cos (K\Lambda_C)]A(K)$$

The sum of the currents of the lines B and C must match the current of the line A at the branch point. Since $I(x,t) = -(1/R) [\partial V(x,t)/\partial x]$, we have

$$A(K) \cos (K\Lambda_A) = B(K) \sin (K\Lambda_B) + C(K) \sin (K\Lambda_C) \quad (6.27)$$

The three relationships determine the acceptable values of K, which are found from

$$F(K) = \tan (K\Lambda_A)[\tan (K\Lambda_B) + \tan (K\Lambda_C)] - 1 = 0 \quad (6.28)$$

The function $F(K)$ is plotted in Figure 6.13, and the locations of the roots are indicated by the filled circles. In the following example we assume that $\Lambda_A = 1$, $\Lambda_B = 2$, and $\Lambda_C = 1$.

The acceptable values of K are

$$K_{4n} = K(n) - \Delta_1$$

$$K_{4n+1} = K(n) - \Delta_2$$

$$K_{4n+2} = K(n) + \Delta_2 \qquad (6.29)$$

$$K_{4n+3} = K(n) + \Delta_1$$

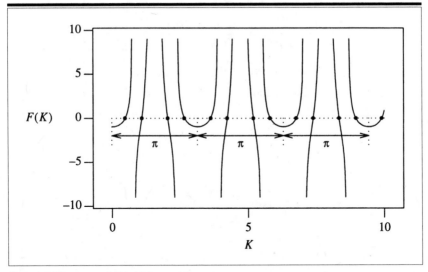

Figure 6.13 Acceptable Values of K

where $K(n) = (\pi/2)(2n + 1)$, $\Delta_1 = 1.093138$, and $\Delta_2 = 0.477658$. The mode voltage profiles for values of K_m from $m = 0$ to $m = 7$ are plotted in Figure 6.14. For $m = 0$ to $m = 3$ the filled circles show the voltage profiles in line A, the open circles in line B, and the open squares in line C. For $m = 4$ to $m = 7$ the solid curves show the voltage profiles in lines A and B, and the dotted curves in line C.

Determination of the expansion coefficients A_m is carried out as follows. The Fourier transform of the function shown in Figure 6.15 is given by

$$G(x) = \frac{1}{\sqrt{2\pi}} \int_{-\infty}^{\infty} \exp(jKx)\, \Gamma(K)\, dK$$

$$\Gamma(K) = \frac{1}{\sqrt{2\pi}} \int_{-\infty}^{\infty} \exp(-jKx)\, G(x)\, dx \qquad (6.30)$$

$$= -\sqrt{\frac{2}{\pi}} \frac{j[1 - \cos(K\Lambda)]}{K}$$

where Λ is an adjustable parameter. If the integral over K is replaced by a sum, we have

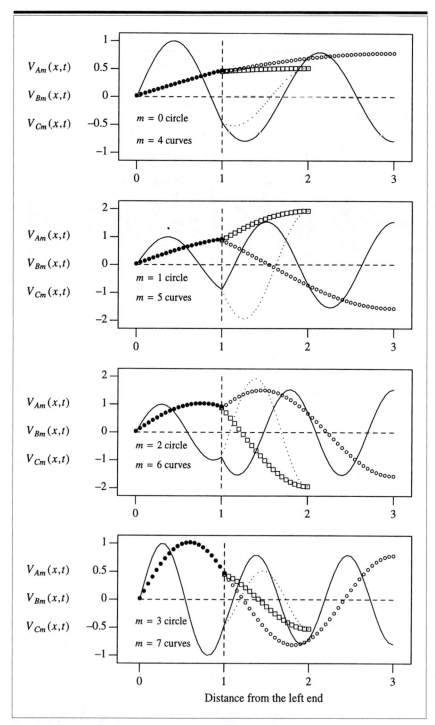

Figure 6.14 First Few Modes of the Branched *RC* Chain

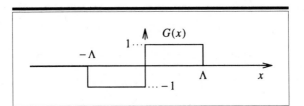

Figure 6.15 Determination of Fourier Expansion Coefficients

$$G(x) = \sum_{m=0}^{\infty} \frac{1 - \cos (K_m \Lambda)}{K_m} \frac{\Delta K_m}{\pi} \sin (K_m x) \qquad (6.31)$$

where ΔK_m is the range of K that is assigned to a single discrete value K_m. Since there are four acceptable values of K in the range having width π as observed from Figure 6.13, we have $\Delta K_m = \pi/4$.

This formula is for only one value of Λ and is therefore not general enough. We generalize it by including more than one value of Λ with weight factor $f(\Lambda)$. We write

$$V_A(x_A, 0) = \sum_{m=0}^{\infty} A_m \sin (K_m x)$$

where $\qquad\qquad\qquad\qquad\qquad\qquad\qquad\qquad\qquad\qquad\qquad\qquad$ (6.32)

$$A_m = \frac{V_{DD}}{4} \sum_{\Lambda} f(\Lambda) \frac{1 - \cos (K_m \Lambda)}{K_m}$$

The sum over Λ should be chosen so that the profiles $V_A(x_A,0)$, $V_B(x_B,0)$, and $V_C(x_C,0)$ are independent of the coordinates x_A, x_B, and x_C. If the sum over Λ is too complicated, the mathematical technique is not useful; but often the sum is quite simple. Success of the procedure depends much on the quality of the guess, which is best obtained by computing and plotting the initial voltage profile for several simple cases. The initial voltage profiles for $\Lambda = 2$ and $\Lambda = 4$ are shown in Figure 6.16. If the two profiles are superposed with weight factors $f(2) = -2$ and $f(4) = 3$, the resultant initial voltage profile is independent of the coordinate except at $x_A = 0$, as shown in Figure 6.17.

Using the expansion coefficient A_m, the time-dependent voltage profile of a branched RC line is given by

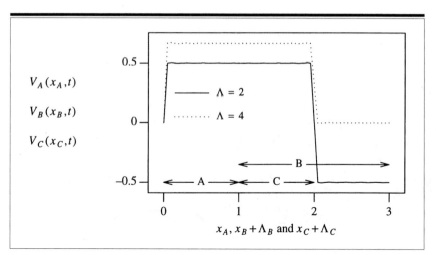

Figure 6.16 Determination of Fourier Expansion Coefficients

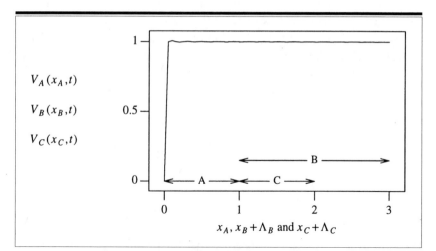

Figure 6.17 Determination of Fourier Expansion Coefficients

$$V_A(x_A,t) = \sum_0^\infty A_m \exp[-(K_m^2/CR)t] \sin(K_m x_A)$$

$$V_B(x_B,t) = \sum_0^\infty B_m \exp[-(K_m^2/CR)t] \cos[K_m(\Lambda_B - x_B)] \qquad (6.33)$$

$$V_C(x_C,t) = \sum_0^\infty C_m \exp[-(K_m^2/CR)t] \cos[K_m(\Lambda_C - x_C)]$$

The time-dependent voltage profiles are shown in Figure 6.18.

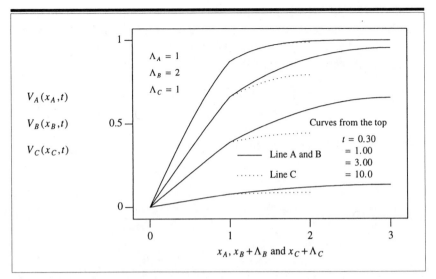

$V_A(x_A,t)$

$V_B(x_B,t)$

$V_C(x_C,t)$

$\Lambda_A = 1$
$\Lambda_B = 2$
$\Lambda_C = 1$

Curves from the top

$t = 0.30$
$= 1.00$
$= 3.00$
$= 10.0$

—— Line A and B

······· Line C

$x_A, x_B + \Lambda_B$ and $x_C + \Lambda_C$

Figure 6.18 Time Dependence of Node Voltage Profiles

6.5 *RC* Transmission Line with Ground Resistance

If an RC transmission line is fabricated on a resistive substrate, the equivalent circuit of the line is as shown in Figure 6.19. R_0 is the resistance of the substrate. The transmission line is infinitely long to the right. The line capacitances are charged to $V_{DD} = 1$, and they are discharged by closing the switch at the left end.

We seek a closed-form solution. If the line voltage profiles are given by $V_0(x,t)$ and $U_0(x,t)$ for the upper node (the line itself) and the lower node (the substrate ground line), respectively, they satisfy

$$C \frac{\partial}{\partial t}[V_0(x,t) - U_0(x,t)] = \frac{1}{R_2} \frac{\partial^2 V_0(x,t)}{\partial x^2}$$

$$C \frac{\partial}{\partial t}[V_0(x,t) - U_0(x,t)] + \frac{1}{R_0} \frac{\partial^2 U_0(x,t)}{\partial x^2} = 0$$

(6.34)

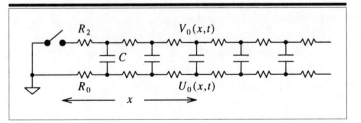

R_2

$V_0(x,t)$

C

R_0

$U_0(x,t)$

x

Figure 6.19 *RC* Transmission Line with Ground Resistance

In the equivalent circuit of Figure 6.19, the currents flow from right to left only through the resistors R_0 and R_2. Then the total current that crosses any section is zero. This is a restriction set by the equivalent circuit structure, but it is often confusing. We have

$$\frac{1}{R_2} \frac{\partial V_0(x,t)}{\partial x} = - \frac{1}{R_0} \frac{\partial U_0(x,t)}{\partial x} \qquad (6.35)$$

Using this relationship, the equations are converted to

$$C(R_0 + R_2) \frac{\partial \phi(x,t)}{\partial t} = \frac{\partial^2 \phi(x,t)}{\partial x^2}$$

and   (6.36)

$$U_0(x,t) = - \frac{R_0}{R_0 + R_2} \phi(x,t)$$

where $\phi(x,t) = V_0(x,t) - U_0(x,t)$ is the voltage developed across the capacitance of the RC line, which carries the information.

The equation of the RC line has solutions of the form $\exp[-(K^2/CR)t]$ $\sin(K_m x)$ that we used before. The solutions, or *modes*, are spatially extended *waves*. The other solution is a spatially localized *particle* of the form

$$\phi(x,t,\xi,\tau) = \frac{A}{\sqrt{t - \tau}} \exp\left[- \frac{C(R_0 + R_2)(x - \xi)^2}{4(t - \tau)} \right] \qquad (6.37)$$

$\phi(x,t,\xi,\tau)$ is the voltage profile developed on the RC line if a pair of charges is generated in the capacitance at location ξ at time τ. A superposition of the localized modes is a solution of the circuit equation as well.

If an infinitely long line is charged to the uniform initial voltage V_{DD} and the line's left end is grounded at $t = 0$, the voltage profile is given by

$$\phi(x,t) = \frac{A}{\sqrt{t}} \int_0^\infty \exp\left[- \frac{C(R_0 + R_2)(x - \xi)^2}{4t} \right] d\xi$$

$$- \frac{A}{\sqrt{t}} \int_{-\infty}^0 \exp\left[- \frac{C(R_0 + R_2)(x - \xi)^2}{4t} \right] d\xi$$

$$= \frac{4A}{\sqrt{C(R_0 + R_2)}} f_E[\sqrt{C(R_0 + R_2)/4t}\, x] \qquad (6.38)$$

where

$$f_E(X) = \int_0^X \exp(-\theta^2) \, d\theta$$

here the integral from $-\infty$ to 0 is the sum over the *image* sources required to maintain $\phi(0,t) = 0$ all the time. Since $f_E(x) \rightarrow \sqrt{\pi}/2$ for $x \rightarrow \infty$, we have $A = \sqrt{C(R_0 + R_2)}\, V_{DD}/2\sqrt{\pi}$. By substituting A back into the expression for $\phi(x,t,\xi,\tau)$ and integrating over x, we find that charge CV_{DD} is generated per unit length at time $\tau = 0$ (at the beginning), and that is reasonable. We then have

$$\phi(x,t) = \frac{2}{\sqrt{\pi}} V_{DD} f_E[\sqrt{C(R_0 + R_2)/4t}\, x]$$

(6.39)

$$U_0(x,t) = -\frac{R_0}{R_0 + R_2}\phi(x,t) \qquad V_0(x,t) = \frac{R_2}{R_0 + R_2}\phi(x,t)$$

Figure 6.20 shows $V_0(x,t)$, $U_0(x,t)$, and $\phi(x,t)$ for $t = 0.01, 0.03, 0.1$, and 0.3, assuming that $R_0 = 0.2$, $R_2 = 1.0$, and $C = 1.0$.

We see that the voltage of the line a long distance away from the left end changes immediately after the discharge starts. This infinite-speed "signal" propagation occurs because no capacitance is assigned directly between the left end and the nodes of the line. The signal on the line is represented by ϕ, the voltage difference between the two lines. The voltage difference is developed across the line capacitance C and therefore is associated with the electrostatic energy of the capacitance, which makes the voltage ϕ measurable. $U_0(x,t)$ and $V_0(x,t)$, referred to the ground at the left end of the line, are not developed across a capacitor, and no energy

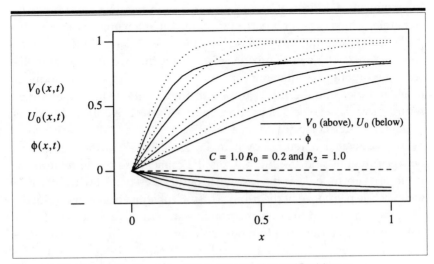

Figure 6.20 Voltage Profile of the *RC* Line Having Ground Resistance

can be drawn to measure them. This is an example of the principle that an electrical signal that is not associated with stored electrostatic energy is unmeasurable, and it cannot be considered as an information-carrying signal.

The analysis of this section is reduced to the special case of a perfectly conducting ground if R_0 is set equal to zero. Then the displacement current of the line capacitance per unit length, $J_0(x,t)$, is given by

$$J_0(x,t) = C \frac{\partial}{\partial t} \phi(x,t)$$

where   (6.40)

$$\phi(x,t) = \frac{2}{\sqrt{\pi}} V_{DD} f_E[\sqrt{CR_2/4t}\, x]$$

and after differentiation with respect to time,

$$J_0(x,t) = -\frac{CV_{DD}}{2\sqrt{\pi}} \sqrt{CR_2/t}(x/t) \exp(-CR_2 x^2/4t) \quad (6.41)$$

This is the current injected into the ground when the RC line discharges.

6.6 Coupled *RC* Transmission Lines

Interconnects of silicon ICs are fabricated on a resistive semiconductor substrate, which is never a perfect ground. The interconnects are densely packed, and their mutual coupling is strong. The signals on different interconnects mix. The signal mixing arises from the two mechanisms: the coupling among the interconnects above the monolithic semiconductor level, and the coupling through the substrate by the voltage the ground return currents develop. Let us consider how the resistive substrate works.

An interconnect fabricated on a resistive p-tub of CMOS is represented by the equivalent circuit shown in Figure 6.21. C_1 is the capacitance between the line and the p-tub, C_0 is the capacitance of the pn junction p-tub-$n+$ substrate, R_1 is the resistance of the interconnect wire, and R_0 is the effective resistance of the p-tub. The $n+$ substrate has high conductivity. This two-layered RC transmission line structure is a generalization of the simple RC ladder used to model an interconnect. The voltage profiles of the two-layered RC line, $V_0(x,t)$ and $V_1(x,t)$, satisfy the circuit equations

$$(C_0 + C_1) \frac{\partial V_0}{\partial t} - C_1 \frac{\partial V_1}{\partial t} = \frac{1}{R_0} \frac{\partial^2 V_0}{\partial x^2}$$

$$-C_1 \frac{\partial V_0}{\partial t} + C_1 \frac{\partial V_1}{\partial t} = \frac{1}{R_1} \frac{\partial^2 V_1}{\partial x^2}$$

(6.42)

We assume an exponential time dependence $\exp(St)$. Then the time derivatives $\partial/\partial t$ can be replaced by S. By eliminating V_0 among the equations we have

$$\frac{\partial^4 V_1}{\partial x^4} - [R_0(C_0 + C_1) + R_1 C_1]S \frac{\partial^2 V_1}{\partial x^2} + (R_0 C_0)(R_1 C_1)S^2 V_1 = 0 \quad (6.43)$$

We seek a special solution of the form

$$V_1 = \exp(St) \sin (Kx) \tag{6.44}$$

The parameter K satisfies the equation

$$K^4 + [R_0(C_0 + C_1) + R_1 C_1]SK^2 + (C_0 R_0)(C_1 R_1)S^2 = 0 \quad (6.45)$$

We write the solutions as

$$K^2 = S/\alpha \quad \text{or} \quad S/\beta \tag{6.46}$$

where

$\alpha(+), \beta(-)$

$$= \frac{2}{-[R_0(C_0 + C_1) + R_1 C_1] \pm \sqrt{[R_0(C_0 + C_1) + R_1 C_1]^2 - 4(C_0 R_0)(C_1 R_1)}}$$

(6.47)

Since $S < 0$ and α and β are both negative, all four solutions of K are real numbers. We assume that the left end of the first-layer line of Figure 6.21 is permanently grounded, the capacitances C_1 are all charged to $V_{DD} = 1$, and the left end of the second layer line is switched to ground at $t = 0$. The right ends of both lines are open-circuited. The solution takes the form

$$V_1(x,t) = \int_0^\infty [A(K) \exp(\alpha K^2 t) + B(K) \exp(\beta K^2 t)] \sin(Kx) \, dK \quad (6.48)$$

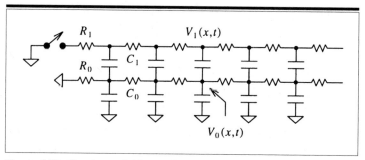

Figure 6.21 Two-Layer *RC* Transmission Line

The set of circuit equations can be solved for $\partial V_0/\partial t$, in a form that contains V_1 only. By substituting V_1 and integrating over t, we obtain

$$V_0(x,t) = \int_0^\infty [[1 + (1/\alpha C_1 R_1)]A(K)\ \exp(\alpha K^2 t)$$
$$+ [1 + (1/\beta C_1 R_1)]B(K)\ \exp(\beta K_2 t)]\ \sin\ (Kx)\ dK$$

$$(6.49)$$

where the integral may contain an arbitrary function of x as the integral constant, which can however be set equal to zero. Since the right end of the line having length $\Lambda = 1$ is open-circuited, the currents must be zero there. We note that this boundary condition does not necessarily reflect the real hardware (as we will see in the later examples), but at the right end of the line, where the voltages depend on the time and coordinate slowly, we may still use the approximation. Then we have

$$K_m = \pi(1 + 2m)/2 \quad \text{where } m = \text{positive integer} \quad (6.50)$$

The voltage profiles are written as

$$V_1(x,t) = \sum_{m=0}^\infty [A_m \exp(\alpha K_m^2 t) + B_m \exp(\beta K_m^2 t)]\ \sin\ (K_m x)$$

$$V_0(x,t) = \sum_{m=0}^\infty [[1 + (1/\alpha C_1 R_1)]A_m \exp(\alpha K_m^2 t) \qquad (6.51)$$

$$+ [1 + (1/\beta C_1 R_1)]B_m \exp(\beta K_m^2 t)]\ \sin\ (K_m x)$$

A_m and B_m are chosen so that $V_1(x,0) = 1$ and $V_0(x,t) = 0$ in the range $0 < x \leq 1$. We have from the second requirement

$$\frac{A_m}{B_m} = -\frac{1 + (1/\beta C_1 R_1)}{1 + (1/\alpha C_1 R_1)} \tag{6.52}$$

and the first requirement is satisfied by choosing the Fourier coefficient as before (Section 6.4):

$$A_m + B_m = (4/\pi)[1/(1 + 2m)] \tag{6.53}$$

We obtain

$$A_m = -\frac{4}{\pi} \frac{\alpha(1 + C_1 R_1 \beta)}{\beta - \alpha} \frac{1}{1 + 2m} \tag{6.54}$$

$$B_m = \frac{4}{\pi} \frac{\beta(1 + C_1 R_1 \alpha)}{\beta - \alpha} \frac{1}{1 + 2m}$$

For the parameter values $C_0 = 0.1$, $C_1 = 1.0$, $R_0 = 0.2$, and $R_1 = 1.0$ the evolution of the voltage profile is shown in Figure 6.22. The evolution of the voltage profile is explained as follows. As the second-layer line discharges, its decreasing voltage couples through C_1 to the first-layer line, and the voltage of the first-layer line is pushed down below ground poten-

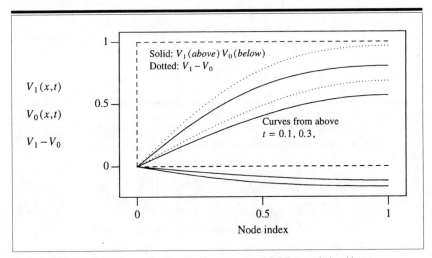

Figure 6.22 Node Voltage Profile of a Two-Layered *RC* Transmission Line

tial. The signal on the first-layer line propagates rapidly to the right, since R_0 and C_0 are much smaller than R_1 and C_1. Then the voltage at the right end of the second-layer line decreases as soon as the signal goes through the first-layer line, as shown in Figure 6.22 by the solid curve. In this interconnect structure the voltage developed between the second-layer line and the common ground at the right end is *not* the information-carrying voltage. Therefore this voltage reduction cannot be interpreted as the indication of the arrival of the signal. The true information-carrying signal is the difference between the line voltages, $V_1(x,t) - V_0(x,t)$, which is plotted by the dotted curves in Figure 6.22. The signal propagates, essentially, through the *RC* ladder structure whose parallel capacitance is C_1 and whose series resistance is $R_0 + R_1$. Some voltages lose their meanings even in RC-only circuits. The reason why the physical meaning of V_1 is lost is that the equivalent circuit of Figure 6.21 dictates that V_1 is developed across a series connection of C_0 and C_1, which is much less than C_1. In the circuit only $V_1 - V_0$ is developed across a large capacitance C_1 and so is measurable and has physical meaning.

The second example is a pair of RC transmission lines that share a common, resistive ground line, through which the two lines interact. In Figure 6.23 the middle line (line 0) is the resistive ground, and the upper and the lower lines, which are symmetrically disposed about the resistive ground, interact through the voltages developed to the common ground resistances. The circuit equations are

$$C_0 \frac{\partial}{\partial t}(V_2 - V_0) = \frac{1}{R_2}\frac{\partial^2 V_2}{\partial x^2} \qquad C_0 \frac{\partial}{\partial t}(V_1 - V_0) = \frac{1}{R_2}\frac{\partial^2 V_1}{\partial x^2}$$

$$\frac{1}{R_0}\frac{\partial V_0}{\partial x} + \frac{1}{R_2}\left[\frac{\partial V_1}{\partial x} + \frac{\partial V_2}{\partial x}\right] = 0$$

(6.55)

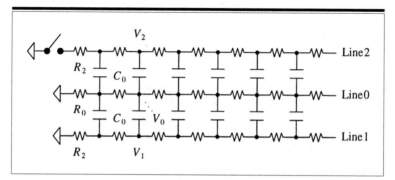

Figure 6.23 *RC* Transmission Line with Resistive Common Ground

We assume an exponential time dependence $\exp(St)$ and replace the time derivative $\partial/\partial t$ by S, and then eliminate V_0 and V_1 among the equations. After some algebra we obtain

$$\frac{1}{R_0 R_2 C_0 S} \frac{\partial^5 V_2}{\partial x^5} - 2\left(\frac{1}{R_0} + \frac{1}{R_2}\right) \frac{\partial^3 V_2}{\partial x^3} + C_0 S\left(2 + \frac{R_2}{R_0}\right) \frac{\partial V_2}{\partial x} = 0 \quad (6.56)$$

If we assume a special solution of the form $V_2 = A \sin(Kx)$ for a line that is grounded at $x = 0$, we obtain an equation satisfied by K as

$$K[K^4 + 2C_0 S(R_0 + R_2)K^2 + C_0^2 S^2 R_2(2R_0 + R_2)] = 0 \quad (6.57)$$

whose solutions are

$$K^2 = -C_0 R_2 S \quad \text{or} \quad -C_0(2R_0 + R_2)S \quad \text{or} \quad 0 \quad (6.58)$$

We write

$$S = \alpha K^2 \text{ or } \beta K^2$$

where $\qquad\qquad\qquad\qquad\qquad\qquad\qquad\qquad\qquad\qquad\qquad (6.59)$

$$\alpha = -1/[C_0(2R_0 + R_2)] \quad \beta = -1/C_0 R_2$$

Suppose that the left ends of line 1 and line 0 are permanently grounded, and line 2 is charged to V_{DD} (=1). The right ends of the lines are all open-circuited (unloaded). The left end of line 2 is grounded at $t = 0$. The solution that describes the potential profile of line 2 is a superposition of the solutions satisfying the boundary conditions:

$$V_2(x,t) = \sum_{m=0}^{\infty} [A_m \exp(\alpha K_m^2 t) + B_m \exp(\beta K_m^2 t)] \sin(K_m x) \quad (6.60)$$

where $K_m = (\pi/2\Lambda)(2m + 1)$, to satisfy the boundary condition at the right end. Since

$$\frac{\partial V_0}{\partial t} = \frac{\partial V_2}{\partial t} - \frac{1}{C_0 R_2} \frac{\partial^2 V_2}{\partial x^2} \quad \text{and} \quad \frac{\partial V_1}{\partial x} = -\frac{\partial V_2}{\partial x} - \frac{R_2}{R_0} \frac{\partial V_0}{\partial x} \quad (6.61)$$

we obtain the solution for V_1 and V_0 as

$$V_1(x,t) = \sum_{m=0}^{\infty} [A_m \exp(\alpha K_m^2 t) - B_m \exp(\beta K_m^2 t)] \sin(K_m x) \quad (6.62)$$

and

$$V_0(x,t) = -(2R_0/R_2) \sum_{m=0}^{\infty} A_m \exp(\alpha K_m^2 t) \sin(K_m x) \quad (6.63)$$

The initial conditions that determine A_m and B_m are the following. Immediately following the pulldown of the left end of line 2, lines 1 and 0 have the same potential, since the capacitances C_0 that connect the two lines have no charge. As we saw in Section 6.3, the sudden voltage change on line 2 is transmitted immediately to lines 1 and 0, and the ratio of the initial voltages is determined by the ratio of R_2 of line 2 and $R_0 R_2/(R_0 + R_2)$, which is the paralleled resistances of lines 1 and 0. We have

$$V_2: \quad A_m + B_m = \frac{R_0 + R_2}{2R_0 + R_2} \frac{4}{\pi} \frac{1}{2m + 1}$$

$$(6.64)$$

$$V_1: \quad A_m - B_m = -\frac{R_0}{2R_0 + R_2} \frac{4}{\pi} \frac{1}{2m + 1}$$

and therefore

$$A_m = (2/\pi)[R_2/(2R_0 + R_2)][1/(2m + 1)]$$

$$(6.65)$$

$$B_m = (2/\pi)[1/(2m + 1)]$$

Figure 6.24 shows V_2, V_0, and $V_2 - V_0$ versus time. V_2 and V_1 are not

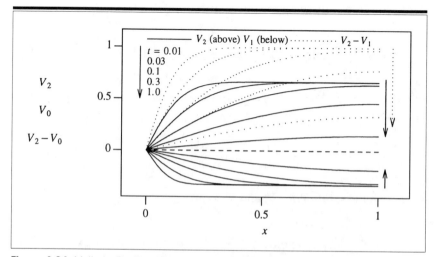

Figure 6.24 Voltage Profile of Resistive-Ground Parallel Transmission Lines

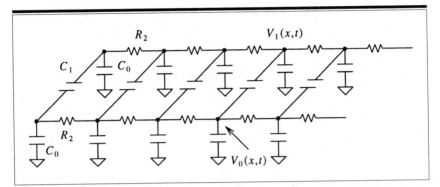

Figure 6.25 Coupled Parallel *RC* Transmission Lines

information-carrying signals. The voltage difference $V_2 - V_1$ carries the digital information.

The second mechanism of coupling between interconnect lines is through capacitance between adjacent lines. The equivalent circuit in this case is shown in Figure 6.25. The potential profiles $V_0(x,t)$ and $V_1(x,t)$ satisfy the circuit equations

$$(C_0 + C_1) \frac{\partial V_0}{\partial t} - C_1 \frac{\partial V_1}{\partial t} = \frac{1}{R_2} \frac{\partial^2 V_0}{\partial x^2}$$

$$-C_1 \frac{\partial V_0}{\partial t} + (C_0 + C_1) \frac{\partial V_1}{\partial t} = \frac{1}{R_2} \frac{\partial^2 V_1}{\partial x^2} \tag{6.66}$$

If the exponential time dependence $\exp(St)$ is assumed, V_1 is eliminated between the two equations and we obtain

$$\frac{\partial^4 V_0}{\partial x^4} - 2R_2(C_0 + C_1)S \frac{\partial^2 V_0}{\partial x^2} + R_2^2(C_0^2 + 2C_0C_1)S^2 V_0 = 0 \tag{6.67}$$

We consider the case that the left end of the first line is grounded at $t = 0$. We seek a special solution of the form

$$V_0(x,t) = A(K)\exp(St)\sin(Kx) \tag{6.68}$$

where K satisfies the equation

$$K^4 + 2R_2(C_0 + C_1)SK^2 + (C_0^2 + 2C_0C_1)R_2^2S^2 = 0 \tag{6.69}$$

whose solutions are

$$K^2 = -R_2(C_0 + 2C_1)S \quad \text{and} \quad K^2 = -R_2C_0S \tag{6.70}$$

If $\alpha = -1/[R_2(C_0 + 2C_1)]$ and $\beta = -1/R_2C_0$, the solutions can be written as

$$V_0(x,t) = \int_0^\infty [A(K)\exp(\alpha K^2 t) + B(K)\exp(\beta K^2 t)]\sin(Kx)\,dK$$

$$V_1(x,t) = \int_0^\infty [-A(K)\exp(\alpha K^2 t) + B(K)\exp(\beta K^2 t)]\sin(Kx)\,dK$$

$$+ \text{(integration constant)} \tag{6.71}$$

The integration constant can be an arbitrary function of x. Let us consider the case where the left end of the second line is permanently grounded. The integration constant is zero, since $V_1(x,t) \to 0$ as $t \to \infty$. The right ends of the lines are open-circuited (or unloaded). From the boundary condition at the right end,

$$K = K_m = \pi(1 + 2m)/2 \tag{6.72}$$

At $t = 0$, we have $V_0(x,0) = 1.0$ ($V_{DD} = 1$) and $V_1(x,0) = 0$ for $[0 < x \le \Lambda$ $(\Lambda = 1)]$. We have

$$A_m + B_m = (4/\pi)[1/(1 + 2m)] \quad \text{and} \quad A_m = B_m \tag{6.73}$$

and

$$V_0(x,t) = \frac{2}{\pi}\sum_{m=0}^\infty \frac{1}{1 + 2m}[\exp(\alpha K_m^2 t) + \exp(\beta K_m^2 t)]\sin(K_m t)$$

$$V_1(x,t) = \frac{2}{\pi}\sum_{m=0}^\infty \frac{1}{1 + 2m}[-\exp(\alpha K_m^2 t) + \exp(\beta K_m^2 t)]\sin(K_m x)$$

$$\tag{6.74}$$

The voltage profile evolution for $C_0 = C_1 = R_2 = 1$ is shown in Figure 6.26. As the first line discharges, a negative voltage is induced on the second line, and a voltage minimum is created on the second line, as shown by the dotted curves for $t = 0.03$ and $t = 0.1$. This is because pulldown of the first line is most active somewhere between the two ends, and to that region the second line injects current through C_1. The left end of the second line is still held at ground potential by C_0.

A mathematically interesting problem is that if $V_0(x,t)$ is expanded in $\sin(Kx)$ as in Equation (4), then $V_1(x,t)$ must also be expressed in terms of

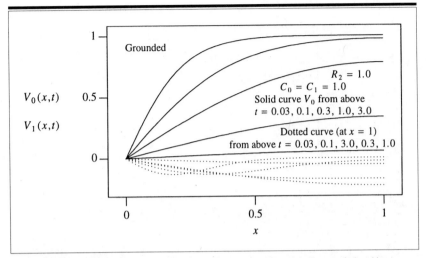

Figure 6.26 Voltage Profile of Capacitively Coupled Double Transmission Line

$\sin(Kx)$ only. This means that it is apparently impossible to apply an open-circuit (unloaded) boundary condition to the second line at $x = 0$ (for the second line to be floating). Suppose that line 2 is charged to $V_{DD} = 1$ and is tristated. In this case the integration constant for $V_1(x,t)$ must be chosen so that

$$V_1(0,+0) = \lim_{t\to\infty} V_1(x,t) = C_0/(C_0 + C_1) \qquad (6.75)$$

and the complete solution is given by further requiring that

$$V_0(0,t) = 0 \qquad V_0(x,+0) = V_1(x,+0) = 1 \quad (0 < x \le \Lambda) \qquad (6.76)$$

We have

$$V_0(x,t) = \frac{2}{\pi} \sum_{m=0}^{\infty} \frac{1}{1 + 2m} [[C_0/(C_0 + C_1)] \exp(\alpha K_m^2 t)$$
$$+ [(C_0 + 2C_1)/(C_0 + C_1)] \exp(\beta K_m^2 t)] \sin(K_m x)$$

$$V_1(x,t) = [C_0/(C_0 + C_1)]$$
$$+ \frac{2}{\pi} \sum_{m=0}^{\infty} \frac{1}{1 + 2m} [-[C_0/(C_0 + C_1)] \exp(\alpha K_m^2 t)$$
$$+ [(C_0 + 2C_1)/(C_0 + C_1)] \exp(\beta K_m^2 t)] \sin(K_m x)$$

$$(6.77)$$

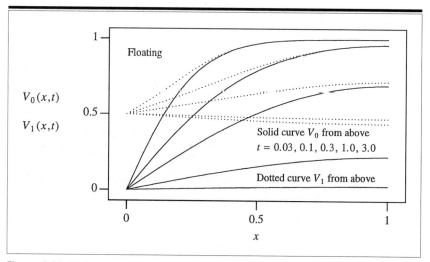

Figure 6.27 Voltage Profile of a Capacitively Coupled Double Transmission Line

The evolution of the potential profile is shown in Figure 6.27 for the same parameter values as before. In this profile $\partial V_1(x,t)/\partial x$ is not zero at $x = 0$ for small t, even if the left end is open-circuited. This is because a large current is injected from line 1 through C_1 at $t = 0$ and creates an anomaly. If we compute line 2 current as

$$I_1(x,0) = -(1/R_2)[\partial V_1(x,0)/\partial x] \tag{6.78}$$

$$= [2C_1/(C_0 + C_1)](1/R_2L) \sum_{m=0}^{\infty} \cos(K_m x)$$

At $x = 0$

$$I_1(0,0) = [2C_1/(C_0 + C_1)](1/R_2L)[1 - 1 + 1 - 1 + 1 - \cdots] \tag{6.79}$$

and this is a diverging infinite series. Closed-form analysis for this case, as well as numerical simulation, requires care in the vicinity of the singularity.

6.7 *LC* Transmission Line

Signal transmission through a lossless transmission line has been discussed in many textbooks, but only for the simplest cases of single, isolated lines. We extend the theory to cover more complex cases that are

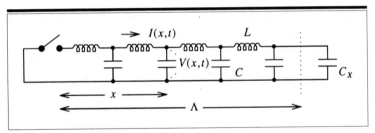

Figure 6.28 *LC* Transmission Line (with Capacitive Termination)

relevant to IC interconnects. The equations satisfied by the current $I(x,t)$ and the voltage $V(x,t)$ defined in Figure 6.28 are

$$\frac{\partial V}{\partial x} = -L\frac{\partial I}{\partial t} \quad \text{and} \quad \frac{\partial I}{\partial x} = -C\frac{\partial V}{\partial t} \quad \text{whence} \quad \frac{\partial^2 V}{\partial x^2} = LC\frac{\partial^2 V}{\partial t^2} \quad (6.80)$$

The general solution of this *wave equation* is given by

$$V(x,t) = F[x - (t/\sqrt{LC})] + B[x + (t/\sqrt{LC})] \quad (6.81)$$

where F and B are arbitrary functions. The first term represents a wave traveling to the right, and the second term, to the left. This solution represents a wave whose amplitude does not change with time, since inductances and capacitances consume no power. We follow the same mathematical techniques we used in the earlier sections.

Suppose that the LC line of Figure 6.28 is charged uniformly to V_{DD} (=1) and the switch at the left end of the line is closed at $t = 0$. The right end of the line is open-circuited (unloaded). The voltage profile in the line subject to this condition is given by a superposition of the modes of the form

$$V_m(x,t) = f_m(t) \sin (K_m x) \quad (6.82)$$

where $K_m = \pi(2m + 1)/2\Lambda$, since the voltage profile must have an extremum at the right end of the line, where Λ is the length of the line. The function $f_m(t)$ satisfies

$$LCf''_m(t) = -K^2_m f_m(t) \quad (6.83)$$

where the primes indicate the time derivative. We have

$$f_m(t) = A_m \cos[(K_m/\sqrt{LC})t] + B_m \sin[(K_m/\sqrt{LC})t] \quad (6.84)$$

The sine term is not needed, since it has only the effect of shifting the origin of time. We have a solution

$$V(x,t) = \sum_{m=0}^{\infty} A_m \cos[(K_m/\sqrt{LC})t] \sin(K_m x)$$

where   (6.85)

$$A_m = (4/\pi)[1/(2m + 1)]$$

that satisfies the initial condition

$$V(x,0) = 0 \quad (x \le 0) \qquad V(x,0) = 1 \quad (0 < x \le \Lambda) \quad (6.86)$$

Figure 6.29 shows evolution of the voltage profile, which is a traveling wave reflected back and forth between the shorted left end and the open-circuited right end. The voltage profiles at $t = 2.2$ and $t = 2.4$ are shown by the filled and the open circles to indicate that the time evolution backtracks and returns to the initial condition with period $t = 4$. The transient may be considered as a standing-wave phenomenon, which repeats unattenuated an infinite number of times.

If the right end of the LC transmission line is terminated by an extra

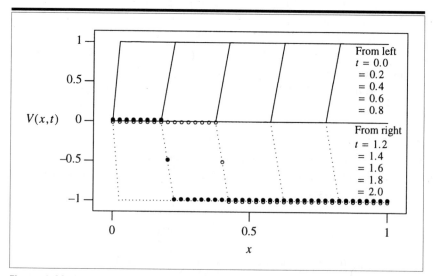

Figure 6.29 Traveling Wave in the Open-Ended *LC* Transmission Line

capacitance C_X, the voltage profile is expressed by a similar formula as a sum of modes:

$$V(x,t) = \sum_{m=0}^{\infty} V_m(x,t)$$

where   (6.87)

$$V_m(x,t) = A_m \cos[(K_m/\sqrt{LC})t] \sin (K_m x)$$

where the K_m's no longer satisfy the simple, harmonic relationship. Since for each mode

$$\partial I_m /\partial x = -C(\partial V_m /\partial t)$$

or   (6.88)

$$I_m(x,t) = -\sqrt{C/L} A_m \sin[(K_m/\sqrt{LC})t] \cos (K_m x)$$

we apply the boundary condition at the right end,

$$I_m(\Lambda,t) = C_X(\partial/\partial t)V_m(x,t)$$ (6.89)

to obtain

$$\tan (\xi) = \gamma/\xi \quad \gamma = C\Lambda/C_X \quad \xi = K\Lambda$$ (6.90)

In the limit as $C_X \to 0$ we have $K_m = \pi(2m + 1)/2\Lambda$. This is the same expression for K_m as with the capacitively terminated RC line in the last section. Since $V(x,0)$ has the same form as the RC transmission line, the expansion coefficients have already been determined. An RC and an LC transmission line have a close mathematical similarity. The evolution of the potential profile is shown in Figure 6.30. At the moment when the wavefront arrives at the right end, C_X cannot be discharged immediately: The right end tries to hold the voltage, and generates a reflected wave having the same amplitude. As the reflected wave travels back to the left, C_X is kept discharging, and the right-end voltage decreases. The reflected wavefront develops a ramp behind the front as it travels back to the left end. When the right-end voltage reaches zero, the discharge does not stop, since the current of the inductance of the LC line continues to draw charge from C_X and the voltage decreases still more.

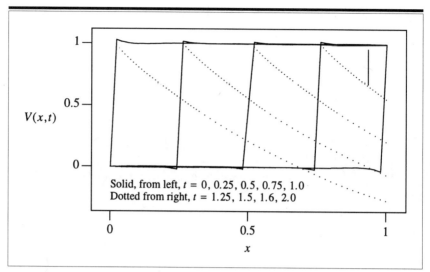

$V(x,t)$

Solid, from left, $t = 0, 0.25, 0.5, 0.75, 1.0$
Dotted from right, $t = 1.25, 1.5, 1.6, 2.0$

Figure 6.30 Evolution of the Voltage Profile
of a Capacitively Terminated *LC* Transmission Line

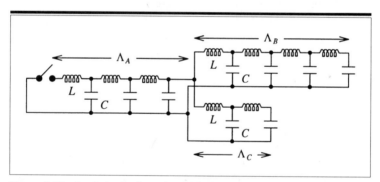

Figure 6.31 Analysis of a Branched *LC* Transmission Line

We consider a branched LC line whose equivalent circuit is shown in Figure 6.31. The line is uniformly charged to V_{DD} (=1), and the switch at the left end of the line A is turned on at $t = 0$ to initiate the discharge. The right ends of the lines B and C are open-circuited. The voltage profiles in the lines A, B, and C are functions of the coordinates x_A, x_B, and x_C, respectively, which are taken along the respective lines with their origins at their left ends. By taking the boundary conditions at $x_A = 0$, $x_B = \Lambda_B$, and $x_C = \Lambda_C$ into consideration, the potential profiles $V_A(x_A,t)$ through $V_C(x_C,t)$ are written as sums of modes:

$$V_A(x_A,t) = \sum_{m=0}^{\infty} V_{Am}(x_A,t)$$

$$V_B(x_B,t) = \sum_{m=0}^{\infty} V_{Bm}(x_B,t) \qquad (6.91)$$

$$V_C(x_C,t) = \sum_{m=0}^{\infty} V_{Cm}(x_C,t)$$

where

$$
\begin{aligned}
V_{Am}(x_A,t) &= [A_{Cm} \cos[(K_m/\sqrt{LC})t] \\
&\quad + A_{Sm} \sin[(K_m/\sqrt{LC})t]] \sin (K_m x) \\
V_{Bm}(x_B,t) &= [B_{Cm} \cos[(K_m/\sqrt{LC})t] \\
&\quad + B_{Sm} \sin[(K_m/\sqrt{LC})t]] \cos[K_m(\Lambda_B - x_B)] \\
V_{Cm}(x_C,t) &= [C_{Cm} \cos[(K_m/\sqrt{LC})t] \\
&\quad + C_{Sm} \sin[(K_m/\sqrt{LC})t]] \cos[K_m(\Lambda_C - x_C)]
\end{aligned}
\qquad (6.92)
$$

where A_{Cm} through C_{Sm} are the expansion coefficients. The currents in the lines are given by $\partial I(x,t)/\partial x = -C[\partial V(x,t)/\partial t]$ as

$$
\begin{aligned}
I_{Am}(x_A,t) &= \sqrt{C/L}[-A_{Cm} \sin[(K_m/\sqrt{LC})t] \\
&\quad + A_{Sm} \cos[(K_m/\sqrt{LC})t]] \cos(K_m x_A) \\
I_{Bm}(x_B,t) &= \sqrt{C/L}[-B_{Cm} \sin[(K_m/\sqrt{LC})t] \\
&\quad + B_{Sm} \cos[(K_m/\sqrt{LC})t]] \sin[K_m(\Lambda_B - x_B)] \\
I_{Cm}(x_C,t) &= \sqrt{C/L}[-C_{Cm} \sin[(K_m/\sqrt{LC})t] \\
&\quad + C_{Sm} \cos[(K_m/\sqrt{LC})t]] \sin[K_m(\Lambda_C - x_C)]
\end{aligned}
\qquad (6.93)
$$

By requiring that the voltages of the three profiles match at the joint, we have that

$$
\begin{aligned}
A_{Cm,Sm} \sin (K_m \Lambda_A) &= B_{Cm,Sm} \cos (K_m \Lambda_B) \\
A_{Cm,Sm} \sin (K_m \Lambda_A) &= C_{Cm,Sm} \cos (K_m \Lambda_C)
\end{aligned}
\qquad (6.94)
$$

and that the sum of the currents in the lines B and C equals the current in the line A:

$$A_{Cm,Sm} \cos (K_m \Lambda_A) = B_{Cm,Sm} \sin (K_m \Lambda_B) + C_{Cm,Sm} \sin (K_m \Lambda_C) \quad (6.95)$$

The terms including $\sin[(K_m/\sqrt{LC})t]$ have only the effect of shifting the origin of the time, and therefore they are not necessary. We obtain the equation that determines the values of K_m as

$$\tan (K_m \Lambda_A)[\tan (K_m \Lambda_B) + \tan (K_m \Lambda_C)] - 1 = 0 \quad (6.96)$$

This is the same equation as for the branched RC transmission line discussed in the last section. Since the mathematical expression of the voltage profiles is the same except for the time dependence, all the results of the last section can be used. For the time dependence the following substitution is made:

$$\exp[-(K_m^2/CR)t] \rightarrow \cos[(K_m/\sqrt{LC})t] \quad (6.97)$$

We assume the same values $\Lambda_A = 1.0$, $\Lambda_B = 2.0$, and $\Lambda_C = 1.0$ as before. The results are shown in Figure 6.32(a)–(j).

We observe that when the downgoing step-function wavefront arrives at the branch point, a reflected wave having amplitude one-third is generated, and the step-function height beyond the branch point in the lines B and C is reduced to two-thirds. This is understood from the discharge mechanisms of the LC line as follows. In Figure 6.33, the LC line having a branch is uniformly charged to V_{DD} before $t = 0$, when the switch is closed at the left end and the discharge begins. We consider a uniform line. The current I_1 of the first loop is the time integral of $V_1(t)$ divided by L_1:

$$I_1 = \frac{1}{L_1} \int_0^t V_1(\theta) \, d\theta \quad (6.98)$$

$V_1(t)$ decreases and reaches zero at $t = t_1$, which is given by $t_1 = \sqrt{L_1 C_1}$. The current established in L_1 *relays* the current from L_2 that is built up at a time $\sqrt{L_2 C_2}$ later without generating voltage across L_1. Therefore $V_1(t) = 0$ for $t > t_1$. The current established in L_1 is given by

$$I_1(t > t_1) = V_{DD}\sqrt{L_1 C_1}/L_1 = \sqrt{C_1/L_1}\, V_{DD} = V_{DD}/Z_0 \quad (6.99)$$

This step repeats for loops 2, 3, . . . , and the downgoing voltage step propagates to the right.

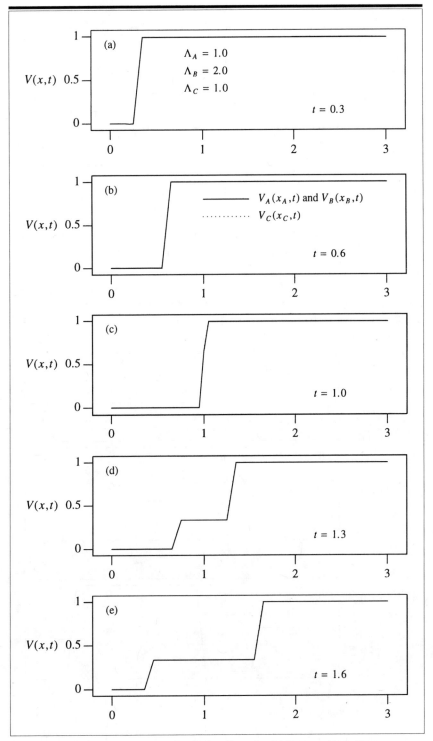

Figure 6.32 Evolution of Voltage Profile in a Branched *LC* Transmission Line

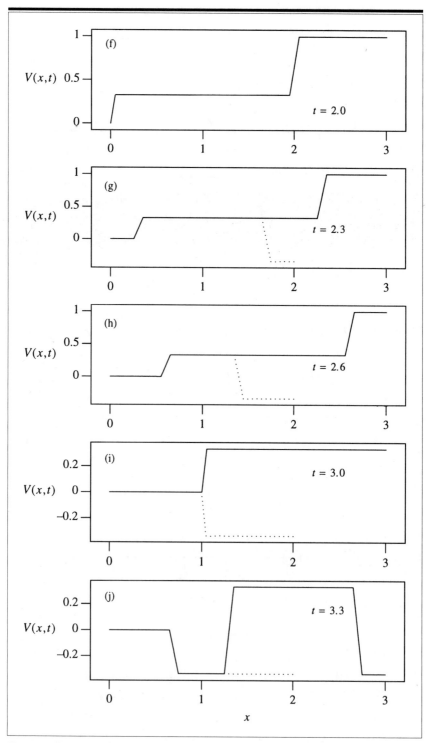

Figure 6.32 Continued

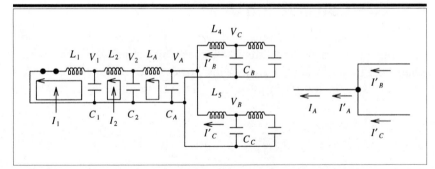

Figure 6.33 Wave Reflection at Branch Point of *LC* Transmission Line

At the branch point, the current I_A established in L_A is $\sqrt{C/L}\,V_{DD}$. At the branch point the downgoing forward wavefronts in the lines B and C, and an upgoing backward wavefront in the line A are launched. The currents carried by the three wavefronts past the branch point are I'_B, I'_C, and I'_A, respectively. We must have

$$I_A + I'_A = I'_B + I'_C, \quad \text{where} \quad I'_B = I'_C \quad (6.100)$$

Here we have used the facts that (1) the lines B and C are symmetrical with respect to A before the wavefronts are reflected back by their respective ends, and (2) the reflected wave in *A increases* the current in the inductor, since the wave carries the charge required to recharge A from the right end as the reflected wavefront travels to the left. Since the height of the voltage step created in the lines A, B, and C (the difference in the line voltage before and after the incidence at the branch point) is proportional to the current,

$$I'_A + I'_B = I_A \quad (6.101)$$

must be satisfied. From Equations (6.100) and (6.101) we find that

$$I'_B = (2/3)I_A \qquad I'_A = (1/3)I_A \quad (6.102)$$

The voltage steps generated in the lines are, from the proportionality to the currents,

$$V_{A\text{step}} = (1/3)V_{DD}$$

$$V_{B\text{step}} = (2/3)V_{DD} \quad (6.103)$$

$$V_{C\text{step}} = (2/3)V_{DD}$$

and this is what we observed in Figure 6.32.

6.8 *LCR* Transmission Lines

If resistance loss is included the analysis becomes significantly more complex, but the mathematics leads to an interesting physical insight, which relates the LC and the RC transmission lines as extreme cases. The circuit equations satisfied by $I(x,t)$ and $V(x,t)$ of the LCR line of Figure 6.34 are

$$\frac{\partial I}{\partial x} = -C\frac{\partial V}{\partial t} \quad \text{and} \quad \frac{\partial V}{\partial x} = -L\frac{\partial I}{\partial t} - RI$$

whence   (6.104)

$$\frac{\partial^2 V}{\partial x^2} = LC\frac{\partial^2 V}{\partial t^2} + RC\frac{\partial V}{\partial t}$$

If we assume the solution of the form $V = A\exp(St)\sin(Kx)$ appropriate for a left-end-grounded LCR line, S satisfies the equation

$$LCS^2 + RCS + K^2 = 0$$
                                                                                          (6.105)
or

$$S = (R/2L)[-1 \pm \sqrt{1 - (4L/CR^2)K^2}]$$

Since the right end of the line is open-circuited (unloaded), $K_m = (\pi/2\Lambda)(2m + 1)$, where $m = 0, 1, 2, \ldots$. If

$$m < m_C = \frac{1}{2}\left[\frac{\Lambda R}{\pi\sqrt{L/C}} - 1\right] \tag{6.106}$$

both roots of the secular equation are real and negative. They are given by

$S_\alpha(K_m)(+), \; S_\beta(K_m)(-)$
$$= (R/2L)[-1 \pm \sqrt{1 - (4L/CR^2)K_m^2}] \quad (6.107)$$

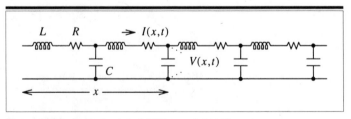

Figure 6.34 Analysis of an *LCR* Transmission Line

If $m > m_C$ we define

$$S_R(K_m) = -R/2L \tag{6.108}$$
$$S_I(K_m) = (R/2L)\sqrt{(4L/CR^2)K_m^2 - 1}$$

For $m < m_C$ a monotonically decaying exponential function of time is the solution, and for $m > m_C$ a decaying sinusoidal function of time is the solution. The two regimes change over at $m = m_C$, where m_C is solved from

$$[\Lambda/(2m_C + 1)]R = \pi\sqrt{L/C} = \pi Z_0 \tag{6.109}$$

where Z_0 is the *characteristic impedance* of the LC transmission line.

The quantity $\Lambda/(2m_C + 1)$ in this equation is the characteristic length for mode m_C: Within that length a significant change of the line voltage takes place. $V(x,t)$ can be written as

$$
\begin{aligned}
V(x,t) = \sum_{m=0}^{m<m_C} & \{A_m \exp[S_\alpha(K_m)t] + B_m \exp[S_\beta(K_m)t]\} \sin(K_m x) \\
& + \sum_{m>m_C}^{\infty} \{A_m \exp[S_R(K_m)t] \cos[S_I(K_m)t] \\
& + B_m \exp[S_R(K_m)t] \sin[S_I(K_m)t]\} \sin(K_m x)
\end{aligned}
\tag{6.110}
$$

The expansion coefficients A_m and B_m are determined from the initial conditions

$$V(x,0) = V_{DD} = 1 \quad I(x,0) = 0 \quad (0 < x \le \Lambda) \tag{6.111}$$

$I(x,t)$ is found from the first of the circuit equations (6.104), by differentiating $V(x,t)$ with respect to t and then integrating the result with respect to x, as

$$
\begin{aligned}
I(x,t) = C \sum_{m=0}^{m<m_C} & (1/K_m)[A_m S_\alpha(K_m) \exp[S_\alpha(K_m)t] \\
& + B_m S_\beta(K_m) \exp[S_\beta(K_m)t]] \cos(K_m x) \\
& + \sum_{m>m_C}^{\infty} (1/K_m)[[A_m S_R(K_m) \\
& + B_m S_I(K_m)] \exp[S_R(K_m)t] \cos[S_I(K_m)t] \\
& + [-A_m S_I(K_m) \\
& + B_m S_R(K_m)] \exp[S_R(K_m)t] \sin[S_I(K_m)t]] \cos(K_m x)
\end{aligned}
\tag{6.112}
$$

At $t = 0$ the initial conditions are imposed to determine A_ms and B_ms. They are

$$S_\alpha(K_m)A_m + S_\beta(K_m)B_m = 0 \quad (m < m_C) \tag{6.113}$$

$$S_R(K_m)A_m + S_I(K_m)B_m = 0 \quad (m > m_C)$$

and

$$A_m + B_m = (4/\pi)[1/(2m + 1)] = X_m \quad (m < m_C) \tag{6.114}$$

$$A_m = X_m \quad (m > m_C)$$

From them we find

$$A_m = \frac{S_\beta(K_m)X_m}{S_\beta(K_m) - S_\alpha(K_m)}$$

$$B_m = -\frac{S_\alpha(K_m)X_m}{S_\beta(K_m) - S_\alpha(K_m)} \quad (m < m_C)$$

and $\hspace{6cm}$ (6.115)

$$A_m = X_m$$

$$B_m = -\frac{S_R(K_m)}{S_I(K_m)}X_m \quad (m > m_C)$$

Figure 6.35(a)–(e) show the evolution of the voltage profile. In the series-resistance-dominated LCR line [Figure 6.35(a) and (b)] the discharge process is similar to that of an RC line, except that a step-function voltage front launched at the left end is superposed on the profile. The step function moves to the right, it is reflected, its amplitude decreases, and finally it disappears altogether. If L is increased, the height of the step function increases. For intermediate inductance values the profile upstream of the step function forms a loosely defined backbone, on which the step function slides back and forth, as is observed in Figure 6.35(c) and (d). This backbone represents the slow discharge process controlled by R and C. The backbone is more clearly defined as L increases. For the large L [Figure 6.35(e)] the backbone is almost quasistable, and it acts as if it were the local voltage reference profile for the step-function traveling waves [1].

The capacitance-terminated LCR line can be analyzed similarly. At the right end of the line the boundary condition

$$C_X(d/dt)V(\Lambda,t) = I(\Lambda,t) \tag{6.116}$$

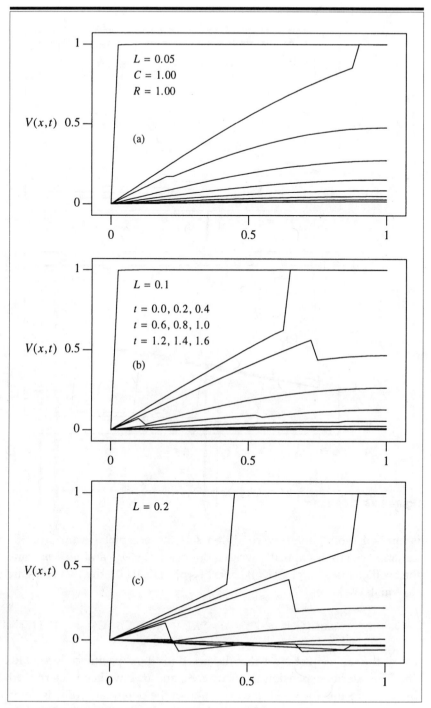

Figure 6.35 Potential Profile in an *LCR* Transmission Line

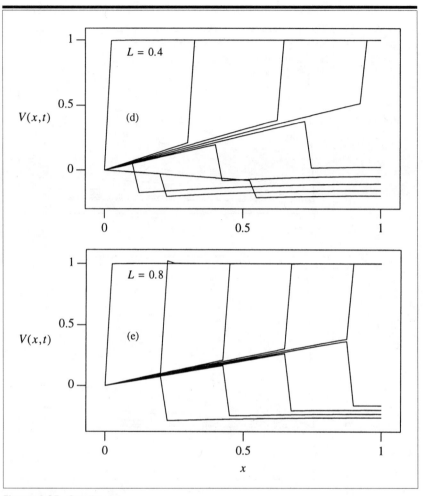

Figure 6.35 Continued

is applied. This boundary condition sets the acceptable values of K_m. Equation (6.116) is explicitly written down for $m < m_C$ and $m > m_C$, and the coefficients of $\exp[S_a(K_m)t]$ and of $\sin[S_t(K_m)t]$ on the two sides are set equal. We have

$$\tan(\xi) = \gamma/\xi \qquad \xi = K_m\Lambda \qquad \gamma = C\Lambda/C_X \qquad (6.117)$$

This is the same equation as that discussed for the capacitance-loaded RC line. The Fourier expansion coefficients A_m and B_m have been determined. Using the results for $\gamma = 1$, the evolution of the potential profile is determined as shown in Figure 6.36(a) and (b). For small L [Figure 6.36(a)] the

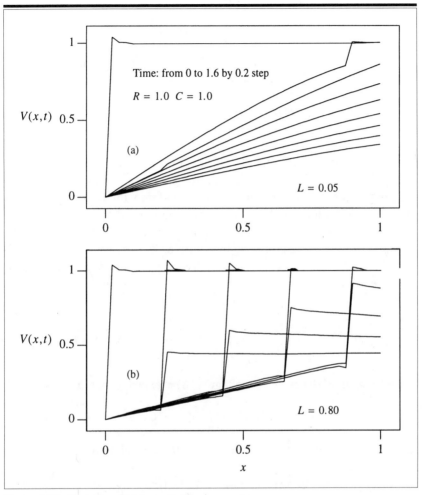

Figure 6.36 Evolution of Voltage Profile
in a Capacitively Terminated *LCR* Transmission Line

discharge process is similar to that of the capacitance-loaded RC line: The termination capacitance slows the discharge process down, and the potential profile in the later phase of the transient is close to a linear ramp. The LCR line works effectively as a resistor discharging the capacitor at the right end. For large L [Figure 6.36(b)] the discharge process is similar to that of an LC line [Figure 6.30], except that the reflected step-function amplitude decreases. The decreasing step function slides along the quasistable backbone of the profile that represents the RC discharge mechanisms of the LCR line.

A practically important problem is to determine the proper value of

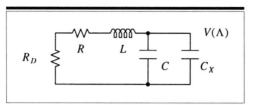

Figure 6.37 Simplified Equivalent Circuit of an
LCR Line

series resistance R of an LCR line for the best digital signal transmission. This problem was discussed in detail by Frye and Chen [50]. Here I summarize the qualitative conclusion. If R is too large, the RC delay becomes excessive, and if too small, the line generates an oscillatory response. The best compromise is to seek the critical damping condition of the line, which is understood as follows. In Figure 6.37 the entire inductance, capacitance, and series resistance of the line are combined into L, C, and R of the equivalent circuit. R_D is the internal resistance of the pulldown driver, and C_X is the load capacitance to the line. The circuit equation is

$$L(C + C_X) \frac{d^2V(\Lambda)}{dt^2} + (R + R_D)(C + C_X)\frac{dV(\Lambda)}{dt} + V(\Lambda) = 0 \quad (6.118)$$

If the time dependence of the voltage $V(\Lambda)$ is exp(St), then S satisfies

$$L(C + C_X)S^2 + (R + R_D)(C + C_X)S + 1 = 0 \quad (6.119)$$

The equation should have double roots at the critical damping. We have

$$(R + R_D)^2(C + C_X)^2 - 4L(C + C_X) = 0$$

or   (6.120)

$$R = \sqrt{4L/(C + C_X)} - R_D$$

This relationship is not exact, since the equivalent circuit was simplified. To improve the approximation the LCR line must be divided into many segments, ultimately into an infinitely large number. It is not convenient to carry out this analysis in closed form. The problem can be simplified, however, by choosing $L = C = 1$ without losing generality. It then contains only three independent variables, R, R_D, and C_X. The problem can be solved by dividing the transmission line into segments, and computing the voltage waveform at the end of the line. If the line is divided into nine segments and typical parameter values are assigned, we obtain numerical results as shown in Figure 6.38.

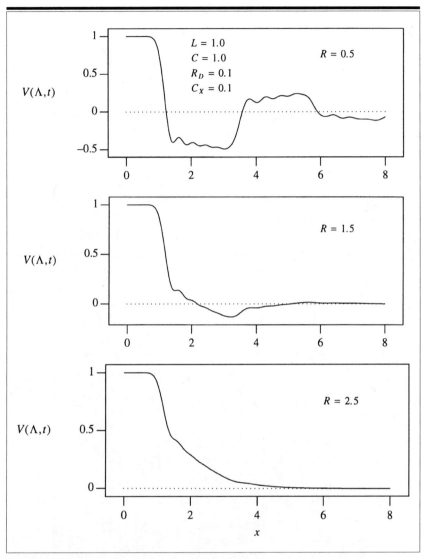

Figure 6.38 Voltage Waveform at End of *LCR* Line

As is observed from the waveforms, it is not a simple matter to determine at what parameter values the critical damping is reached, so that the best waveform for digital signal transmission is available. It is especially difficult to set up a criterion for computerization. It is an easy matter, however, to determine the desirability of a waveform by inspecting simulation results. This is an example of the difficulty of using a computer in a problem that involves a value judgement. Positive assurance of the best waveform can never be achieved.

In this problem, the following criteria can be checked. If $V(\Lambda, t)$ becomes negative at certain t, the waveform will show ringing, and the LCR line-driver-load capacitance combination is in the underdamping regime. This is a clear criterion, so it should be checked first. The circuit may go into the underdamping regime, however, without satisfying the condition. If at time t we have $V(\Lambda, t) > V(\Lambda, t - \Delta t)$, where Δt is the positive integration step, the LCR combination is weakly underdamped, which creates waveform distortion. This criterion can be checked second. The period of oscillation is given by

$$\text{period} = 2\pi \sqrt{L(C + C_X) - (R + R_D)^2(C + C_X)^2} \quad (6.121)$$

and the voltage is expected to go through a minumum and a maximum within it. If the waveform is observed over several times the period and if it is guaranteed that the voltage is positive and is decreasing monotonically with time, we may say that the combination is not strongly underdamped. Contour maps created by this criterion are shown in Figures 6.39(a) and (b). The two criteria discussed above were applied in the given order. The overdamping regime is represented by open circles,

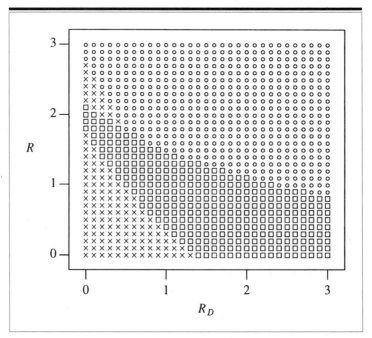

Figure 6.39(a) R and R_D Versus Damping of the Line, Observed at the Right End ($C_X = 0$)

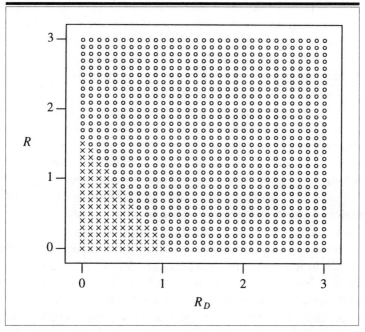

Figure 6.39(b) R and R_D Versus Damping of the Line, Observed at the
Right End ($C_X = 1$)

the stronger underdamping condition by $\times$, and the weaker underdamping condition by open squares. To reach a practical design decision we need to allow slight underdamping, since a negative swing less than a certain limit is not harmful for circuit operation. It has been established that the driver (or the termination) resistance adds to the total resistance of the line to determine the state of damping of the transmission line [50]. This conclusion is simple and beautiful, but it is not easy to derive directly from the fundamental principles.

6.9 Future Directions for Integrated Circuit Theory

Interconnects in present integrated circuits are characterized as resistance-dominated, overdamped transmission lines as discussed in this section. The RC delay is still many times longer than the intrinsic LC delay. Many efforts are being directed to reducing the RC delay. The most fundamental solution is high-temperature superconductor interconnect. I consider this trend inevitable, at the least for the highest-performance systems, so that device performance gained by scaledown is not wasted. When the ultimate LC delay limit is reached, we will see a strange new world, which is anticipated in this work. The reader will agree that voltageless electronic circuits and backward flow of time in circuits are quite futuristic. Yet the theory I have developed here stands entirely on the classical works by Ampère, Faraday, and Maxwell; all that I have added is new interpretation. For me, this is very surprising, perhaps more so than the many curious conclusions presented in this book.

Electronic circuits span the wide gap between simplicity and complexity, and between physical science and information science. Electronic circuits are so simple that they have some features of elementary particles, such as the indistinguishability of excitations. Still, they can be integrated into a complex system that ultimately has high functions such as intelligence. I favor the idea that human intelligence emerges in the process of organizing the quantum physical world into the classical physical world [51]. I see a possibility of integrating the simple and complex extremes of natural science with electronic circuit theory as the medium. For this integration, information science will play a crucial role in circuit theory. This is the reason I expect a prosperous future for electronic circuit theory.

Fundamental to this work is the urge to *understand* the meanings of the term "circuit operation." The problem is on the boundary between the electronic circuit and the human cognitive function. What I have attempted to do in this book is to solve some of the problems on the circuit side: More precisely, I have singled out and studied the simple principles of circuit theory that can be integrated into complexity. The rest of the problems are left unsolved. This scientific area is wide open to investigation.

REFERENCES

1. M. Shoji, "Theory of CMOS digital circuits and circuit failures," Princeton University Press, Princeton, N.J., 1992.

2. K. Hara, "High temperature superconducting electronics," IOS Press, Amsterdam, 1993.

3. A. W. Wieder, "Renaissance of bipolar technology," 18th Conference on Solid-State Devices and Material, Tokyo, pp. 261–265, 1986.

4. R. N. Bracewell, "The Fourier transform and its applications," McGraw-Hill, New York, 1978.

5. R. K. Moore, "Travelling wave engineering," McGraw-Hill, New York, 1960.

6. M. Shoji and R. M. Rolfe, "Negative capacitance bus terminator for improving the switching speed of a microcomputer data bus," IEEE J. Solid-State Circuits, vol. SC-20, pp. 828–832, August 1985.

7. M. Shoji, "CMOS digital circuit technology," Prentice-Hall, Englewood Cliffs, N.J., 1987.

8. H. Hasegawa, M. Furukawa, and H. Yanai, "Properties of microstrip line on Si-SiO$_2$ system," IEEE Trans. Microwave Theory and Techniques, vol. MTT-19, pp. 869–881, November 1971.

9. J. Laramee, M. J. Aubin, and J. D. N. Cheeke, "Behavior of CMOS inverters at cryogenic temperatures," Solid-State Electronics, vol. 28, pp. 453–456, 1985.

10. W. H. Nottage, "The calculation and measurement of inductance and capacity," Wireless Press, London, 1916.

11. J. A. Stratton, "Electromagnetic theory," McGraw-Hill, New York, 1941.

12. M. E. van Valkenburg, "Introduction to modern network synthesis," Wiley, New York, 1960.

13. E. H. Frost-Smith, "The theory and design of magnetic amplifiers," Chapman & Hall, London, 1966.

14. W. Shockley, "Electrons and holes in semiconductors," van Nostrand, Princeton, N.J., 1950.

15. J. Sparks, "Junction transistors," Pergamon, Oxford, England, 1966.

16. J. R. Brews, "Physics of the MOS transistor," Applied solid state science, Supplement 2A, Academic Press, New York, 1981.

17. S. M. Sze, "Semiconductor devices, physics and technology," Wiley, New York, 1985.

18. K. K. Ng, "Complete guide to semiconductor devices," McGraw-Hill, New York, 1995.

19. J. Jeans, "The dynamic theory of gases," Dover, New York, 1954.

20. R. F. Probstein, "Physicochemical hydrodynamics," Wiley, New York, 1994.

21. A. van der Ziel, "Noise," Prentice-Hall, Englewood Cliffs, N.J., 1954.

22. C. Y. Wu, K. N. Lai, and C. Y. Wu, "Generalized theory and realization of a voltage-controlled negative resistance MOS device (lambda MOSFET)," Solid-State Electronics, vol. 23, pp. 1–7, January 1980.

23. K. Seeger, "Semiconductor physics," Springer-Verlag, New York, 1973.

24. L. Esaki, "New phenomenon in narrow germanium p-n junctions," Phys. Rev., vol. 109, pp. 603–604, February 1958.

25. B. L. van der Waerden, "Modern algebra," Frederick Unger, New York, 1931.

26. B. Alder, S. Fernbach, and M. Rotenberg, ed., "Methods in computational physics," Academic Press, New York, 1963.

27. W. C. Elmore, "The transient response of damped linear network with particular regard to wideband amplifiers," J. Appl. Phys., vol. 19, pp. 55–63, January 1948.

28. T. J. Chaney and C. E. Molnar, "Anomalous behavior of synchronizer and arbiter circuits," IEEE Trans. Computers, vol. C-22, pp. 421–422, April 1973.

29. H. J. M. Veendrick, "The behavior of flip-flops used as synchronizers and prediction of their failure rate," IEEE J. Solid-State Circuits, vol. SC-19, pp. 169–176, April 1980.

30. M. Shoji, "Analysis of high frequency thermal noise of enhancement mode MOS field-effect transistors," IEEE Trans. Electron Devices, vol. ED-13, pp. 520–524, June 1966.

31. A. van der Ziel, "Noise in VLSI," VLSI handbook, N. G. Einspruch, ed., Academic Press, Orlando, Fla., 1985.

32. E. D. Sunde, "Earth conduction effects in transmission systems," van Nostrand, New York, 1949.

33. L. Brillouin, "Relativity reexamined," Academic Press, New York, 1970.

34. W. L. Cassell, "Linear electric circuits," Wiley, New York, 1964.

35. E. Bodewig, "Matrix calculus," North-Holland, Amsterdam, 1959.

36. L. A. Pipes, "Applied mathematics for engineers and physicists," McGraw-Hill, New York, 1946.

37. L. C. Parrillo, R. S. Payne, R. E. Davis, G. W. Reutlinger, and R. L. Field, "Twin-tub CMOS—a technology for VLSI circuits," IEDM80 Digest, pp. 752–755, December 1980.

38. I. Prigogine, "Thermodynamics of irreversible processes," Wiley-Interscience, New York, 1955.

39. J. D. Cole, "Perturbation methods in applied mathematics," Blaisdel, Waltham, Mass., 1968.

40. M. Van Dyke, "Perturbation methods in fluid mechanics," Academic Press, New York, 1964.

41. Z. Nehari, "Complex analysis," Allyn and Bacon, Boston, 1968.

42. R. Schlegel, "Time and the physical world," Michigan State University Press, 1961.

43. T. Gold and D. L. Schumacher, ed., "The nature of time," Cornell University Press, Ithaca, N.Y., 1967.

44. C. L. Seitz, A. H. Frey, S. Mattisson, S. D. Ravin, D. A. Speck, J. L. van de Snepscheut, "Hot-clock nMOS," 1985 Chapel Hill Conference on VLSI, pp. 1–17.

45. G. Metzger and J. P. Vabre, "Transmission lines with pulse excitation," Academic Press, New York, 1969.

46. P. Grivet and P. W. Hawkes, "The physics of transmission lines at high and very high frequencies," Academic Press, New York, 1970.

47. L. M. Magid, "Electromagnetic fields, energy, and waves," Wiley, New York, 1972.

48. B. M. Budak, A. A. Samarskii, and A. N. Tikhonov, "A collection of problems on mathematical physics," Macmillan, New York, 1964.

49. E. T. Whittaker and G. N. Watson, "Modern analysis," Cambridge University Press, Cambridge, 1927.

50. R. C. Frye and H. Z. Chen, "High speed interconnections using self-damped lossy transmission lines," IEEE Symposium on High Density Integration in Communication and Computer Systems, Waltham, Mass., 17–18, October 1991.

51. H. P. Stapp, "Mind, matter and quantum mechanics," Springer-Verlag, Berlin, 1993.

INDEX